Endohedral Metallofullerenes

BASICS AND APPLICATIONS

Endohedral Metallofullerenes

BASICS AND APPLICATIONS

Edited by

Xing Lu • Luis Echegoyen • Alan L. Balch
Shigeru Nagase • Takeshi Akasaka

CRC Press
Taylor & Francis Group
Boca Raton London New York

CRC Press is an imprint of the
Taylor & Francis Group, an **informa** business

CRC Press
Taylor & Francis Group
6000 Broken Sound Parkway NW, Suite 300
Boca Raton, FL 33487-2742

First issued in paperback 2019

© 2015 by Taylor & Francis Group, LLC
CRC Press is an imprint of Taylor & Francis Group, an Informa business

No claim to original U.S. Government works

ISBN-13: 978-1-4665-9394-7 (hbk)
ISBN-13: 978-0-367-37800-4 (pbk)

Visit the Taylor & Francis Web site at
http://www.taylorandfrancis.com

and the CRC Press Web site at
http://www.crcpress.com

Contents

Foreword

Without doubt, the most salient examples of nanomaterials and nano-sized chemistry are the endohedral metallofullerenes (EMFs).* In the pages that follow, the editors, Xing Lu, Luis Echegoyen, Alan L. Balch, Shigeru Nagase, and Takeshi Akasaka, have collected an impressive amount of information regarding this family of a truly new form of matter, *sui generis*, to the cage nature of the fullerenes, especially the larger cages. In some respects this will not be an easy read, but for those interested in exotic molecular and condensed matter structure as well as their electronic, optical, and optoelectronic properties, reading this text will be a rewarding effort.

The bulk preparation of endohedrals was an imposing challenge in the early days following their discovery, which occurred almost simultaneously with the advent of the empty fullerene cages. The reader will learn that in the more recent past the *preparation*, *separation*, and *isolation*, though still somewhat challenging, no longer require the heroic efforts of the 1990s. The reader will also learn how theory has advanced at the same pace to illuminate the subtle effects of cage structure and geometry as well as metal–metal and metal–cage interactions that influence EMF preparation and properties.

Most importantly, the reader will learn how the difficult X-ray structure solution of these solids was accomplished using unusual techniques, particularly in this case with the use of co-crystallization.

Indeed this novel *hybrid* allotrope of carbon is poised to provide important developments in the applications of their nano-size to materials science based on the unique physicochemical phenomenon of intramolecular charge transfer that will be reflected in device physics such as interaction of light with matter (photovoltaics, optical sensors and light emitters) and electronics (transistors, diodes).

Fred Wudl
University of California

---

* A very unfortunate acronym that clashes with the fundamentals of batteries, electromotive force (EMF).

Preface

I first learned of fullerenes from a lecture given by Harry Kroto in our small conference room here in the Chemistry Building at U. C. Davis in 1987 or so. His enthusiastic lecturing style drew me into the room as I was passing by. I had no intention of attending a seminar that day; too much to do. But I did attend and learned about the infancy of fullerenes. Then, I considered them as chicken wire in the sky, gas-phase molecules with intriguing structures but nothing I would ever become involved with. And now here I am ~27 years later hastily trying to write a preface for a book on endohedral metallofullerenes.

The fullerenes, those carbon cages that Kroto, Heath, O'Brien, Curl, and Smalley discovered in 1985, have now been isolated as molecular species that a chemist can manipulate and modify. They are available in ton quantities and have been examined for many possible applications including as electron acceptors in organic photo-devices, as components in photodynamic therapy in tumor management, and as radical traps. These same carbon cages can also trap atoms, molecules, and atomic clusters inside with the resulting formation of endohedral fullerenes. Indeed, the second publication on fullerenes from the team at Rice led by Kroto, Curl, and Smalley describes the formation of "Lanthanum complexes of spheroidal carbon shells", the first examples of what we now know as endohedral metallofullerenes.

This book involves an important subset of endohedral fullerenes, the ones termed metallofullerenes in which an electropositive metal atom or a collection of metal atoms is mechanically trapped inside a closed carbon cage. In constructing this book, the editors have brought together a set of informed scientists who are actively engaged in various aspects of fullerene research to present an up-to-date account of the field. Our volume begins with a general introduction to the field by Professor Xing Lu and it ends with an overall perspective from him. The formation of endohedral metallofullerenes in sufficient quantity for further study is an important issue in the development of the field. Professor Steven Stevenson provides us with an authoritative chapter that concerns practical issues involving metallofullerene formation and purification. Understanding the stability and structures of various endohedral metallofullerenes is a challenging task, one that has required sophisticated computational studies to advance our understanding. Alexey Popov has been an active participant and pioneer in such studies. He has written an informative chapter for us on relevant computational studies. Structural characterization of endohedral metallofullerenes has relied heavily on two major techniques: single crystal X-ray diffraction and nuclear magnetic resonance (NMR) spectroscopy of fullerenes in solution. Wenjun Zhang, Muqing Chen, Lipiao Bao, Michio Yamada, and Xing Lu have written on NMR studies, while Marilyn Olmstead, Kamran Ghiassi, and I report on the crystallographic identification of endohedral metallofullerenes.

The fundamental physical properties (absorption spectroscopy in the UV-vis-NIR region, electrochemical behavior, luminescence and nonlinear optical behavior, and magnetism) of various endohedral metallofullerenes are the subject of a chapter by Song Wang, Taishan Wang, Tao Wei, Chunru Wang, and Shangfeng Yang. Endohedral metallofullerenes are neutral molecules with an all carbon external surface that is readily available for chemical reactions and the addition of various functional groups. Muqing Chen and Xing Lu contribute a section on the chemical modification of the exterior of metallofullerenes containing one or two metal atoms, while Maira R. Cerón, Marta Izquierdo, and Luis Echegoyen write about chemical reactions on the outside of metallofullerenes that contain clusters, such as the popular M_3N unit. Metallofullerenes are being examined for a number of applications. Charge-transfer processes are important for their use in electro-optical devices. This area is discussed here by Marcus Lederer, Marc Rudolf, Maximilian Wolf, and Dirk M. Guldi. The biomedical applications, which rely greatly on the properties of the encapsulated metal atoms, are described by Shasha Zhao and Xing Lu. Finally, Lai Feng considers endohedral metallofullerenes in the context of electronic nanomaterials.

Certain related topics have been omitted from this volume. Endohedral fullerenes containing molecules such as molecular hydrogen or carbon monoxide form a different subset and have not been discussed. Likewise, endohedral fullerenes containing the inert gasses and atomic nitrogen have not been considered. Separate reviews of both of these types of endohedral fullerenes are available elsewhere.

Alan L. Balch
University of California

Editors

Xing Lu is currently a full professor at Huazhong University of Science and Technology supported by The National Thousand Talents Program of China. His research interests lie in the rational design and facile generation of novel hybrid carbon materials with applications in energy storage/conversion and biology. He was born in 1975 and received a PhD from Peking University in 2004. Then he worked as a COE postdoctoral fellow at Nagoya University, Japan. From 2006 to 2011, he has been a Senior Scientist at the University of Tsukuba, Japan. In September 2011, he joined his current institution to found the Center of Carbon Materials. He has published more than 60 scientific papers including more than 20 at JACS and ACIE. He is the recipient of the Ambassador Award from the Chinese embassy in Japan (2009) and the Osawa Award from the Fullerenes and Nanotubes Research Society of Japan (2011).

Luis Echegoyen has been the Robert A. Welch Chair Professor of Chemistry at the University of Texas at El Paso since August 2010. He was the director of the Chemistry Division at the National Science Foundation from August 2006 until August 2010, where he was instrumental in establishing new funding programs and research centers. He was simultaneously a professor of chemistry at Clemson University in South Carolina, where he maintained a very active research program with interests in fullerene electrochemistry, monolayer films, supramolecular chemistry, and spectroscopy; endohedral fullerene chemistry and electrochemistry; and carbon nano-onions, synthesis, derivatization, and fractionation. He served as chair for the Department of Chemistry at Clemson from 2002 until his NSF appointment. Luis has published around 340 research articles and 43 book chapters. He was elected fellow of the American Association for the Advancement of Science in 2003 and has been the recipient of many awards, including the 1996 Florida ACS Award, the 1997 University of Miami Provost Award for Excellence in Research, the 2007 Herty Medal Award from the ACS Georgia Section, the 2007 Clemson University Presidential Award for Excellence in Research, and the 2007 University of Puerto Rico Distinguished Alumnus Award. He was also selected as an ACS fellow in 2011 and was the first recipient of the ACS Award for Recognizing Underrepresented Minorities in Chemistry for Excellence in Research & Development, also in 2011. Luis is a coveted speaker who has to his record over 370 scientific invited lectures and presentations.

Alan L. Balch is currently Distinguished Professor of Chemistry at the University of California, Davis. He earned a BA in chemistry from Cornell University in 1962 and a PhD from Harvard University in 1967. In 1966, he was appointed as an assistant professor at the University of California, Davis. In 2011 he received the F. Albert Cotton Award in Synthetic Inorganic Chemistry from the American Chemical Society. He is a Fellow of the American Chemical Society and the American Association for the

Advancement of Science. His current research interests include synthetic and structural studies of fullerenes and endohedral fullerenes, examination of metal–metal bonding and its effects on luminescence of transition metal complexes, and supramolecular chemistry in crystals.

Shigeru Nagase received a PhD degree from Osaka University in 1975. After doing research work as a postdoctoral fellow from 1976 to 1979 at the University of Rochester and Ohio State University, he joined the Institute for Molecular Science in 1979. He moved to Yokohama National University as an associate professor in 1980 and became a professor in 1991. As a professor, he moved to Tokyo Metropolitan University in 1995 and returned to the Institute for Molecular Science in 2001. After he retired from the Institute for Molecular Science in 2012, he moved to the Fukui Institute for Fundamental Chemistry at Kyoto University as a senior research fellow. He has been a member of the International Academy of Quantum Molecular Science since 2008. He received the Commendation for Science and Technology (Research Category) from the MEXT in 2011, the Fukui Medal from the Asia-Pacific Association of Theoretical and Computational Chemists in 2012, and the Chemical Society of Japan Award in 2012. He has great interest in developing new molecules and reactions through close interplay between theoretical predictions and experimental tests.

Takeshi Akasaka was born in Kyoto and grew up in Osaka, Japan. He received a PhD degree from the University of Tsukuba in 1979. After working as a postdoctoral fellow at Brookhaven National Laboratory in the United States, he returned to the University of Tsukuba in 1981. In 1996, he moved to Niigata University as a professor. From 2001 until 2013, he was a professor of the TARA Center and Department of Chemistry, University of Tsukuba. After his retirement from the University of Tsukuba, he moved to the Foundation for Advancement of International Science as a senior researcher. His research interests include the organic chemistry of nanocarbons (fullerenes, endohedral metallofullerenes, carbon nanotubes and so forth). He received the Chemical Society of Japan Award for Creative Work for 2001, the Commendation for Science and Technology (Research Category) from the Ministry of Education, Culture, Sports, Science and Technology of Japan in 2011, and the Chemical Society of Japan Award for 2013.

Contributors

Lipiao Bao
School of Materials Science and
Engineering
Huazhong University of Science and
Technology (HUST)
Wuhan, China

Maira R. Cerón
Department of Chemistry
University of Texas at El Paso
El Paso, Texas

Muqing Chen
School of Materials Science and
Engineering
Huazhong University of Science and
Technology (HUST)
Wuhan, China

Luis Echegoyen
Department of Chemistry
University of Texas at El Paso
El Paso, Texas

Lai Feng
Jiangsu Key Laboratory of Thin Films
and School of Energy
Soochow University
Suzhou, China

Kamran B. Ghiassi
Department of Chemistry
University of California, Davis
Davis, California

Dirk M. Guldi
Department of Chemistry
and Pharmacy Interdisciplinary
Center for Molecular Materials
(ICMM)
Friedrich-Alexander-Universitaet
Erlangen-Nuernberg
Erlangen, Germany

Marta Izquierdo
Department of Chemistry
University of Texas at El Paso
El Paso, Texas

Marcus Lederer
Department of Chemistry
and Pharmacy Interdisciplinary
Center for Molecular Materials
(ICMM)
Friedrich-Alexander-Universitaet
Erlangen-Nuernberg
Erlangen, Germany

Xing Lu
School of Materials Science and
Engineering
Huazhong University of Science and
Technology (HUST)
Wuhan, China

Marilyn M. Olmstead
Department of Chemistry
University of California, Davis
Davis, California

Alexey A. Popov
Leibniz Institute for Solid State and
  Materials Research
Dresden, Germany

Marc Rudolf
Department of Chemistry and
  Pharmacy Interdisciplinary
  Center for Molecular Materials
  (ICMM)
Friedrich-Alexander-Universitaet
  Erlangen-Nuernberg
Erlangen, Germany

Steven Stevenson
Indiana-Purdue University
  at Fort Wayne
Fort Wayne, Indiana

Chunru Wang
Beijing National Laboratory for
  Molecular Sciences
Laboratory of Molecular
  Nanostructure and Nanotechnology
Institute of Chemistry
Beijing, China

Song Wang
Hefei National Laboratory for Physical
  Science at Microscale
CAS Key Laboratory of Materials
  for Energy Conversion
  & Department of Materials Science
  and Engineering
University of Science and Technology
  of China (USTC)
Hefei, China

Taishan Wang
Beijing National Laboratory for
  Molecular Sciences
Laboratory of Molecular Nanostructure
  and Nanotechnology
Institute of Chemistry
Beijing, China

Tao Wei
Hefei National Laboratory for Physical
  Science at Microscale
CAS Key Laboratory of Materials for
  Energy Conversion & Department of
  Materials Science and Engineering
University of Science and Technology
  of China (USTC)
Hefei, China

Maximilian Wolf
Department of Chemistry and
  Pharmacy Interdisciplinary Center
  for Molecular Materials (ICMM)
Friedrich-Alexander-Universitaet
  Erlangen-Nuernberg
Erlangen, Germany

Michio Yamada
Department of Chemistry
Tokyo Gakugei University
Koganei, Tokyo, Japan

Shangfeng Yang
Hefei National Laboratory for Physical
  Science at Microscale
CAS Key Laboratory of Materials for
  Energy Conversion & Department of
  Materials Science and Engineering
University of Science and Technology
  of China (USTC)
Hefei, China

Wenjun Zhang
School of Materials Science and
  Engineering
Huazhong University of Science and
  Technology (HUST)
Wuhan, China

Shasha Zhao
School of Materials Science and
  Engineering
Huazhong University of Science and
  Technology (HUST)
Wuhan, China

1 Introduction to Endohedral Metallofullerenes

Xing Lu

CONTENTS

From the point of view of a chemist, carbon is surely the most intriguing element among the elements that have been discovered to date. As the *core* element of thousands of organic compounds, carbon is associated closely with the origin of life.[1] The discovery of fullerenes, a collection of spherical molecules consisting of pure carbon atoms, has not only added a new form of elemental carbon but also opened an avenue for research into molecular architecture.[2] Soon after the proposal of the soccer-ball-like molecular structure of C_{60}, the most abundant fullerene isomer, the pioneering work by Smalley and coworkers predicted the possibility of fullerenes for encapsulating metal atoms (or other molecules) inside hollow-cage cavities based on the observation of the molecular ion peak of LaC_{60} in mass spectrum.[3] This new class of metal–carbon hybrid molecules is referred to as endohedral metallofullerenes (EMFs).[4] In a later report on the solvent extraction of La-containing EMFs, the authors suggested that the symbol @ can be used to indicate the endohedral

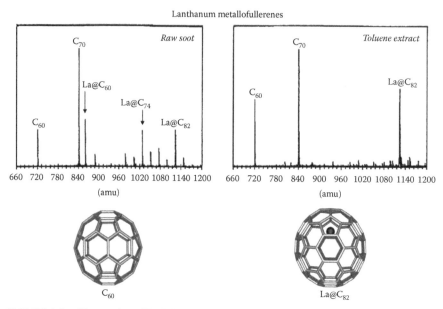

FIGURE 1.1   Observation of La@C_{2n} EMFs with mass spectroscopy. The lower panel shows the molecular structures of C_{60} and La@C_{82}. (Reprinted with permission from Chai, Y. et al., *J. Phys. Chem.* 95, 7564–7568, 1991. Copyright 1991 American Chemical Society.)

characteristics of the metal atom(s) inside a fullerene cage.[5] It is noteworthy that although La@C_{2n} ($2n$ = 60, 70, 74–84) are observable in raw soot, only La@C_{82} can be extracted by toluene (Figure 1.1). In this regard, La@C_{82} is viewed as a prototype of EMFs.[6]

However, research on EMFs did not receive wide attention following their discovery because of their low availability. That relative obscurity changed with the development of the arc-discharge method, which, initially invented for high-efficiency fullerene production, was used to produce EMFs.[7] During recent years, research on EMFs has advanced considerably, yielding many meaningful results.[8–10] It is now acknowledged that the structures and inherent properties of EMFs differ explicitly from those of empty fullerenes because of the presence of the metallic species and their strong interactions with cage carbons.[11–16]

An intriguing feature of EMFs is intramolecular electron transfer, which results in novel structures, unique properties, and potential applications of these hybrid molecules.[17] Although the formation mechanisms of EMFs remain somewhat unclear,[18] it is commonly accepted that the endohedral metallic cluster and the surrounding cage are mutually dependent because of their strong electrostatic interactions. Figure 1.1 illustrates a graph showing the electron transfer from the internal metallic species to the carbon cage in EMFs. It is noteworthy that the number of electrons transferred from metal to cage depends strongly on the cluster properties.

As the research on EMFs proceeded rapidly, it was interesting to find that only these metal elements centered in Groups 2, 3, and 4 can form EMFs in isolable

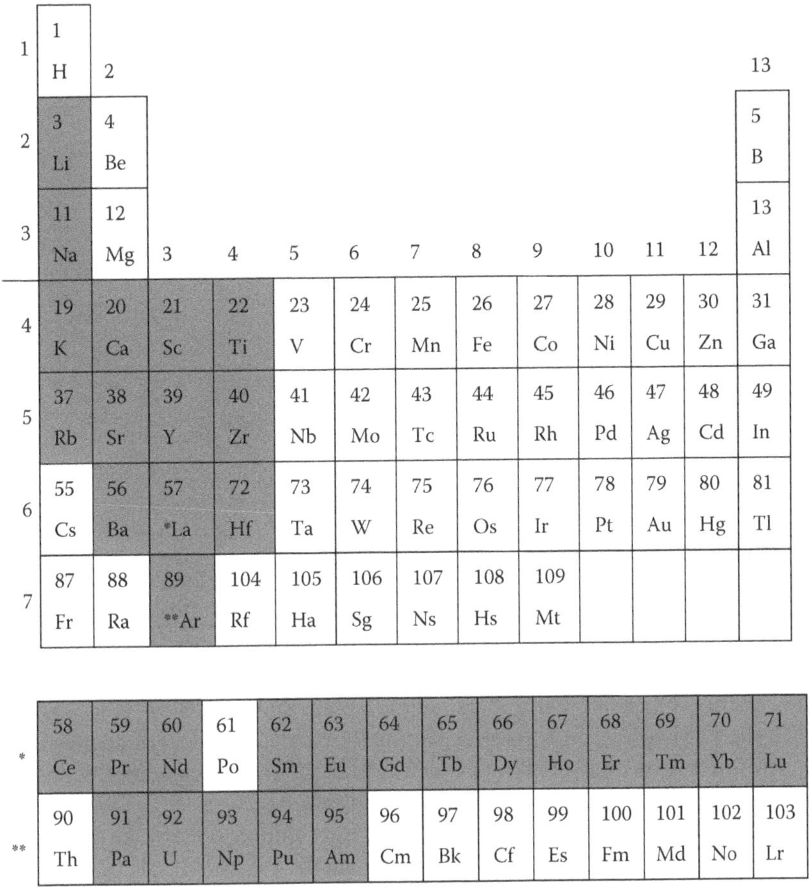

1	2	3	4	5	6	7	8	9	10	11	12	13
1 H												
3 Li	4 Be											5 B
11 Na	12 Mg											13 Al
19 K	20 Ca	21 Sc	22 Ti	23 V	24 Cr	25 Mn	26 Fe	27 Co	28 Ni	29 Cu	30 Zn	31 Ga
37 Rb	38 Sr	39 Y	40 Zr	41 Nb	42 Mo	43 Tc	44 Ru	45 Rh	46 Pd	47 Ag	48 Cd	49 In
55 Cs	56 Ba	57 *La	72 Hf	73 Ta	74 W	75 Re	76 Os	77 Ir	78 Pt	79 Au	80 Hg	81 Tl
87 Fr	88 Ra	89 **Ar	104 Rf	105 Ha	106 Sg	107 Ns	108 Hs	109 Mt				

58 Ce	59 Pr	60 Nd	61 Po	62 Sm	63 Eu	64 Gd	65 Tb	66 Dy	67 Ho	68 Er	69 Tm	70 Yb	71 Lu
90 Th	91 Pa	92 U	93 Np	94 Pu	95 Am	96 Cm	97 Bk	98 Cf	99 Es	100 Fm	101 Md	102 No	103 Lr

FIGURE 1.2 Metal elements that reportedly form EMFs.

amounts.[4] In contrast, other metallic elements, especially transition metals and noble metals showing remarkable catalytic abilities in many chemical transformations, do not form EMFs in reasonable amounts (Figure 1.2).

1.1 CLASSIFICATION OF EMFs

According to the number and types of encapsulated metal species, EMFs are classifiable into several categories.

1.1.1 Mono-EMFs

Early in EMF research, mono-EMFs were observed frequently.[19–22] As the simplest prototypes of EMFs that encapsulate single metal atoms inside fullerene cages of various sizes, mono-EMFs are particularly suitable for elucidating metal–cage interactions.

As summarized in Table 1.1, almost all metals in Groups 2, 3, and 4 reportedly form mono-EMFs. For these compounds, either two or three electrons are transferred

TABLE 1.1

Summary of Identified EMFs Showing Diversity of Both the Internal Metallic Species and the Cage Size/Structures

Metallic Species	Carbon Cage
Li^+, Na^+, K^+, Rb^+	C_{60}, C_{70}
Ca, Sr, Ba, Ti, Sc, Y, La-Yb, Zr, Hf, Ac, Th, U, Np, Pu, Am	C_{28}, C_{72}, C_{74}, C_{80-84}, C_{94}
M_2 (M = Sc, Y, La-Lu Hf, U)	C_{80-84}, C_{100}, C_{104}
M_3 (M = Y, Sm)	C_{80}
M_3N (M = Sc, Y, La-Lu)	C_{68}, C_{78}, $C_{79}N$, C_{80-84}, C_{88}, C_{96}
$M_xSc_{3-x}N$ (M = Ti, Y, La-Lu)	C_{80-84}
M_2C_2 (M = Ti, Sc, Y, Er,); M_3C_2 (M = Sc, Lu); Sc_4C_2	C_{78}, C_{80}, C_{82}, C_{84}, C_{92}
Sc_2O, Sc_4O_2, Sc_4O_3	C_{80}, C_{82}
M_2S (Sc, Y, Dy, Lu)	C_{72}, C_{82}
Sc_3CN, Sc_3HC, YCN	C_{78}, C_{80}, C_{82}

from the internal metal atom to the fullerene cage. Therefore, mono-EMFs are classified into two sub-branches: trivalent mono-EMFs and divalent mono-EMFs. It is commonly accepted that, when forming mono-EMFs, alkali-earth metal elements and Sm, Eu, Tm, and Yb[23–30] normally transfer two electrons to the fullerene cages, whereas other rare-earth metal atoms donate three of their valence electrons to the fullerene cage.[31–35] It is noteworthy that the production yield of divalent EMFs is much lower than that of trivalent ones, as described in numerous reports.[36,37] Furthermore, alkali-metal ions can be inserted into the cages of C_{60} and C_{70} using synthetic methods other than the arc-discharge technique, thereby forming mono-EMFs featuring no intramolecular electron transfer.[38–41]

Although only one metal atom is encapsulated in mono-EMFs, the sizes and structures of the surrounding cages are diverse. For example, M@C_{28} (M = Ti, Zr, U) has been identified as robust in the gas phase, represented as the smallest mono-EMFs that have ever been reported, although their structures remain unclear.[42,43] Among the experimentally identified mono-EMFs, the smallest cage is C_{72}, which contains a pair of fused pentagons, violating the so-called isolated pentagon rule (IPR): La@C_2(10612)-C_{72}.[44] The largest ones are M@C_{3v}(134)-C_{94} (M = Ca, Tm, Sm) as determined with single crystal X-ray diffraction spectroscopy.[24] Among all mono-EMFs obtained to date, M@C_{2v}(9)-C_{82} (M = Sc, Y, La, Ce, etc.) species are generally more abundant than others. Accordingly, they have been investigated more intensively. In these compounds, the single metal cation interacts strongly with the cage carbons, resulting in an off-center-positioned metal ion inside the cages.

1.1.2 Di-EMFs

EMFs containing two metal atoms, di-EMFs are among the earliest discovered species, but they are generally less abundant than the corresponding mono-EMFs encapsulating the same type of metals.[45–48] Moreover, EMFs encapsulating two divalent metal atoms have rarely been reported. One example with a clear molecular structure is $Sm_2@C_{104}$, which also represents the largest fullerene cage that has been structurally identified.[49]

In addition, cage structures encapsulating two metal atoms differ completely from those of mono-EMFs because the number of electrons accepted by the cages is different. In reality, the most common di-EMF species are $M_2@I_h(7)-C_{80}$ (M = La, Ce, etc.). Emergence of some small species such as $Sc_2@C_{66}$ and $La_2@C_{72}$, both bearing two pairs of fused pentagons violating the IPR, suggests that the metal ions interact strongly with the cage carbons, thereby making the highly reactive pentalene carbons more stable.[50,51] Larger cages such as $Sm_2@D_{3d}(822)-C_{104}$ and $La_2@D_5(450)-C_{100}$ have also been identified recently.[52]

The location of the metal atoms inside the fullerene cages presents another interesting issue for di-EMFs. Previous studies of $La_2@C_{80}$ and $Er_2@C_{82}$ demonstrated that the two metal atoms tend to move inside the cages, but recent results have suggested that the metal location and motion are highly sensitive to the cage geometry and the size of the metal atoms.[53,54] For example, the two La^{3+} ions coordinate strongly with the pentalene units in $La_2@D_2(10611)-C_{72}$,[55,56] but they more likely tend to move inside $La_2@C_s(17490)-C_{76}$. In the large tubular cages like $La_2@C_{100}$, the two metal ions are always separated to the greatest degree to avoid electrostatic repulsion.[52]

1.1.3 TRI-EMFs

EMFs with more than two pure metals have rarely been encountered. The first tri-EMF, $Sc_3@C_{3v}(7)-C_{82}$, was proposed in 1999,[57] but it was recently identified as a carbide cluster EMF, $Sc_3C_2@I_h(7)-C_{80}$.[58] Yet another example is $Y_3@I_h(7)-C_{80}$, which was characterized solely by theoretical calculations.[59] The only experimentally confirmed tri-EMF is $Sm_3@I_h(7)-C_{80}$, in which the three divalent Sm^{2+} ions transfer totally six electrons to the cage.[60] X-ray crystallographic results reveal that the three metal ions move rapidly inside the cage, which is expected to be a result of the highly symmetric-cage structure.

The situation that tri-EMFs have rarely been found can be attributed to the strong repulsions among three or more cations, each of which takes a valence higher than 2, inside a confined space in a fullerene cage with a normal size (C_{60}–C_{100}). Accordingly, when more than three metal atoms are encapsulated inside a fullerene cage forming EMFs, one or more nonmetallic atoms are generally necessary for stabilization of the whole cluster, for example, to partially neutralize the positive charges of the metal cations.[61]

1.1.4 CLUSTER EMFs

Results have increasingly showed that cluster formation is an effective way for the generation of novel EMFs containing various metallic clusters. The strong coordination ability of the metal elements and strong metal–cage interactions are primary driving forces.[62] In fact, examples of cluster EMFs are far more common than those containing merely metal atoms. In some cases, mono-EMFs and di-EMFs are called conventional EMFs, whereas cluster EMFs (also tri-EMFs) are regarded as novel EMFs.

1.1.4.1   Nitride Cluster EMFs

In 1999, Dorn and coworkers reported the detection of a series of nitride cluster EMFs with the formula $M_3N@C_{2n}$: the so-called trimetallic nitride template (TNT) family.[63] This discovery opened a new era of research on EMFs. Importantly, one of the TNT members, $Sc_3N@I_h(7)-C_{80}$, was found to be the most abundant EMF compound, for which the production yield is even higher than that of C_{84}. More importantly, these TNT species are more stable than EMFs of other types, which may lead to their higher potential in future applications.[64–66] What is interesting is that cage size depends strongly on the metal cation radius, which determines the size of the encapsulated metal cluster. Such small atoms as Sc, Y and from Gd to Lu, C_{80} is always the most suitable one.[67] C_{88} is found to be better for M_3N, where M represents Nd, Pr, or Ce[68,69]; the large La_3N cluster prefers C_{96}.[70] The larger the fullerene cage, the lower the production yields of the TNT species. No divalent metal ions such as alkali-earth metal ions or Sm^{2+}, Eu^{2+}, and Yb^{2+} have been found in nitride cluster EMFs, which strongly supports the ionic model of TNT EMFs, that is, $M_3N^{6+}@C_{2n}{}^{6-}$.[71]

In addition, $A_xB_{3-x}N@C_{2n}$-type ($x = 0$–3; A and B are different rare-earth metal elements) nitride cluster EMFs have also been reported.[72–78] An interesting example is $ScYErN@C_{80}$, which contains four elements of different types.[79] Recently, titanium was also encapsulated inside the fullerene cages together with scandium in the form of mixed metal nitride cluster. $Sc_2TiN@C_{80}$ and $Y_2TiN@C_{80}$ have been isolated and fully characterized.[80,81] Accordingly, it is expected that other transition metal elements such as V and Fe/Co/Ni can be trapped inside fullerenes via cluster formation.

1.1.4.2   Carbide Cluster EMFs

Carbide cluster EMFs have the formula $M_xC_2@C_{2n}$ (M = Sc, Y, Er, Gd, Ti, etc.; $x = 2, 3, 4$). Actually, several carbide cluster EMFs have been believed to be di-EMFs or tri-EMFs in previous reports.[82] Particularly, most of the scandium-containing EMFs have been confirmed as carbide cluster EMFs. Such examples include $Sc_4C_2@I_h(7)-C_{80}$, $Sc_3C_2@I_h(7)-C_{80}$ instead of $Sc_3@C_{3v}(7)-C_{82}$, $Sc_2C_2@C_{2v}(5)-C_{80}$ instead of $Sc_2@C_{82}$, $Sc_2C_2@C_s(6)-C_{82}$ instead of $Sc_2@C_s(10)-C_{84}$, $Sc_2C_2@C_{2v}(9)-C_{82}$ instead of $Sc_2@C_{2v}(17)-C_{84}$, and $Sc_2C_2@C_{3v}(8)-C_{82}$ instead of $Sc_2@D_{2d}(23)-C_{84}$.[83–87] Additionally, such elements as Ti, Y, Er, Lu, and Gd have been found to form carbide cluster fullerenes effectively.[88,89] At present, these carbide cluster EMFs contain at least two metal cations and exactly two carbon atoms, but it is anticipated that EMFs containing a monometallic carbide cluster, such as MC_2, will be discovered in the near future because $YCN@C_{82}$ has been reported.[90]

1.1.4.3   Oxide Cluster EMFs

Although oxygen has been reported to compress the formation of conventional EMFs, novel species containing a cluster of metal oxide (Sc_2O, Sc_4O_2, Sc_4O_3) have been synthesized by introducing a small amount of oxygen (e.g., air) into the discharge reactor, as reported by Stevenson and collaborators.[91–93] All these metal oxide cluster EMFs were structurally characterized with single-crystal X-ray crystallography. Considering the electronic configuration of oxide cluster EMFs, it is commonly accepted that six electrons are transferred from the internal cluster to the fullerene cage, in spite of the cluster type.

1.1.4.4 Sulfide Cluster EMFs

EMFs containing a sulfide cluster M_2S (M = Sc, Y, Dy, and Lu) have been successfully synthesized,[94,95] either by doping the graphite rods with a solid sulfur source or by use of SO_2 gas inside the reactor. First, Dunsch and coworkers isolated $Sc_2S@C_{3v}(8)-C_{82}$ and characterized it using a combination of spectroscopic measurements and theoretical calculations. A recent X-ray crystallographic study of $Sc_2S@C_{3v}(8)-C_{82}$ and $Sc_2S@C_s(6)-C_{82}$ in combination with computational works proposed that the cluster is particularly suitable to be encapsulated in these two cages.[96] Sc_2S can also be trapped inside either a C_{70} or a C_{72} cage, both violating the IPR, featuring strong metal–cage interactions.[97,98] In sulfide cluster EMFs, it is believed that four electrons are transferred from the cluster to the cage. Accordingly, both the cluster composition and the cage structure of sulfide cluster EMFs differ markedly from those of oxide EMFs, which are characterized by six-electron transfer, although sulfur and oxygen are both Group 6 elements. Interestingly, the cage structures and cluster orientation of isolated sulfide cluster EMFs closely resemble those of some abundantly produced carbide cluster EMFs, which suggests both may have similar physicochemical properties.

1.1.4.5 Cyanide Cluster EMFs

Metal cyanide cluster EMFs were formed simultaneously with nitride cluster EMFs because nitrogen is added into the chamber. An example of cyanide cluster EMFs is $Sc_3NC@I_h(7)-C_{80}$.[99] Crystallographic results reveal that the Sc_3NC cluster is on the same plane as the N atom in the center of the cage, with the three Sc atoms and the C atom surrounding it. $Sc_3NC@C_{78}$ has also been reported.[100] A recent cyanide cluster EMF $YNC@C_s(6)-C_{80}$ breaks the dogma that cluster EMFs always require two or more metal atoms.[90] The triangular structure of the YCN cluster indicates that the interactions among the three atoms inside the confined space of the fullerene cage differ greatly from those in common organo-metallic species. In addition, $Sc_3HC@C_{80}$ has been isolated but its molecular structure remains unclear.[101]

1.1.5 Aza-EMFs

Aza-fullerene $C_{59}N$ has been widely investigated.[102] Aza-EMFs were first observed in the mass spectrometric characterization of the EMF derivatives such as $La@C_{82}NCH_2Ph$ and $La_2@C_{80}NCH_2Ph$. According to computational results, neutral $La@C_{81}N$ and $La_2@C_{79}N$ can exist as monomers because the unpaired electron from the nitrogen is localized on the internal metal atom(s).[103] $M_2@C_{79}N$ (M = Y, Gd, Tb) compounds have recently been identified using crystallography.[104] Although the doping position of nitrogen is difficult to distinguish experimentally, calculations reveal that N-substitution of a [665]-junction-cage carbon is more stable. It has been demonstrated that the unpaired electron of the nitrogen atom has moved to the center of the two internal metal ions, producing a long M—M bonding.[104] Similarly, $La_3N@C_{79}N$ has been reported, in which the nitrogen is believed to occupy a [665]-junction position.[105]

1.2   ELECTRONIC CONFIGURATIONS OF EMFs

The electronic features of EMFs are more complicated because of electron transfer from the internal metallic species to the fullerene cage (Figure 1.3). In this regard, the bonding character between the encapsulated metal ions and the cage carbons is mainly ionic. Accordingly, the EMF molecules are regarded as a kind of salt that is not dissociated upon solvation.

For mono-EMFs, either two or three electrons are transferred from the metal atom to cage. They are called *divalent EMFs* and *trivalent EMFs*, respectively. These two categories differ markedly in terms of their cage structures and intrinsic properties. For instance, $M^{3+}@C_{2n}^{3-}$ type trivalent EMFs are all paramagnetic species because of the existence of an unpaired electron on the lowest unoccupied molecular orbital (LUMO). They accordingly feature radical characters in most cases.[106-109]

Electronic structures of di-EMFs are simpler. The most frequently encountered $M_2@C_{80}$ (M = La, Ce, Pr, etc.) species have the electronic structure of $(M^{3+})_2@C_{2n}^{6-}$, so that they are all diamagnetic species. Therefore, each metal atom transfers three electrons to the carbon cage. The metal atoms in other di-EMFs, such as $M_2@C_{2n}$ (M = La, Ce; $2n$ = 72, 78), also take a +3 valence. Recently, $Sm_2@C_{104}$ and $Sc_2@C_{82}$ were reported as new di-EMFs in which the metal ions are more likely to be divalent.[110] In fact, the number of transferred electrons cannot be measured accurately. Sometimes calculation results are helpful. For example, the electronic states of Sc in different EMFs vary among cluster types: in $Sc_3N@C_{80}$ all Sc atoms take a +3 valence but they are believed to be divalent ions in $Sc@C_{84}$.[111]

For cluster EMFs, the internal cluster is commonly viewed as an intact unit that transfers a certain number of electrons to the cage. For nitride cluster EMFs, the cages invariably receive six electrons from the cluster. Consequently, the metal ions are all trivalent (N takes a −3 valence). The same applies to cyanide cluster EMFs such as $Sc_3NC@C_{80}$.[99] However, for the monometallic $YNC@C_{82}$, only a two-electron process has been reported.[90]

M_2C_2-type carbide clusters are believed to provide four electrons to the fullerene cages, although different electronic configurations have been proposed for carbide cluster EMFs. In contrast, the electronic structures of $Sc_xC_2@C_{80}$ (x = 3 or 4) should be described as $(Sc_xC_2)^{6+}@C_{80}^{6-}$.[112,113]

The electronic structures of metal oxide cluster EMFs are also sensitive to the cluster type. Sc_2O transfers four electrons, whereas both Sc_4O_2 and Sc_4O_3 transfer six electrons to the fullerene cages. Consequently, the covalent valences of oxygen must change to adopt the corresponding configurations. For the newly emerged sulfide

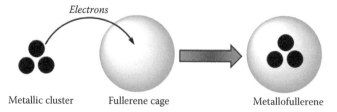

Electrons

Metallic cluster   Fullerene cage   Metallofullerene

FIGURE 1.3   Schematic of EMF formation emphasizing intramolecular electron transfer.

cluster EMFs, M_2S is believed to give four electrons to the cages, a case that is similar to that of M_2C_2.[96] It must be pointed out that the electronic configurations of the internal clusters strongly influence the cage structures (vide infra).

1.3　CAGE STRUCTURES AND METAL–CAGE RELATIONS

Fullerene cages are acknowledged to contain exactly 12 pentagonal rings and a variable number of hexagons, although heptagon-containing fullerene derivatives have also been discovered recently.[114] Fullerene structures faithfully fulfill the IPR, which states that all pentagons should be separated by hexagons because fused-pentagon carbons bear higher local strains and thus are very reactive.[115] During EMF research, all the isolated species were believed to comply with the IPR. For example, the two isomers of $M@C_{82}$ (M = La, Ce, Pr, etc.), respectively, have IPR-satisfying cages with C_{2v} and C_s symmetries. Furthermore, the I_h-symmetric C_{80} cage is particularly suitable for encapsulation of a M_3N cluster (M = Sc, Y, Lu, etc.) or a dimetallic M_2 cluster (M = Y, La, Ce, Pr, etc.) because it becomes particularly stable after accepting six electrons.[116–119]

It is noteworthy that EMFs sometimes adopt carbon cages violating the IPR. Non-IPR EMF was first predicted for $Ca@C_{72}$ theoretically in 1997.[120] Soon after, $Sc_2@C_{66}$ and $Sc_3N@C_{68}$ were reported as the first examples of non-IPR fullerenes, in 2000.[121] Examples of non-IPR structures abound, with structures such as $Sc_2C_2@C_{68}$, $Sc_3N@C_{70}$, $La@C_{72}$, $La_2@C_{72}$, $Sc_2S@C_{72}$, $La_2@C_{72}$, $Gd_3N@C_{78}$, $Tb_3N@C_{82}$, and $Tb_3N@C_{84}$ $(Tm_3N@C_{84}, Gd_3N@C_{84})$.[122–125] In all these species, only fused-pentagon pairs, instead of multiply fused pentagons, have been observed. It is believed that metal–pentalene interactions are the main reason for the existence of these non-IPR EMFs. Accordingly, the IPR is no longer a regulation for EMF structures.

1.4　STRUCTURES AND MOTIONS
　　OF THE ENCAGED METALLIC SPECIES

In mono-EMFs, the metal ion does not stay at the center of the fullerene cage. For frequently obtained $M@C_{2v}(9)$-C_{82} (M = Sc, Y, La, Ce, etc.), a hexagonal ring along the C_2 axis is found to be suitable for keeping the La^{3+} ion. Although anomalous structures have been reported for $Gd@C_{82}$ and $Eu@C_{82}$, in which the metal atom was found to stay closely to a [6,6]-junction opposite the hexagon, recent crystallographic results confirmed that the Gd^{3+} ion in $Gd@C_{82}$ should also sit under a six-member ring, as found in other $M@C_{82}$ species.[126] Actually, the off-center location of the metal ion in mono-EMFs results in an anisotropic distribution of the charges on the cage surface, which in turn causes the aggregation of such molecules in solid state or on clean surfaces, and which is responsible for the different chemical reactivity of the cage carbons at different positions.

For di-EMFs, the strong metal–metal repulsion is an additional factor for the location of the metal cations, in addition to specific metal–cage interactions. Sophisticated studies of $La_2@C_{80}$[127] and $Er_2@C_{82}$[53,54] have nicely demonstrated the circular motions of the metal ions. However, when the cage shrinks, the two metal ions have to be confined to a certain place inside the cage.[128] Also in large cages, the two metal ions are trapped into the limited spaces of the two ends of the cages and interact strongly with the cage carbon atoms, and thus, they do not want to move.

The M_2C_2-type carbide clusters generally adopt a bent structure when they are encapsulated inside a fullerene cage smaller than C_{90}.[129] Similar configurations have also been found in M_2S clusters in sulfide cluster EMFs. An X-ray study of $Gd_2C_2@C_{92}$[130] revealed that the cluster is nearly planar, although Gd^{3+} is much larger than Sc^{3+}. In sharp contrast, the more crowded Sc_3C_2 cluster is planar inside C_{80}, indicating that for controlling the cluster structures the configuration of the cluster itself is more important than cage size.

A planar M_3N cluster is common in nitride cluster EMFs because this configuration can best minimize repulsions among the three trivalent metal ions. However, if the cage shrinks or the cluster size increases, the pyramidalized M_3N cluster has been observed. For instance, a distorted structure of the Gd_3N cluster with the N atom displaced 0.5 Å above the Gd_3 plane has been found in $Gd_3N@C_{80}$.[131]

In this book, we address all aspects of EMFs, including the synthesis, isolation, structures, chemical properties, and theoretic considerations and their applications in related fields.

REFERENCES

1. Hirsch, A. The era of carbon allotropes. *Nat. Mater.* **2010**, *9*, 868–871.
2. Kroto, H. W.; Heath, J. R.; Obrien, S. C.; Curl, R. F.; Smalley, R. E. C_{60}—Buckminsterfullerene. *Nature* **1985**, *318*, 162–163.
3. Heath, J. R.; Obrien, S. C.; Zhang, Q.; Liu, Y.; Curl, R. F.; Kroto, H. W.; Tittel, F. K.; Smalley, R. E. Lanthanum complexes of spheroidal carbon shells. *J. Am. Chem. Soc.* **1985**, *107*, 7779–7780.
4. Lu, X.; Feng, L.; Akasaka, T.; Nagase, S. Current status and future developments of endohedral metallofullerenes. *Chem. Soc. Rev.* **2012**, *41*, 7723–7760.
5. Chai, Y.; Guo, T.; Jin, C. M.; Haufler, R. E.; Chibante, L. P. F.; Fure, J.; Wang, L. H.; Alford, J. M.; Smalley, R. E. Fullerenes with metals inside. *J. Phys. Chem.* **1991**, *95*, 7564–7568.
6. Akasaka, T.; Wakahara, T.; Nagase, S.; Kobayashi, K.; Waelchli, M.; Yamamoto, K.; Kondo, M. et al. La@C_{82} anion. An unusually stable metallofullerene. *J. Am. Chem. Soc.* **2000**, *122*, 9316–9317.
7. Kratschmer, W.; Lamb, L. D.; Fostiropoulos, K.; Huffman, D. R. Solid C_{60}—a new form of carbon. *Nature* **1990**, *347*, 354–358.
8. Chaur, M. N.; Melin, F.; Ortiz, A. L.; Echegoyen, L. Chemical, electrochemical, and structural properties of endohedral metallofullerenes. *Angew. Chem. Int. Ed.* **2009**, *48*, 7514–7538.
9. Dunsch, L.; Yang, S. F. Endohedral clusterfullerenes—playing with cluster and cage sizes. *Phys. Chem. Chem. Phys.* **2007**, *9*, 3067–3081.
10. Dunsch, L.; Yang, S. Metal nitride cluster fullerenes: Their current state and future prospects. *Small* **2007**, *3*, 1298–1320.
11. Akasaka, T.; Nagase, S. *Endofullerenes: A New Family of Carbon Clusters*; Kluwer: Dordrecht, the Netherlands, 2002.
12. Akasaka, T.; Wudl, F.; Nagase, S. *Chemisty of Nanocarbons*; Wiley: Chichester, UK, 2010.
13. Lu, X.; Akasaka, T.; Nagase, S. Rare earths trapped inside fullerenes—endohedral metallofullerenes. In *Rare Earth Coordination Chemistry: Fundamentals and Applications*; Huang, C. H., Ed.; Wiley-Blackwell: Singapore, **2009**, 273–308.

14. Yang, S. F. Synthesis, separation, and molecular structures of endohedral fullerenes. *Curr. Org. Chem.* **2012**, *16*, 1079–1094.
15. Yang, S. F.; Liu, F. P.; Chen, C. B.; Jiao, M. Z.; Wei, T. Fullerenes encaging metal clusters—clusterfullerenes. *Chem. Commun.* **2011**, *47*, 11822–11839.
16. Lu, X.; Akasaka, T.; Nagase, S. Chemistry of endohedral metallofullerenes: The role of metals. *Chem. Commun.* **2011**, *47*, 5942–5957.
17. Popov, A.; Yang, S.; Dunsch, L. Endohedral fullerenes. *Chem. Rev.* **2013**, *113*, 5989–6113, DOI:10.1021/cr300297r.
18. Dunk, P. W.; Kaiser, N. K.; Hendrickson, C. L.; Quinn, J. P.; Ewels, C. P.; Nakanishi, Y.; Sasaki, T.; Shinohara, H.; Marshall, A. G.; Kroto, H. W. Closed network growth of fullerenes. *Nat. Commun.* **2012**, *3*, 855.
19. Hoinkis, M.; Yannoni, C. S.; Bethune, D. S.; Salem, J. R.; Johnson, R. D.; Crowder, M. S.; Devries, M. S. Multiple species of La@C82 and Y@C82—mass spectroscopic and solution EPR studies. *Chem. Phys. Lett.* **1992**, *198*, 461–465.
20. Soderholm, L.; Wurz, P.; Lykke, K. R.; Parker, D. H.; Lytle, F. W. An EXAFS study of the metallofullerene YC82—is the yttrium inside the cage. *J. Phys. Chem.* **1992**, *96*, 7153–7156.
21. Wang, L. S.; Alford, J. M.; Chai, Y.; Diener, M.; Smalley, R. E. Photoelectron-spectroscopy and electronic structure of Ca@C60. *Z. Phys. D.* **1993**, *26*, S297–S299.
22. Wang, L. S.; Alford, J. M.; Chai, Y.; Diener, M.; Zhang, J.; McClure, S. M.; Guo, T.; Scuseria, G. E.; Smalley, R. E. The electronic structure of Ca@C60. *Chem. Phys. Lett.* **1993**, *207*, 354–359.
23. Xu, Z. D.; Nakane, T.; Shinohara, H. Production and isolation of Ca@C82 (I-IV) and Ca@C84 (I,II) metallofullerenes. *J. Am. Chem. Soc.* **1996**, *118*, 11309–11310.
24. Che, Y. L.; Yang, H.; Wang, Z. M.; Jin, H. X.; Liu, Z. Y.; Lu, C. X.; Zuo, T. M. et al. Isolation and structural characterization of two very large, and largely empty, endohedral fullerenes: Tm@C-3v-C(94) and Ca@C-3v-C(94). *Inorg. Chem.* **2009**, *48*, 6004–6010.
25. Kubozono, Y.; Ohta, T.; Hayashibara, T.; Maeda, H.; Ishida, H.; Kashino, S.; Oshima, K.; Yamazaki, H.; Ukita, S.; Sogabe, T. Preparation and extraction of Ca@C60. *Chem. Lett.* **1995**, 457–458.
26. Zhang, Y.; Xu, J. X.; Hao, C.; Shi, Z. J.; Gu, Z. N. Synthesis, isolation, spectroscopic and electrochemical characterization of some calcium-containing metallofullerenes. *Carbon* **2006**, *44*, 475–479.
27. Xu, J. X.; Li, M. X.; Shi, Z. J.; Gu, Z. N. Electrochemical survey: The effect of the cage size and structure on the electronic structures of a series of ytterbium metallofullerenes. *Chem. Eur. J.* **2006**, *12*, 562–567.
28. Xu, J. X.; Lu, X.; Zhou, X. H.; He, X. R.; Shi, Z. J.; Gu, Z. N. Synthesis, isolation, and spectroscopic characterization of ytterbium-containing metallofullerenes. *Chem. Mater.* **2004**, *16*, 2959–2964.
29. Lu, X.; Lian, Y.; Beavers, C. M.; Mizorogi, N.; Slanina, Z.; Nagase, S.; Akasaka, T. Crystallographic X-ray analyses of Yb@C2v(3)-C80 reveal a feasible rule that governs the location of a rare earth metal inside a medium-sized fullerene. *J. Am. Chem. Soc.* **2011**, *133*, 10772–10775.
30. Lu, X.; Slanina, Z.; Akasaka, T.; Tsuchiya, T.; Mizorogi, N.; Nagase, S. Yb@C-2n (n = 40, 41, 42): New fullerene allotropes with unexplored electrochemical properties. *J. Am. Chem. Soc.* **2010**, *132*, 5896–5905.
31. Akasaka, T.; Wakahara, T.; Nagase, S.; Kobayashi, K.; Waelchli, M.; Yamamoto, K.; Kondo, M. et al. Structural determination of the La@C82 isomer. *J. Phys. Chem. B* **2001**, *105*, 2971–2974.
32. Wakahara, T.; Kobayashi, J.; Yamada, M.; Maeda, Y.; Tsuchiya, T.; Okamura, M.; Akasaka, T. et al. Characterization of Ce@C82 and its anion. *J. Am. Chem. Soc.* **2004**, *126*, 4883–4887.

33. Akasaka, T.; Okubo, S.; Kondo, M.; Maeda, Y.; Wakahara, T.; Kato, T.; Suzuki, T.; Yamamoto, K.; Kobayashi, K.; Nagase, S. Isolation and characterization of two Pr@ C82 isomers. *Chem. Phys. Lett.* **2000**, *319*, 153–156.

34. Kubozono, Y.; Maeda, H.; Takabayashi, Y.; Hiraoka, K.; Nakai, T.; Kashino, S.; Emura, S.; Ukita, S.; Sogabe, T. Extractions of Y@C60, Ba@C60, La@C60, Ce@C60, Pr@C60, Nd@C60 and Gd@C60 with aniline. *J. Am. Chem. Soc.* **1996**, *118*, 6998–6999.

35. Yamaoka, H.; Sugiyama, H.; Kubozono, Y.; Kotani, A.; Nouchi, R.; Vlaicu, A. M.; Oohashi, H.; Tochio, T.; Ito, Y.; Yoshikawa, H. Charge transfer satellite in Pr@C-82 metallofullerene observed using resonant x-ray emission spectroscopy. *Phys. Rev. B* **2009**, *80*, 205403.

36. Huang, H. J.; Yang, S. H. Relative yields of endohedral lanthanide metallofullerenes by arc synthesis and their correlation with the elution behavior. *J. Phys. Chem. B* **1998**, *102*, 10196–10200.

37. Lian, Y. F.; Shi, Z. J.; Zhou, X. H.; Gu, Z. N. Different extraction behaviors between divalent and trivalent endohedral metallofullerenes. *Chem. Mater.* **2004**, *16*, 1704–1714.

38. Aoyagi, S.; Nishibori, E.; Sawa, H.; Sugimoto, K.; Takata, M.; Miyata, Y.; Kitaura, R. et al. A layered ionic crystal of polar Li@C-60 superatoms. *Nat. Chem.* **2010**, *2*, 678–683.

39. Campbell, E. E. B.; Couris, S.; Fanti, M.; Koudoumas, E.; Krawez, N.; Zerbetto, F. Third-order susceptibility of Li@C60. *Adv. Mater.* **1999**, *11*, 405–408.

40. Krawez, N.; Gromov, A.; Buttke, K.; Campbell, E. E. B. Thermal stability of Li@C60. *Eur. Phys. J. D* **1999**, *9*, 345–349.

41. Tellgmann, R.; Krawez, N.; Lin, S. H.; Hertel, I. V.; Campbell, E. E. B. Endohedral fullerene production. *Nature* **1996**, *382*, 407–408.

42. Guo, T.; Diener, M. D.; Chai, Y.; Alford, M. J.; Haufler, R. E.; McClure, S. M.; Ohno, T.; Weaver, J. H.; Scuseria, G. E.; Smalley, R. E. Uranium stabilization of C28—a tetravalent fullerene. *Science* **1992**, *257*, 1661–1664.

43. Dunk, P. W.; Kaiser, N. K.; Mulet-Gas, M.; Rodriguez-Fortea, A.; Poblet, J. M.; Shinohara, H.; Hendrickson, C. L.; Marshall, A. G.; Kroto, H. The smallest stable fullerene, M@C28 (M = Ti, Zr, U): Stabilization and growth from Carbon vapor. *J. Am. Chem. Soc.* **2012**, *134*, 9380–9389.

44. Nikawa, H.; Kikuchi, T.; Wakahara, T.; Nakahodo, T.; Tsuchiya, T.; Rahman, G. M. A.; Akasaka, T. et al. Missing metallofullerene La@C74. *J. Am. Chem. Soc.* **2005**, *127*, 9684–9685.

45. Alvarez, M. M.; Gillan, E. G.; Holczer, K.; Kaner, R. B.; Min, K. S.; Whetten, R. L. La2C80—a soluble dimetallofullerene. *J. Phys. Chem.* **1991**, *95*, 10561–10563.

46. Yeretzian, C.; Hansen, K.; Alvarez, M. M.; Min, K. S.; Gillan, E. G.; Holczer, K.; Kaner, R. B.; Whetten, R. L. Collisional probes and possible structure of La2C80. *Chem. Phys. Lett.* **1992**, *196*, 337–342.

47. Beyers, R.; Kiang, C. H.; Johnson, R. D.; Salem, J. R.; Devries, M. S.; Yannoni, C. S.; Bethune, D. S. et al. Preparation and structure of crystals of the metallofullerene Sc2C84. *Nature* **1994**, *370*, 196–199.

48. Akasaka, T.; Nagase, S.; Kobayashi, K.; Suzuki, T.; Kato, T.; Kikuchi, K.; Achiba, Y.; Yamamoto, K.; Funasaka, H.; Takahashi, T. Syntheis of the first adducts of the dimetallofullerenes La2C80 and Sc2C84 by addition of a disilirane. *Angew. Chem. Int. Ed.* **1995**, *34*, 2139–2141.

49. Mercado, B. Q.; Jiang, A.; Yang, H.; Wang, Z. M.; Jin, H. X.; Liu, Z. Y.; Olmstead, M. M.; Balch, A. L. Isolation and structural characterization of the molecular nanocapsule Sm-2@D-3d(822)-C-104. *Angew. Chem. Int. Ed.* **2009**, *48*, 9114–9116.

50. Wang, C. R.; Kai, T.; Tomiyama, T.; Yoshida, T.; Kobayashi, Y.; Nishibori, E.; Takata, M.; Sakata, M.; Shinohara, H. Materials science—C66 fullerene encaging a scandium dimer. *Nature* **2000**, *408*, 426–427.

51. Kato, H.; Taninaka, A.; Sugai, T.; Shinohara, H. Structure of a missing-caged metallofullerene: La2@C72. *J. Am. Chem. Soc.* **2003**, *125*, 7782–7783.

52. Beavers, C. M.; Jin, H.; Yang, H.; Wang, Z. M.; Wang, X.; Ge, H.; Liu, Z.; Mercado, B. Q.; Olmstead, M. H.; Balch, A. L. Very large, soluble endohedral fullerenes in the series La2C90 to La2C138: Isolation and crystallographic characterization of La2@ D5(450)-C100. *J. Am. Chem. Soc.* **2011**, *133*, 15338–15341.

53. Olmstead, M. M.; de Bettencourt-Dias, A.; Stevenson, S.; Dorn, H. C.; Balch, A. L. Crystallographic characterization of the structure of the endohedral fullerene {Er-2@ C-82 Isomer I} with C-s cage symmetry and multiple sites for erbium along a band of ten contiguous hexagons. *J. Am. Chem. Soc.* **2002**, *124*, 4172–4173.

54. Olmstead, M. M.; Lee, H. M.; Stevenson, S.; Dorn, H. C.; Balch, A. L. Crystallographic characterization of Isomer 2 of Er-2@C-82 and comparison with Isomer 1 of Er-2@C-82. *Chem. Commun.* **2002**, 2688–2689.

55. Lu, X.; Nikawa, H.; Nakahodo, T.; Tsuchiya, T.; Ishitsuka, M. O.; Maeda, Y.; Akasaka, T. et al. Chemical understanding of a non-IPR metallofullerene: Stabilization of encaged metals on fused-pentagon bonds in La2@C72. *J. Am. Chem. Soc.* **2008**, *130*, 9129–9136.

56. Lu, X.; Nikawa, H.; Tsuchiya, T.; Maeda, Y.; Ishitsuka, M. O.; Akasaka, T.; Toki, M. et al. Bis-Carbene adducts of non-IPR La2@C72: Localization of high reactivity around fused Pentagons and electrochemical properties. *Angew. Chem. Int. Ed.* **2008**, *47*, 8642–8645.

57. Takata, M.; Nishibori, E.; Sakata, M.; Inakuma, M.; Yamamoto, E.; Shinohara, H. Triangle scandium cluster imprisoned in a fullerene cage. *Phys. Rev. Lett.* **1999**, *83*, 2214–2217.

58. Iiduka, Y.; Wakahara, T.; Nakahodo, T.; Tsuchiya, T.; Sakuraba, A.; Maeda, Y.; Akasaka, T. et al. Structural determination of metallofullerene Sc_3C_{82} revisited: A surprising finding. *J. Am. Chem. Soc.* **2005**, *127*, 12500–12501.

59. Popov, A.; Zhang, L.; Dunsch, L. A Pseudoatom in a cage: Trimetallofullerene Y3@ C80 mimics Y3N@C80 with Nitrogen substituted by a Pseudoatom. *Acs Nano* **2010**, *4*, 795–802.

60. Xu, W.; Feng, L.; Calvaresi, M.; Liu, J.; Liu, Y.; Niu, B.; Shi, Z. J.; Lian, Y. F.; Zerbetto, F. An experimentally observed Trimetallofullerene Sm-3@I-h-C-80: Encapsulation of three metal atoms in a cage without a nonmetallic mediator. *J. Am. Chem. Soc.* **2013**, *135*, 4187–4190.

61. Rivera-Nazario, D. M.; Pinzon, J. R.; Stevenson, S.; Echegoyen, L. A. Buckyball maracas: Exploring the inside and outside properties of endohedral fullerenes. *J. Phys. Org. Chem.* **2013**, *26*, 194–205.

62. Akasaka, T.; Lu, X. Structural and electronic properties of endohedral metallofullerenes. *Chem. Rec.* **2012**, *12*, 256–269.

63. Stevenson, S.; Rice, G.; Glass, T.; Harich, K.; Cromer, F.; Jordan, M. R.; Craft, J. et al. Small-bandgap endohedral metallofullerenes in high yield and purity. *Nature* **1999**, *401*, 55–57.

64. Slanina, Z.; Uhlik, F.; Sheu, J. H.; Lee, S. L.; Adamowicz, L.; Nagase, S. Stabilities of fullerenes: Illustration on C-80. *MATCH Commun. Math. Comput. Chem.* **2008**, *59*, 225–238.

65. Popov, A. A.; Dunsch, L. Structure, stability, and cluster-cage interactions in nitride clusterfullerenes $M_3N@C_{2n}$ (M = Sc, Y; 2n = 68–98): A density functional theory study. *J. Am. Chem. Soc.* **2007**, *129*, 11835–11849.

66. Maki, S.; Nishibori, E.; Terauchi, I.; Ishihara, M.; Aoyagi, S.; Sakata, M.; Takata, M.; Umemoto, H.; Inoue, T.; Shinohara, H. A structural diagnostics diagram for metallofullerenes encapsulating metal carbides and nitrides. *J. Am. Chem. Soc.* **2013**, *135*, 918–923.

67. Chaur, M. N.; Melin, F.; Athans, A. J.; Elliott, B.; Walker, K.; Holloway, B. C.; Echegoyen, L. The influence of cage size on the reactivity of trimetallic nitride metallofullerenes: A mono- and bis-methanoadduct of Gd3N@C80 and a monoadduct of Gd3N@C84. *Chem. Commun.* **2008**, 2665–2667.

68. Chaur, M. N.; Melin, F.; Elliott, B.; Athans, A. J.; Walker, K.; Holloway, B. C.; Echegoyen, L. Gd3N@C2n (n = 40, 42, and 44): Remarkably low HOMO-LUMO gap and unusual electrochemical reversibility of Gd3N@C88. *J. Am. Chem. Soc.* **2007**, *129*, 14826–14829.

69. Chaur, M. N.; Melin, F.; Elliott, B.; Kumbhar, A.; Athans, A. J.; Echegoyen, L. New M3N@C2n endohedral metallofullerene families (M = Nd, Pr, Ce; n = 40–53): Expanding the preferential templating of the C88 cage and approaching the C96 cage. *Chem. Eur. J.* **2008**, *14*, 4594–4599.

70. Chaur, M. N.; Melin, F.; Ashby, J.; Elliott, B.; Kumbhar, A.; Rao, A. M.; Echegoyen, L. Lanthanum nitride endohedral fullerenes La$_3$N@C$_{2n}$ (43 <= n<= 55): Preferential formation of La$_3$N@C$_{96}$. *Chem. Eur. J.* **2008**, *14*, 8213–8219.

71. Chaur, M. N.; Valencia, R.; Rodriguez-Fortea, A.; Poblet, J. M.; Echegoyen, L. Trimetallic nitride endohedral fullerenes: Experimental and theoretical evidence for the M3N6+@C-2n(6-) model. *Angew. Chem. Int. Ed.* **2009**, *48*, 1425–1428.

72. Zhang, Y.; Popov, A. A.; Schiemenz, S.; Dunsch, L. Synthesis, isolation, and spectroscopic characterization of holmium-vased mixed-metal nitride clusterfullerenes: HoxSc3-xN@C-80 (x = 1, 2). *Chem. Eur. J.* **2012**, *18*, 9691–9698.

73. Yang, S.; Liu, F.; Chen, C.; Zhang, W. Novel endohedral mixed metal nitride clusterfullerenes: Synthesis, structure and property. *Prog. Chem.* **2010**, *22*, 1869–1881.

74. Yang, S.; Popov, A. A.; Chen, C.; Dunsch, L. Mixed metal nitride clusterfullerenes in cage isomers: LuxSc3-xN@C-80 (x = 1, 2) as compared with MxSc3-xN@C-80 (M = Er, Dy, Gd, Nd). *J. Phys. Chem. C* **2009**, *113*, 7616–7623.

75. Tarabek, J.; Yang, S. F.; Dunsch, L. Redox properties of mixed Lutetium/Yttrium nitride clusterfullerenes: Endohedral LuxY3-xN@C-80(I) (x = 0–3) compounds. *Chemphyschem* **2009**, *10*, 1037–1043.

76. Yang, S. F.; Popov, A. A.; Dunsch, L. Large mixed metal nitride clusters encapsulated in a small cage: The confinement of the C68-based clusterfullerenes. *Chem. Commun.* **2008**, 2885–2887.

77. Yang, S. F.; Popov, A. A.; Dunsch, L. Carbon pyramidalization in fullerene cages induced by the endohedral cluster: Non-scandium mixed metal nitride clusterfullerenes. *Angew. Chem. Int. Ed.* **2008**, *47*, 8196–8200.

78. Stevenson, S.; Chancellor, C. J.; Lee, H. M.; Olmstead, M. M.; Balch, A. L. Internal and external factors in the structural organization in cocrystals of the mixed-metal endohedrals (GdSc2N@I—C-0, Gd2ScN@I—C-0, and TbSc2N@I—C-0) and nickel(II) octaethylporphyrin. *Inorg. Chem.* **2008**, *47*, 1420–1427.

79. Chen, N.; Zhang, E. Y.; Wang, C. R. C80 encaging four different atoms: The synthesis, isolation, and characterizations of ScYErN@C80. *J. Phys. Chem. B* **2006**, *110*, 13322–13325.

80. Chen, C. B.; Liu, F. P.; Li, S. J.; Wang, N.; Popov, A.; Jiao, M. Z.; Wei, T.; Li, Q.; Dunsch, L.; Yang, S. F. Titanium/Yttrium mixed metal nitride clusterfullerene TiY2N@C80: Synthesis, isolation and effect of the Group-III Metal. *Inorg. Chem.* **2012**, *51*, 3039–3045.

81. Yang, S.; Chen, C.; Popov, A. A.; Zhang, W.; Liu, F.; Dunsch, L. An endohedral titanium(III) in a clusterfullerene: Putting a non-group-III metal nitride into the C-80-I-h fullerene cage. *Chem. Commun.* **2009**, 6391–6393.

82. Lu, X.; Akasaka, T.; Nagase, S. Carbide cluster metallofullerenes: Structure, properties, and possible origin. *Acc. Chem. Res.* **2013**, *46*, 1627–1635, DOI:10.1021/ar4000086.

83. Kurihara, H.; Lu, X.; Iiduka, Y.; Mizorogi, N.; Slanina, Z.; Tsuchiya, T.; Akasaka, T.; Nagase, S. Sc$_2$C$_2$@C$_{80}$ rather than Sc$_2$@C$_{82}$: Templated formation of unexpected C$_{2v}$(5)-C$_{80}$ and temperature-dependent dynamic motion of internal Sc$_2$C$_2$ cluster. *J. Am. Chem. Soc.* **2011**, *133*, 2382–2385.

84. Kurihara, H.; Lu, X.; Iiduka, Y.; Nikawa, H.; Mizorogi, N.; Slanina, Z.; Tsuchiya, T.; Nagase, S.; Akasaka, T. Chemical understanding of carbide cluster metallofullerenes: A case study on $Sc_2C_2@C_{2v}(5)$-C_{80} with complete X-ray crystallographic characterizations. *J. Am. Chem. Soc.* **2012**, *134*, 3139–3144.

85. Lu, X.; Nakajima, K.; Iiduka, Y.; Nikawa, H.; Mizorogi, N.; Slanina, Z.; Tsuchiya, T.; Nagase, S.; Akasaka, T. Structural elucidation and regioselective functionalization of an unexplored carbide cluster metallofullerene $Sc_2C_2@C_s(6)$-C_{82}. *J. Am. Chem. Soc.* **2011**, *133*, 19553–19558.

86. Lu, X.; Nakajima, K.; Iiduka, Y.; Nikawa, H.; Tsuchiya, T.; Mizorogi, N.; Slanina, Z.; Nagase, S.; Akasaka, T. The long-believed $Sc_2@C_{2v}(17)$-C_{84} is actually $Sc_2C_2@$ $C_{2v}(9)$-C_{82}: Unambigous structure assignment and chemical functionalization. *Angew. Chem. Int. Ed.* **2012**, *51*, 5889–5892.

87. Iiduka, Y.; Wakahara, T.; Nakajima, K.; Nakahodo, T.; Tsuchiya, T.; Maeda, Y.; Akasaka, T. et al. Experimental and theoretical studies of the scandium carbide endohedral metallofullerene $Sc_2C_2@C_{82}$ and its carbene derivative. *Angew. Chem. Int. Ed.* **2007**, *46*, 5562–5564.

88. Zhang, J.; Fuhrer, T.; Fu, W. J.; Ge, Z.; Bearden, D. W.; Dallas, J.; Duchamp, J. et al. Nanoscale fullerene compression of an yttrium carbide cluster. *J. Am. Chem. Soc.* **2012**, *134*, 8487–8493.

89. Tan, K.; Lu, X. Ti_2C_{80} is more likely a titanium carbide endohedral metallofullerene $Ti_2C_2@C_{78}$. *Chem. Commun.* **2005**, *41*, 4444–4446.

90. Yang, S. F.; Chen, C. B.; Liu, F. P.; Xie, Y. P.; Li, F. Y.; Jiao, M. Z.; Suzuki, M. et al. An improbable monometallic cluster entrapped in a popular fullerene cage: YCN@C-s(6)-C-82. *Sci. Rep.* **2013**, 1487.

91. Stevenson, S.; Mackey, M. A.; Stuart, M. A.; Phillips, J. P.; Easterling, M. L.; Chancellor, C. J.; Olmstead, M. M.; Balch, A. L. A distorted tetrahedral metal oxide cluster inside an Icosahedral carbon cage. Synthesis, isolation, and structural characterization of Sc4(mu(3)-O)(2)@Ih-C80. *J. Am. Chem. Soc.* **2008**, *130*, 11844–11845.

92. Mercado, B. Q.; Stuart, M. A.; Mackey, M. A.; Pickens, J. E.; Confait, B. S.; Stevenson, S.; Easterling, M. L. et al. Sc-2(mu(2)-O) trapped in a fullerene cage: The isolation and structural characterization of Sc-2(mu(2)-O)@C-s(6)-C-82 and the relevance of the thermal and entropic effects in fullerene isomer selection. *J. Am. Chem. Soc.* **2010**, *132*, 12098–12105.

93. Mercado, B. Q.; Olmstead, M. M.; Beavers, C. M.; Easterling, M. L.; Stevenson, S.; Mackey, M. A.; Coumbe, C. E. et al. A seven atom cluster in a carbon cage, the crystallographically determined structure of Sc-4(mu(3)-O)(3)@I-h-C-80. *Chem. Commun.* **2010**, *46*, 279–281.

94. Dunsch, L.; Yang, S.; Zhang, L.; Svitova, A.; Oswald, S.; Popov, A. A. Metal sulfide in a C-82 fullerene cage: A new form of endohedral clusterfullerenes. *J. Am. Chem. Soc.* **2010**, *132*, 5413–5421.

95. Chen, N.; Chaur, M. N.; Moore, C.; Pinzon, J. R.; Valencia, R.; Rodriguez-Fortea, A.; Poblet, J. M.; Echegoyen, L. Synthesis of a new endohedral fullerene family, Sc2S@C-2n (n = 40–50) by the introduction of SO2. *Chem. Commun.* **2010**, *46*, 4818–4820.

96. Mercado, B. Q.; Chen, N.; Rodriguez-Fortea, A.; Mackey, M. A.; Stevenson, S.; Echegoyen, L.; Poblet, J. M.; Olmstead, M. H.; Balch, A. L. The shape of the Sc2(μ2-S) unit trapped in C82: Crystallographic, computational, and electrochemical studies of the isomers, Sc2(μ2-S)@Cs(6)-C82 and Sc2(μ2-S)@C3v(8)-C82. *J. Am. Chem. Soc.* **2011**, *133*, 6752–6760.

97. Chen, N.; Beavers, C. M.; Mulet-Gas, M.; Rodriguez-Fortea, A.; Munoz, E. J.; Li, Y. Y.; Olmstead, M. M.; Balch, A. L.; Poblet, J. M.; Echegoyen, L. Sc2S@C-s(10528)-C-72: A dimetallic sulfide endohedral fullerene with a non isolated Pentagon rule cage. *J. Am. Chem. Soc.* **2012**, *134*, 7851–7860.

98. Chen, N.; Mulet-Gas, M.; Li, Y. Y.; Stene, R. E.; Atherton, C. W.; Rodriguez-Fortea, A.; Poblet, J. M.; Echegoyen, L. Sc2S@C-2(7892)-C-70: A metallic sulfide cluster inside a non-IPR C-70 cage. *Chem. Sci.* **2013**, *4*, 180–186.

99. Wang, T. S.; Feng, L.; Wu, J. Y.; Xu, W.; Xiang, J. F.; Tan, K.; Ma, Y. H. et al. Planar quinary cluster inside a fullerene cage: Synthesis and structural characterizations of Sc3NC@C-80-I-h. *J. Am. Chem. Soc.* **2010**, *132*, 16362–16364.

100. Wu, J. Y.; Wang, T. S.; Ma, Y. H.; Jiang, L.; Shu, C. Y.; Wang, C. R. Synthesis, isolation, characterization, and theoretical studies of Sc3NC@C78-C2. *J. Phys. Chem. C* **2011**, *115*, 23755–23759.

101. Krause, M.; Ziegs, F.; Popov, A. A.; Dunsch, L. Entrapped bonded hydrogen in a fullerene: The five-atom cluster Sc3CH in C-80. *Chemphyschem* **2007**, *8*, 537–540.

102. Vostrowsky, O.; Hirsch, A. Heterofullerenes. *Chem. Rev.* **2006**, *106*, 5191–5207.

103. Akasaka, T.; Okubo, S.; Wakahara, T.; Yamamoto, K.; Kobayashi, K.; Nagase, S.; Kato, T. et al. Endohedrally metal-doped heterofullerenes: La@C81N and La2@C79N. *Chem. Lett.* **1999**, 945–946.

104. Zuo, T. M.; Xu, L. S.; Beavers, C. M.; Olmstead, M. M.; Fu, W. J.; Crawford, D.; Balch, A. L.; Dorn, H. C. M2@C79N (M = Y, Tb): Isolation and characterization of stable endohedral metallofullerenes exhibiting M-M bonding interactions inside aza[80]fullerene cages. *J. Am. Chem. Soc.* **2008**, *130*, 12992–12997.

105. Stevenson, S.; Ling, Y.; Coumbe, C. E.; Mackey, M. A.; Confait, B. S.; Phillips, J. P.; Dorn, H. C.; Zhang, Y. Preferential encapsulation and stability of La3N cluster in 80 atom cages: Experimental synthesis and computational investigation of La3N@C79N. *J. Am. Chem. Soc.* **2009**, *131*, 17780–17782.

106. Takano, Y.; Ishitsuka, M. O.; Tsuchiya, T.; Akasaka, T.; Kato, T.; Nagase, S. Retroreaction of singly bonded La@C-82 derivatives. *Chem. Commun.* **2010**, *46*, 8035–8036.

107. Takano, Y.; Yomogida, A.; Nikawa, H.; Yamada, M.; Wakahara, T.; Tsuchiya, T.; Ishitsuka, M. O. et al. Radical coupling reaction of paramagnetic endohedral metallofullerene La@C82. *J. Am. Chem. Soc.* **2008**, *130*, 16224–16230.

108. Lu, X.; Nikawa, H.; Kikuchi, K.; Mizorogi, N.; Slanina, Z.; Tsuchiya, T.; Akasaka, T.; Nagase, S. Radical derivatives of Insoluble La@C74 : X-ray structures, metal positions, and isomerization. *Angew. Chem. Int. Ed.* **2011**, *50*, 6356–6359.

109. Lu, X.; Nikawa, H.; Tsuchiya, T.; Akasaka, T.; Toki, M.; Sawa, H.; Mizorogi, N.; Nagase, S. Nitrated Benzyne derivatives of La@C-82: Addition of NO2 and its positional directing effect on the subsequent addition of Benzynes. *Angew. Chem. Int. Ed.* **2010**, *49*, 594–597.

110. Kurihara, H.; Lu, X.; Iiduka, Y.; Mizorogi, N.; Slanina, Z.; Tsuchiya, T.; Nagase, S.; Akasaka, T. $Sc_2@C_{3v}(8)$-C_{82} vs. $Sc_2C_2@C_{3v}(8)$-C_{82}: Drastic effect of C_2 capture on the redox poperties of scandium metallofullerenes. *Chem. Commun.* **2012**, 1290–1292.

111. Takahashi, T.; Ito, A.; Inakuma, M.; Shinohara, H. Divalent scandium atoms in the cage of C84. *Phys. Rev. B* **1995**, *52*, 13812–13814.

112. Tan, K.; Lu, X. Electronic structure and Redox properties of the open-shell metal-carbide Endofullerene Sc3C2@C80: A density functional theory investigation. *J. Phys. Chem. A* **2006**, *110*, 1171–1176.

113. Tan, K.; Lu, X.; Wang, C. R. Unprecedented mu(4)-C2(6-) anion in Sc4C2@C80. *J. Phys. Chem. B* **2006**, *110*, 11098–11102.

114. Tan, Y. Z.; Chen, R. T.; Liao, Z. J.; Li, J.; Zhu, F.; Lu, X.; Xie, S. Y.; Li, J.; Huang, R. B.; Zheng, L. S. Carbon arc production of heptagon-containing fullerene[68]. *Nat. Commun.* **2012**, *2*, 420.

115. Kroto, H. W. The stability of the fullerenes C24, C28, C32, C36, C50, C60 and C70. *Nature* **1987**, *329*, 529–531.

116. Yamada, M.; Minowa, M.; Sato, S.; Kako, M.; Slanina, Z.; Mizorogi, N.; Tsuchiya, T.; Maeda, Y.; Nagase, S.; Akasaka, T. Thermal carbosilylation of endohedral dimetallofullerene La-2@I-h-C-80 with silirane. *J. Am. Chem. Soc.* **2010**, *132*, 17953–17960.

117. Yamada, M.; Minowa, M.; Sato, S.; Slanina, Z.; Tsuchiya, T.; Maeda, Y.; Nagase, S.; Akasaka, T. Regioselective cycloaddition of La-2@/(h)-C-80 with tetracyanoethylene oxide: Formation of an endohedral dimetallofullerene adduct featuring enhanced electron-accepting character. *J. Am. Chem. Soc.* **2011**, *133*, 3796–3799.

118. Yamada, M.; Okamura, M.; Sato, S.; Someya, C. I.; Mizorogi, N.; Tsuchiya, T.; Akasaka, T.; Kato, T.; Nagase, S. Two regioisomers of endohedral pyrrolidinodimetallofullerenes M2@I-h-C-80(CH2)(2)NTrt (M = La, Ce; Trt = trityl): Control of metal atom positions by addition positions. *Chem. Eur. J.* **2009**, *15*, 10533–10542.

119. Yamada, M.; Someya, C. I.; Nakahodo, T.; Maeda, Y.; Tsuchiya, T.; Akasaka, T. Synthesis of endohedral metallofullerene glycoconjugates by carbene addition. *Molecules* **2011**, *16*, 9495–9504.

120. Kobayashi, K.; Nagase, S.; Yoshida, M.; Osawa, E. Endohedral metallofullerenes. Are the isolated pentagon rule and fullerene structures always satisfied? *J. Am. Chem. Soc.* **1997**, *119*, 12693–12694.

121. Stevenson, S.; Fowler, P. W.; Heine, T.; Duchamp, J. C.; Rice, G.; Glass, T.; Harich, K.; Hajdu, E.; Bible, R.; Dorn, H. C. Materials science—a stable non-classical metallofullerene family. *Nature* **2000**, *408*, 427–428.

122. Shi, Z. Q.; Wu, X.; Wang, C. R.; Lu, X.; Shinohara, H. Isolation and characterization of $Sc_2C_2@C_{68}$: A metal-carbide endofullerene with a non-IPR carbon cage. *Angew. Chem. Int. Ed.* **2006**, *45*, 2107–2111.

123. Zuo, T.; Walker, K.; Olmstead, M. M.; Melin, F.; Holloway, B. C.; Echegoyen, L.; Dorn, H. C. et al. New egg-shaped fullerenes: Non-isolated pentagon structures of Tm3N@Cs(51365)C84 and Gd3N@Cs(51365)C84. *Chem. Commun.* **2008**, *44*, 1067–1069.

124. Beavers, C. M.; Chaur, M. N.; Olmstead, M. M.; Echegoyen, L.; Balch, A. L. Large metal ions in a relatively small fullerene cage: The structure of Gd3N@C-2(22010)-C-78 departs from the isolated Pentagon rule. *J. Am. Chem. Soc.* **2009**, *131*, 11519–11524.

125. Beavers, C. M.; Zuo, T. M.; Duchamp, J. C.; Harich, K.; Dorn, H. C.; Olmstead, M. M.; Balch, A. L. Tb3N@C84: An improbable, egg-shaped endohedral fullerene that violates the isolated pentagon rule. *J. Am. Chem. Soc.* **2006**, *128*, 11352–11353.

126. Suzuki, M.; Lu, X.; Sato, S.; Nikawa, H.; Mizorogi, N.; Slanina, Z.; Tsuchiya, T.; Nagase, S.; Akasaka, T. Where does the metal cation stay in Gd@C(2v)(9)-C(82)? A single-crystal X-ray diffraction study. *Inorg. Chem.* **2012**, *51*, 5270–5273.

127. Akasaka, T.; Nagase, S.; Kobayashi, K.; Walchli, M.; Yamamoto, K.; Funasaka, H.; Kako, M.; Hoshino, T.; Erata, T. ^{13}C and [139]La NMR studies of $La_2@C_{80}$: First evidence for circular motion of metal atoms in endohedral dimetallofullerenes. *Angew. Chem. Int. Ed.* **1997**, *36*, 1643–1645.

128. Yamada, M.; Wakahara, T.; Tsuchiya, T.; Maeda, Y.; Kako, M.; Akasaka, T.; Yoza, K.; Horn, E.; Mizorogi, N.; Nagase, S. Location of the metal atoms in Ce2@C78 and its bis-silylated derivative. *Chem. Commun.* **2008**, *44*, 558–560.

129. Kurihara, H.; Lu, X.; Iiduka, Y.; Nikawa, H.; Hachiya, M.; Mizorogi, N.; Slanina, Z.; Tsuchiya, T.; Nagase, S.; Akasaka, T. X-ray structures of $Sc_2C_2@C_{2n}$ (n = 40–42): In-depth understanding of the core–shell interplay in carbide cluster metallofullerenes. *Inorg. Chem.* **2012**, *51*, 746–750.

130. Yang, H.; Lu, C.; Liu, Z.; Jin, H.; Che, Y.; Olmstead, M. M.; Balch, A. L. Detection of a family of gadolinium-containing endohedral fullerenes and the isolation and crystallographic characterization of one member as a metal-carbide encapsulated inside a large fullerene cage. *J. Am. Chem. Soc.* **2008**, *130*, 17296–17300.

131. Stevenson, S.; Phillips, J. P.; Reid, J. E.; Olmstead, M. M.; Rath, S. P.; Balch, A. L. Pyramidalization of Gd3N inside a C80 cage. The synthesis and structure of Gd3N@C80. *Chem. Commun.* **2004**, *40*, 2814–2815.

2 Preparation and Purification of Endohedral Metallofullerenes

Steven Stevenson

CONTENTS

2.1   INTRODUCTION

The period 1990–1995 represented the early years of metallofullerene studies. The abundance of endohedrals in soot was low. It was a time of poor synthetic yields and inefficient separation methods. It was challenging to perform seminal experiments with only micrograms of isolated samples. Naturally, research shifted to improving their abundance and yield. In the 2000s, great improvements were made to *tune* reactor conditions to selectively produce metallofullerenes. In this chapter, we discuss a variety of synthetic approaches, such as the influence of reactive gases, effect of solid additives, and manipulation of reactor conditions.

Beyond metallofullerene synthesis, another challenge was to develop efficient separation methods. Purification approaches in the 1990s relied heavily on high performance liquid chromatography (HPLC) fraction collection. Subsequent advances during the 2000s included the isolation of endohedrals from empty-cage fullerenes based on differences in their physical, oxidative, reductive, or chemical properties (e.g., manipulating cage reactivities). Since the 2010s, improved separation methods are permitting increased amounts of isolated metallofullerenes, which can now be purified from extracts with complex product distributions.

Without the synthesis step, there is no sample to purify. Without a sample to purify, there is no field of separation science. Without purified samples from separation science, there is no characterization and no application development (Figure 2.1). Hence, this chapter represents a vital foundation to the overall field of endohedral metallofullerenes.

Metallofullerene research is 25 years old, and numerous reviews are available.[1–23] The objective of this chapter is not to provide a detailed report of every publication on endohedrals, but rather to provide a historical context for the synthesis, extraction, and separation of endohedrals.

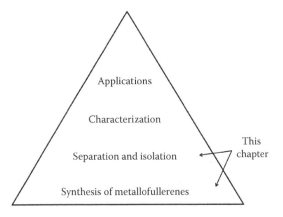

FIGURE 2.1   Synthesis and isolation is the foundation for application development.

2.2   SYNTHESIS

2.2.1   LASER VAPORIZATION

The hypothesis that a metal atom could be entrapped within the hollow cavity of a fullerene cage was suggested in 1985 by Kroto et al.[24] The first experimental evidence for synthesizing an endohedral was reported in 1985 by Heath et al.,[25] who produced a series of lanthanum complexes of spheroidal carbon shells. La@C_{60} was the dominant endohedral. As a source of endohedral metal, a graphite disk was impregnated with lanthanum by refluxing it in a saturated aqueous solution of $LaCl_3$.[25] Subsequent vaporization by a Nd:YAG laser at 532 nm led to a sample and mass spectrum showing the presence of La@C_{60} (Figure 2.2). The debate whether the metal atoms were located on the inside (i.e., endohedral) or outside (i.e., exohedral) of the fullerene cage was addressed experimentally in the early 1990s.

In 1991, Chai et al.[26] performed a laser vaporization of a La_2O_3/graphite composite rod in an argon flow at 1200°C in a tube furnace. They expanded the family of lanthanum endohedrals to include not only La@C_{60}, La@C_{70}, and La@C_{74} but also La@C_{82}. The mass spectrum of a sample obtained from their hot toluene extract is shown in Figure 2.3, which shows La@C_{82} to be a dominant species.[26]

Other endohedrals have also been created by using laser vaporization. In 1992, McElvany[27] used a Nd:YAG laser to vaporize a pressed pellet consisting of Y_2O_3, graphite, and fullerenes. Mass spectral analysis of their product indicated a family of yttrium–fullerene cations, of which Y@C_{60}^+ was a dominant species. The endohedral placement of the yttrium atom was based on its oxidative stability (i.e., the Y@C_{60} species not reacting with N_2O in the gas phase).[27]

2.2.2   ELECTRIC ARC (INERT ATMOSPHERE)

2.2.2.1   Discovery

A severe limitation to the laser ablation approach was the trace quantities of endohedrals being produced during the early days of research on endohedrals (1985–1991). In 1990, the breakthrough needed for synthesizing a much higher throughput of

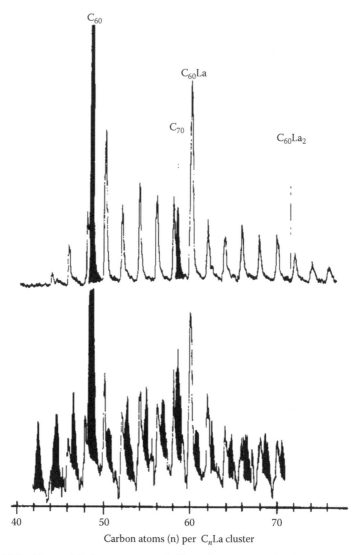

FIGURE 2.2 Time-of-flight mass spectra indicate the presence of LaC_{2n} endohedrals. Mass spectra were obtained at higher (top) or lower (bottom) fluences of the ionizing laser pulse. (Reprinted with permission from Heath, J. R. et al., *J. Am. Chem. Soc.*, 107, 7779–7780, 1985. Copyright 1985 American Chemical Society.)

metallofullerenes was the electric-arc method reported by Kratschmer–Huffman,[28,29] who used resistive heating to place carbon in the vapor phase for subsequent condensation into fullerenes.

The Kratschmer–Huffman approach[28,29] could be modified to introduce metal atoms into the plasma for endohedral metallofullerene production. In the early to mid-1990s with the electric-arc approach, one could vaporize (1) graphite rods soaked in a solution containing metal ions[30] or (2) a cored graphite rod packed with

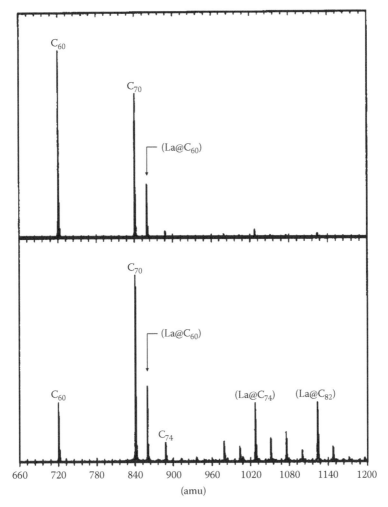

FIGURE 2.3 FT-ICR mass spectra of sublimed extract, which shows the presence of fullerenes and metallofullerenes. The pulsed decelerator was optimized for the C_{60}-C_{70} region (top) or the C_{84} region (bottom). (Reprinted with permission from Chai, Y. et al., *J. Phys. Chem.*, 95, 7564–7568, 1991. Copyright 1991 American Chemical Society.)

the desired metal or metal oxide.[31–37] A modified Kratschmer–Huffman apparatus for generating metallofullerenes is shown in Figure 2.4.

2.2.2.2 Early Studies

In 1991, Alvarez et al.[37] used the Kratschmer–Huffman approach to vaporize a mixture of La_2O_3 and graphite. Analysis of the produced soot extract demonstrated that the electric-arc method can generate an array of lanthanum-based endohedrals, of which $La_2@C_{80}$ was especially dominant. Another key outcome was demonstrating that multiple metals beyond two atoms (e.g., La_3C_{106}) could be encapsulated.[37]

 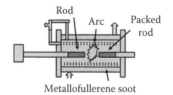 

FIGURE 2.4   Our electric-arc reactor (left), arc-plasma between the packed and solid graphite rod (right), and diagram showing the collected metallofullerene soot on the chamber walls (middle).

By 1992, Ross et al.[36] showed that the electric-arc production could also be used to encapsulate other metals when they added Y_2O_3, in addition to La_2O_3 and graphite, to a cored rod for vaporization. Their approach yielded extracts containing mixed-metal (e.g., $YLa@C_{80}$) as well as the already reported lanthanum-based endohedrals (e.g., $La@C_{82}$).[36] Not introducing La into the plasma resulted in extracts containing only yttrium endohedrals, as reported in 1992 by Shinohara et al.[35] and Weaver et al.[38]

Between 1992 and 1993, the synthesis of endohedral clusters was expanded to include Sc-based metallofullerenes (e.g., $Sc_2@C_{74}$) by using annealed, composite rods containing a mixture of Sc_2O_3 and graphite.[32–34] For lanthanum endohedrals, Bandow et al.[39] reported in 1992 the importance of anaerobic handling of endohedral extracts. In 1993, Bandow et al.[31] introduced the method of *back-burning* (i.e., switching the polarity of the anode and cathode) and improved the endohedral yield in soot extracts by vaporizing lanthanum carbide (LaC_2)-enriched rods.

2.2.2.3   Versatility

From 1993–Present, the electric-arc process for endohedral production has been overwhelmingly the preferred technique due to its (1) adaptability (i.e., ease of introducing solid and gas additives to the plasma), (2) ability to generate gram quantities of metallofullerene-containing soot extract per day, and (3) broad range of types of metallofullerenes produced. Under typical *inert* atmospheres (i.e., He, He/N_2 buffer gas), the electric-arc process creates mono-metal ($M@C_{2n}$), di-metal ($M_2@C_{2n}$), tri-metal ($M_3@C_{2n}$), carbide ($M_2C_2@C_{2n}$, $M_3C_2@C_{2n}$, $M_4C_2@C_{2n}$), cyano ($M_3NC@C_{2n}$), and metallic nitride ($M_3N@C_{2n}$) endohedral clusters.

2.2.2.4   Effect of Catalyst Additives

Despite use of the electric-arc method, the yield of endohedrals in soot extract during the early 1990s was still poor. Experiments were performed to evaluate the influence of *solid additives* into cored graphite rods. The idea was to search for an additive that would exhibit a catalytic effect on endohedral yields. In 2000, additives such as cobalt oxide were used in the synthesis of $Sc_3N@C_{2n}$, mixed-metal $ErSc_2N@C_{2n}$, and rare-earth $Er_3N@C_{2n}$ species.[40] In 2001, cobalt oxide was added to packed rods to improve the yield of $Sc_3N@C_{78}$[41] and again in 2002 to improve yields of $Sc_3N@I_h\text{-}C_{80}$ and $Lu_3N@I_h\text{-}C_{80}$.[42]

In 2000, Gu's research group[43] reported the use of MNi_2 alloys to increase the yield of metallofullerenes. In 2002, Yang's group demonstrated the catalytic effect of a $LaNi_2$ alloy additive to the packed rods as reported by Lian et al.[44] For these experiments, the selection of $LaNi_2$ instead of La_2O_3 was to exclude oxygen, known to decrease yield, from the electric-arc process. This $LaNi_2$ additive was noted to influence the ratio of the *minor* isomer of $La@C_{82}$ to the *major* isomer of $La@C_{82}$.[44] As shown in Figure 2.5, the HPLC data compare soot extracts prepared by vaporizing rods packed with La_2O_3 versus $LaNi_2$ additive, which improves the synthesis of $La@C_{82}$ isomers. In 2007, Stevenson introduced copper metal as an additive to increase the yield of $M_3N@C_{2n}$ species.[45]

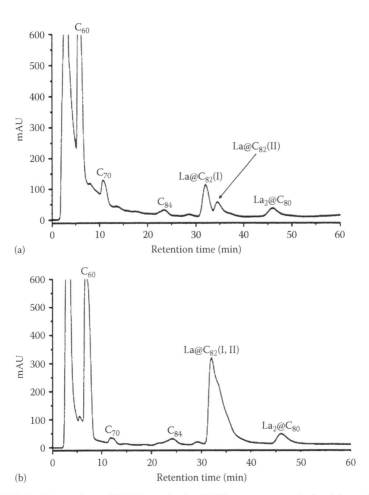

FIGURE 2.5 Comparison of HPLC results for DMF soot extracts obtained from electric-arc-vaporizing packed rods with (a) La_2O_3 or (b) $LaNi_2$ additives. (Reprinted with permission from Lian, Y. F. et al., *J. Phys. Chem. B*, 106, 3112–3117, 2002. Copyright 2002 American Chemical Society.)

2.2.3 ELECTRIC ARC (REACTIVE ATMOSPHERE)

2.2.3.1 Motivation

During the period of 1990–2004, the electric-arc process had two fundamental problems: (1) low yields of endohedrals (typically <1%) in soot extracts and (2) mixtures of more than 50 types of fullerene and metallofullerenes. These synthetic issues often made subsequent purification difficult. Even today, depending on reaction conditions, vaporizing Sc-packed rods can produce complex extracts containing the homologous series of empty cages (e.g., C_{60}-C_{140}) mixed with a diverse array of endohedral clusters, such as metal-only ($Sc@C_{2n}$), carbides ($Sc_2C_2@C_{2n}$), nitrides ($Sc_3N@C_{2n}$), and oxides ($Sc_4O_2@C_{80}$, $Sc_2O@C_{2n}$).[46,47] How this *separation nightmare* was overcome will be discussed in Section 2.3.

Figure 2.6 shows a typical matrix assisted laser desorption ionization (MALDI) mass spectrum of soot extract and indicates the presence of several types of Sc-based metallofullerenes. This mixture of endohedrals was created in our reactor under mild chemically adjusting plasma temperature, energy, and reactivity (CAPTEAR)[48] conditions. Section 2.2.3.6 discusses in more detail the CAPTEAR process and the effect of introducing chemically reactive species to the electric-arc process. Note that several types of metallofullerenes (i.e., carbide, nitride, and oxide clusters) were created from vaporizing a single packed rod in our electric-arc reactor.

Because soot extracts often contained too many species, scientists in the 2000s were increasingly motivated to pursue alternative synthetic approaches to selectively make endohedrals. New methods were sought to minimize the formation of empty-cage fullerenes, which dominated soot extracts from 1990 to 2004. The goal in selective synthesis is to avoid producing contaminant fullerenes or undesirable metallofullerenes in the reactor. The hypothesis involved creation of a harsh environment in the electric-arc reactor so that, if produced, the *weaker* fullerenes and more unstable endohedrals would

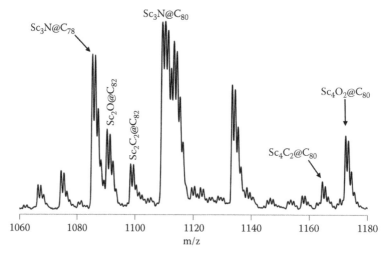

FIGURE 2.6 MALDI mass spectrum of our soot extract obtained under mild CAPTEAR conditions. Several families of endohedral metallofullerenes (e.g., carbides, nitrides, and oxides) are created.

react with a chemical additive and be rendered insoluble prior to soot extraction. These "survival of the fittest" conditions in the reactor ultimately led to simpler extracts with fewer types of fullerenes and fewer endohedrals for subsequent separation.

2.2.3.2   Effect of O_2 on Yield

From 1990 to 2004, the conventional wisdom in electric-arc generators was exclusion of O_2 and other reactive gases. During these early years, care was taken to seal the chambers and introduce only inert buffer gas (e.g., He). There was penalty of decreasing fullerene and endohedral yield if reactive species or air was added to the reactor.[49] Although the reduction in yield was a real consequence, the fear of using a harsh, reactive atmosphere in the electric-arc chamber during 1990–2004 delayed the discovery of new oxide clusters and the selective synthesis[48,50] of $M_3N@C_{2n}$ endohedrals.

2.2.3.3   Advantages

The advantages of replacing a typical reactor environment (e.g., He) with a more chemically reactive, reducing (NH_3, CH_4) or oxidizing atmosphere (NO_x) are discussed below. The impact on the field of endohedral synthesis is significant as these reactive atmospheres have permitted the synthesis and discovery of new clusters containing elements and species not yet encapsulated. Note the creation of new endohedrals has led to the necessity of novel separation schemes required for their purification (Section 2.3). A summary of the different reactor atmospheres and types of molecules formed, isolated, and purified is provided in Figure 2.7.

2.2.3.4   Ammonia, Dunsch Method for $M_3N@C_{2n}$

In 2004, Dunsch et al.[50] made two key discoveries during their search for an alternative synthesis of metallic nitride cluster endohedrals. Ammonia was added in small amounts (10–20 mbar) to the inert gas (He, 200 mbar). In these proportions, some fullerene and metallofullerene formation could still occur, despite the chemically harsh reactor conditions. Although the amount of soot extract decreased, the percentage of metallofullerenes increased dramatically from 3% to 5%[51] to as high as 90%[50] of endohedral species, as shown in Figure 2.8. With a much simpler extract, the problem of co-eluting fullerenes during chromatography was mitigated. This highly enriched endohedral sample would permit a more facile separation of $M_3N@C_{2n}$ endohedrals, which can be purified with only a single HPLC pass.[50]

Another outcome of the Dunsch work[50] was adding the nitrogen source to the packed rod. Until 2004, N_2 was introduced into the reactor from a gas cylinder. The typical additive to a cored graphite rod was a mixture of the metal or metal oxide,

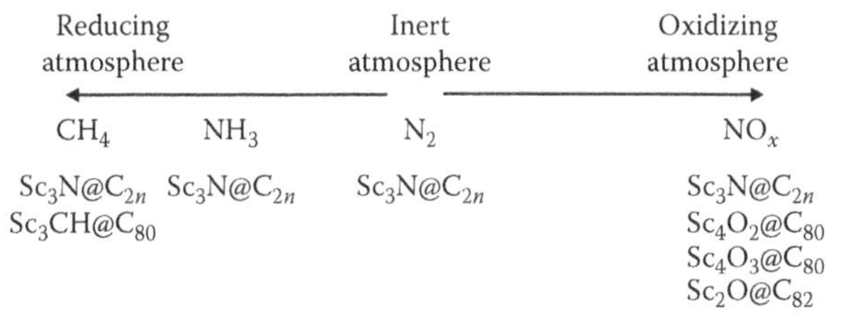

Reducing atmosphere		Inert atmosphere	Oxidizing atmosphere
CH_4	NH_3	N_2	NO_x
$Sc_3N@C_{2n}$	$Sc_3N@C_{2n}$	$Sc_3N@C_{2n}$	$Sc_3N@C_{2n}$
$Sc_3CH@C_{80}$			$Sc_4O_2@C_{80}$
			$Sc_4O_3@C_{80}$
			$Sc_2O@C_{82}$

FIGURE 2.7   Relationship between endohedrals produced versus type of reactor atmosphere.

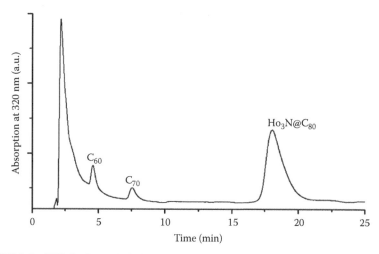

FIGURE 2.8 HPLC of metallofullerene soot extract obtained after vaporizing Ho-packed rods in a *reactive gas* (e.g., NH_3) atmosphere. (Reprinted from *J. Phys. Chem. Solids*, 65, Dunsch, L. et al., Endohedral nitride cluster fullerenes—formation and spectroscopic analysis of $L_{3-x}M_xN@C_{2n}$ ($0 <= x <= 3$; N = 39,40), 309–315, Copyright 2004, with permission from Elsevier.)

which contained the desired metal to be encapsulated. In 2004, the experiment by Dunsch et al.[50] to add calcium cyanamide to the packed rod represented a paradigm shift. Namely, the intent of the cyanamide additive was not to encapsulate Ca, but to use the *in situ* decomposition products of the cyanamide as the alternative source of nitrogen needed to create the $M_3N@C_{2n}$ species.[50] Studies had not yet been done to determine if yields of $M_3N@C_{2n}$ species could be improved with an alternative means of introducing N_2. With calcium cyanamide as the source of nitrogen, Dunsch improved the percentage yield of $Sc_3N@C_{80}$ in soot extract by at least a factor of ten.[50]

2.2.3.5 Selective Organic Solid Method for $M_3N@C_{2n}$

Since 2004, further experiments have investigated the efficacy of other nitrogen additives. In 2010, Dunsch et al.[52] showed that guanidinium salts packed in cored rods permitted a selective organic solid (SOS) approach to prepare $M_3N@C_{2n}$ species, including Sc, Y, Gd, Dy, and Dy/Sc mixed-metal endohedrals. The $M_3N@C_{2n}$ yields obtained from the SOS method were comparable to the reactive gas atmosphere approach with NH_3.[52] Shortly thereafter, improved yields of $Sc_3N@C_{2n}$ were demonstrated with urea, a cheaper source of nitrogen, as reported in 2012 by Jiao et al.[53] In 2013, Yang's group, as reported by Liu et al.,[54] performed a systematic study of 12 nitrogen-containing, inorganic solid compounds that were packed into cored rods containing Sc_2O_3. These nitrogen-containing additives varied in their oxidation states. As shown in Figure 2.9, the highest yield of $Sc_3N@C_{80}$ endohedral was obtained with NH_4SCN.[54]

2.2.3.6 NO_x, Stevenson Method (CAPTEAR) for $M_3N@C_{2n}$ and $M_xO_y@C_{2n}$

With the first $M_3N@C_{2n}$ synthesis performed in 1999[51] under *inert* conditions (i.e., He/N_2) and later in 2004[50] by Dunsch's group under a reducing, reactive environment

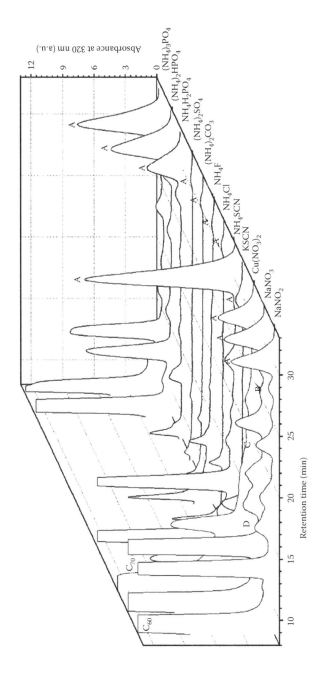

FIGURE 2.9 HPLC comparison of Sc metallofullerene soot extracts from vaporizing cored rods packed with a variety of inorganic additives containing N_2 at different oxidation states. (Reprinted with permission from Liu, F. P. et al., *Inorg. Chem.*, 52, 3814–3822, 2013. Copyright 2013 American Chemical Society.)

(i.e., He/NH_3), the question remained whether $M_3N@C_{2n}$ species could also form in an oxidizing, reactive atmosphere. If yes, what would be the $M_3N@C_{2n}$ yield and percentage composition in soot extracts?

To address these questions, Stevenson's group in 2007 reported the CAPTEAR method[48] for synthesizing endohedrals in NO_x oxidizing atmospheres. The CAPTEAR approach utilized an inorganic compound (e.g., NH_4NO_3 or copper nitrate hydrate), which was mixed with the desired metal or metal oxide (e.g., Sc_2O_3) to be encapsulated and was subsequently packed into a cored graphite rod for electric-arc vaporization. With no rod annealing necessary, the $Cu(NO_3)_2$ hydrate would decompose into NO_x gases, N_2, and O_2 for their *in situ* release into the electric-arc reactor. If NH_4NO_3 was used, NH_3 would also be introduced as a byproduct and plasma additive. With strategic amounts of added copper nitrate hydrate (i.e., 80%), one could produce $Sc_3N@C_{80}$ at 96% purity of the raw soot extract (Figure 2.10).[48] The higher purity but lower yield (i.e., milligrams of extract) was a compromise.

Another outcome of the CAPTEAR method and its oxidizing environment was the serendipitous discovery of a method to entrap oxygen inside the fullerene cage. Within the newly created family of scandium oxide endohedrals, the dominant metallic oxide is $Sc_4O_2@I_h-C_{80}$,[55] followed by $Sc_2O@C_s-C_{82}$,[56] with the lowest-yielding family member being $Sc_4O_3@I_h-C_{80}$.[57] An interesting feature of $Sc_4O_2@I_h-C_{80}$ is the mixed oxidation states of Sc^{2+} and Sc^{3+}.[55,58] The $Sc_4O_3@I_h-C_{80}$ set the world record for the most atoms (seven) in an entrapped fullerene cluster.[57] The CAPTEAR[48] approach also provides the versatility to synthesize many families of endohedrals in a single extract, such as metallic nitride aza-fullerenes (e.g., $M_3N@C_{79}N$),[59,60] in addition to $M_3N@C_{2n}$, mixed-metal $A_xB_y@C_{2n}$,[61] and $M_xO_y@C_{2n}$[55] endohedrals.

The oxygen source can be introduced into the plasma from a solid (e.g., inorganic salt) additive to a cored rod and/or from a gas (e.g., air that is mixed with He/N_2 buffer gas). There are now two book chapters[62,63] describing metal oxide EMFs. One is a 2011 review for $Sc_xO_y@C_{2n}$ species.[62] The second and more recent chapter (in 2014) contains the latest results for $Sc_xO_y@C_{2n}$ experiments and the discovery of the newest members of the metal oxide cluster fullerene family (e.g., transition metal Group IIIB, $Y_2O@C_{2n}$, and rare-earth, $Lu_2O@C_{2n}$).[63]

2.2.3.7 Methane for $M_3CH@C_{2n}$

In 2007, Krause et al.[64] explored additional reactive atmospheres to NH_3, which Dunsch's group had been using for several years. With a goal of observing the effect of methane on the type of metallofullerene produced, they analyzed their resulting soot extracts to determine if new molecules were created. By introducing CH_4 (10%) to He (90%) to create a reactive atmosphere, a new molecule, consistent with a formula $Sc_3CH@C_{80}$, was detected in their mass spectra, isolated, and characterized by HPLC (Figure 2.11).[64]

2.2.3.8 SO_2, Echegoyen's Method for $M_2S@C_{2n}$

In 2010, two research groups reported the first examples of putting sulfur inside a fullerene cage. Dunsch's group[65] introduced the source of sulfur by adding guanidinium thiocyanate ($CH_5N_3 \bullet HSCN$), which decomposed in the electric-arc reactor to create new $M_2S@C_{2n}$ molecules, along with $M_3N@C_{2n}$ endohedrals. Using this

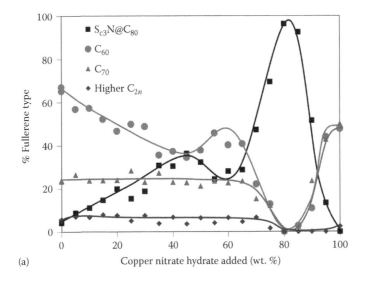

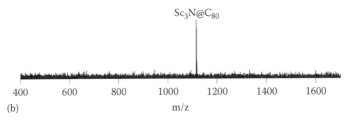

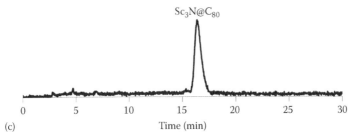

FIGURE 2.10 (a) Comparison of the type of fullerene produced (%) versus the percentage of copper nitrate hydrate in the packing material, (b) HPLC, and (c) MALDI mass spectrum indicating the presence and purity of $Sc_3N@C_{80}$ to be 96% of the soluble CS_2 soot extract. (Reprinted with permission from Stevenson, S. et al., *J. Am. Chem. Soc.*, 129, 16257–16262, 2007. Copyright 2007 American Chemical Society.)

method of solid organic additives, $Sc_2S@C_{3v}(8)-C_{82}$ was created and isolated.[65] Also in 2010, Echegoyen's group[66] used SO_2 in He as the source of sulfur needed to make $M_2S@C_{2n}$ (Figure 2.12).

Not only were metallic sulfide endohedrals produced via the SO_2 approach, but there were two structural isomers of $Sc_2S@C_{82}$ isolated when Chen et al.[66] reported the creation and purification of an additional isomer, $Sc_2S@C_s(6)-C_{82}$. Addition of

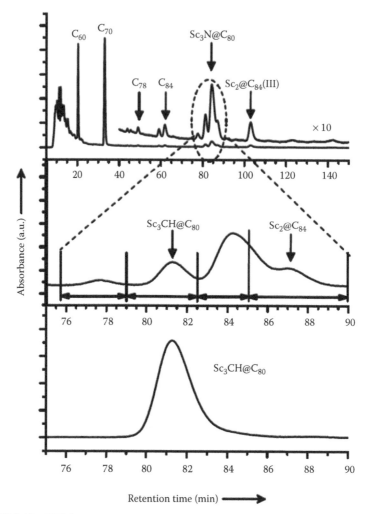

FIGURE 2.11 HPLC of soot extract obtained by vaporizing a Sc-packed rod in a reactive atmosphere of CH_4 (top), expansion of the region containing $Sc_3CH@C_{80}$ (middle), and chromatogram showing purified $Sc_3CH@C_{80}$ after fraction collection (bottom). (From Krause, M. et al., *Chemphyschem*, 2007, 8. 537–540. Copyright Wiley-VCH Verlag GmbH & Co. KGaA. Reproduced with permission.)

SO_2 gas into the reactor resulted in extracts containing significant quantities of C_{60} and C_{70}, in contrast to the Dunsch method of added NH_3,[50] which led to extracts with minimal presence of empty-cage fullerenes.

2.2.4 OTHER METHODS OF SYNTHESIS

2.2.4.1 Radio Frequency Furnace

Other attempts to produce endohedrals include a radio frequency (RF) furnace. Twenty years after the electric-arc method was introduced, in 2010, Krokos[67]

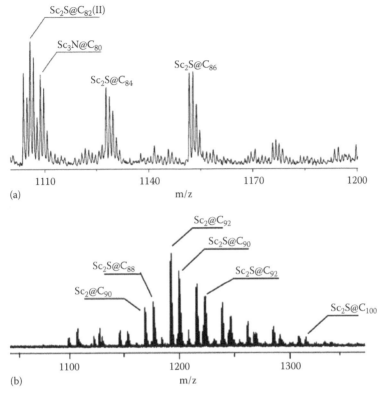

FIGURE 2.12 MALDI mass spectra of (a) 3rd and (b) 4th HPLC fractions, obtained from separating soot extract prepared by addition of SO_2 into an electric-arc reactor. (From Chen, N. et al., *Chem. Commun.*, 46: pp. 4818–4820. Copyright 2010 The Royal Society of Chemistry, reproduced with permission.)

introduced a synthetic approach using a plasma coupled radio frequency furnace (Figure 2.13). In this study, elements such as Ba, La, and Sr possess lower first ionization energies of 5.2, 5.6, and 5.7 eV, respectively, and were encapsulated at lower temperatures of 500–1000°C.[67]

The RF furnace alone had insufficient energy to thermally ionize elements with higher first ionization energies. For this reason, plasma-coupled RF fields were necessary to successfully create Sc-based endohedrals, such as Sc_mC_{2n} and $Sc_3N@C_{2n}$.[67] Under optimized conditions, scandium metallofullerenes were synthesized with suppression of higher empty-cage fullerenes. A dominant species produced and isolated using the plasma-coupled RF furnace was Sc_4C_{82},[67] which may be the carbide species, $Sc_4C_2@C_{80}$, whose X-ray crystal structure had been reported by Wang[68] one year earlier, in 2009.

2.2.4.2 Ion Bombardment

Ion bombardment, though rarely used, is another method of endohedral synthesis. This technique was used in 1996 to encapsulate nonmetal endohedrals, such as

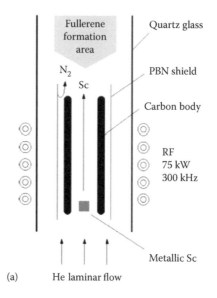

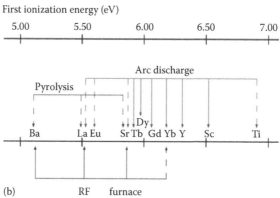

FIGURE 2.13 (a) Design of the production chamber of the RF furnace and (b) correlation of the first ionization energy, desired metal, and proposed synthetic approach for making endohedrals. (Reprinted with permission from Krokos, E., *J. Phys. Chem. C*, 114, 7626–7630, 2010. Copyright 2010 American Chemical Society.)

N@C$_{60}$, by bombarding C$_{60}$ with nitrogen ions from a plasma discharge ion source.[69] More than a decade later, the ion bombardment approach was utilized by Aoyagi et al.[70] in 2010 to produce a polar cationic Li@C$_{60}$ by irradiating Li^{+} ions, generated by contact ionization, onto a target plate while C$_{60}$ vapor was deposited. This approach represents an intriguing new way to synthesize small-cage (e.g., C$_{60}$) endohedrals containing metals with small ionic radii.

2.3   EXTRACTION AND SEPARATION

2.3.1   Pre-Cleanup and Enrichment Methods

2.3.1.1   Solubility and Extraction

Extensive solubility data for C_{60} in diverse solvents was reported in 1993 by Ruoff et al.[71] These results provided the background for subsequent efforts to purify empty-cage fullerenes (e.g., C_{60}) by solubility differences, temperature, and solvent selection.[72] Manipulating these parameters can serve as a pre-cleanup method to provide enriched metallofullerene samples, which are ready for subsequent separation (e.g., HPLC, selective chemical reduction).

Regardless of the synthesis method, metallofullerenes and fullerenes in soot must be extracted. Which solvent is best for extracting endohedrals? During the early 1990s, the solvents used for extracting empty-cage fullerenes were also selected for endohedral-containing soot. Common choices for metallofullerene extractions were aromatic solvents and carbon disulfide. Because these solvents had a high solubility for empty-cage fullerenes, the idea was to extend their use for endohedral extractions. Additional solubility data, specifically obtained for endohedrals, became available as time progressed. For example, dimethylformamide (DMF) was shown to be optimal for soot extractions of mono-metal endohedrals (i.e., $A@C_{2n}$), which were polar and suitable for DMF.[73-75]

Early extractions of empty-cage fullerenes from soot were reported in 1993 by Creasy et al.,[76] who evaluated an array of excellent solvents with high solubility for fullerenes. Extractions at high temperatures (i.e., reflux) and high pressures were performed for o-xylene, 1,2,5-trimethylbenzene, and 1-methylnaphthalene.[76] These solvents have since been used for extractions of metallofullerene soot. In 1996, Ding et al.[74] reported the use of DMF in a new and efficient method to obtain soot extracts enriched in mono-metallofullerenes (i.e., $Ce@C_{82}$) and with discrimination against solubilizing empty-cage fullerenes, C_{60}, C_{70}, and higher empty-cage fullerenes from soot.[74]

Work with DMF for metallofullerene soot extractions includes a 2002 study by Sun et al.,[73] who reported high-temperature extraction with DMF (Figure 2.14) to provide a sample with 15 times more metallofullerene (i.e., $Gd@C_{82}$) than would be obtained by conventional Soxhlet extraction. Similar results were obtained using high-temperature DMF extraction for other Gd-metallofullerenes and for extracting soot containing Tb, Y, and Sm endohedrals.[73]

2.3.1.2   Selective Extraction

Selective extraction of metallofullerenes was demonstrated in 2004 by Lian et al.,[75] who used a four-step extraction procedure using CS_2, followed by DMF, pyridine, and aniline. A variety of different endohedral soots containing Y, La, Nd, Sm, Gd, Tb, Yb, Ca, and Ba metallofullerenes were explored. Extracted samples, obtained from different solvents, showed an increased abundance of endohedral content. The differences in extracting ability among solvents were explained by back-donation from carbon cages to the metal(s) and by the number of electrons transferred to the cage.[75] Differences in the extraction behavior of divalent versus trivalent endohedrals were found with these solvents.[75]

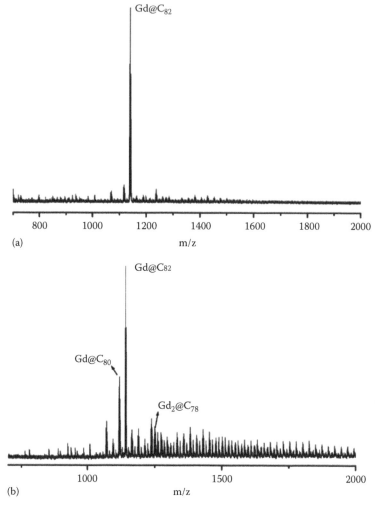

FIGURE 2.14 MALDI mass spectra of soot extracted with (a) toluene compared to (b) a high-temperature DMF extraction. (Reprinted from *Carbon*, 40, Sun, B. Y. et al., Improved extraction of metallofullerenes with DMF at high temperature, 1591–1595, Copyright 2002, with permission from Elsevier.)

In the early 1990s, 1,2,4-trichlorobenzene was found to possess high solubility for empty-cage fullerenes, and it was also used for extracting metallofullerene soot. Of particular interest, Akasaka et al.[23] in 2004 began using 1,2,4-trichlorobenzene (TCB) to extract empty-cage fullerenes and metallofullerenes from soot to perform a *reactive extraction*. TCB had previously been used in endohedral soot extractions due to its excellent ability to dissolve endohedrals. Akasaka et al.[23] manipulated the presence of dichlorophenyl radicals from TCB during their extraction conditions. In this way, several of the more reactive and *missing endohedrals* (i.e., C_{72}–C_{80} cages) were co-extracted and purified as their functionalized derivatives.[23]

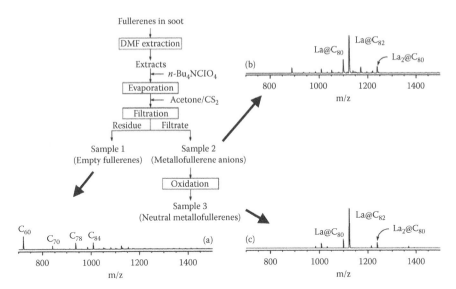

FIGURE 2.15 Separation scheme based on selective extraction and oxidation. LDI mass spectra correspond to (a) sample 1, (b) sample 2, and (c) sample 3 of the flowchart. (Reprinted with permission from Tsuchiya, T. et al., *J. Phys. Chem. B*, 110, 22517–22520, 2006. Copyright 2006 American Chemical Society.)

The concept of using selective extraction as a pre-cleanup stage has also been reported in 2006 by Tsuchiya et al.[77] In the first step, DMF was used to obtain a La@C_{2n}-enriched extract by manipulating solubility differences between empty-cage fullerenes and endohedrals. Subsequent, selective chemical reduction and oxidation of endohedral anions, followed by a single HPLC pass, led to purified La-metallofullerene samples, such as La@C_{82} isomers and La$_2$@C_{80}.[77] A summary of their separation scheme and laser desorption ionization (LDI) mass spectra corresponding to samples 1, 2, and 3 is shown in Figure 2.15.

2.3.1.3   Sublimation Enrichment

Another pre-cleanup approach for enriched samples of endohedrals is sublimation. This technique represents one of the earliest separation methods attempted for metallofullerenes. Seminal work, reported in 1993 by Yeretzian et al.,[78] exploited differences in sublimation temperatures of La endohedrals (e.g., La@C_{74} and La@C_{82}) versus empty-cage fullerenes. An advantage of sublimation is the ability to avoid the use and subsequent removal of extraction solvents.[78,79] This solvent-free approach eliminated the worry and exposure of endohedrals to nucleophilic, reactive solvents, such as amines.[79] A comparison of the mass spectra of soot extract, before and after sublimation, is shown in Figure 2.16.

Sublimation can also be used as the first step in a multistage isolation method. In 2008, Raebiger et al.[79] developed a separation approach that used sublimation as a pre-cleanup step for Gd-based endohedrals. The second stage was a selective extraction with o-dichlorobenzene (ODCB) to obtain a soluble portion, containing Gd@C_{82}

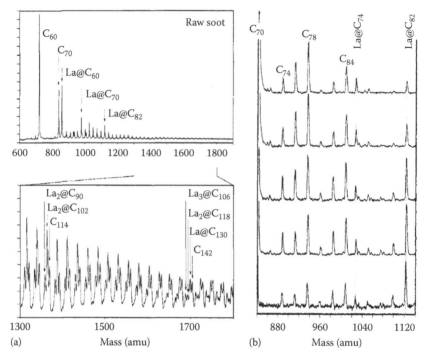

FIGURE 2.16   Laser desorption time-of-flight mass spectra of (a) soot extract versus (b) sublimed material. The spectra from top to bottom (b) represent colder to hotter regions where the sublimate was collected. (Reprinted with permission from Yeretzian, C. et al., *J. Phys. Chem.*, 97, 10097–10101, 1993. Copyright 1993 American Chemical Society.)

and C_{2n}.[79] The insoluble portion was enriched in lower-cage mono-metallofullerenes (e.g., $Gd@C_{74}$) and C_{74}. Both soluble and *insoluble* fractions were subject to selective chemical oxidation for obtaining enriched endohedrals.[79] A flowchart of their optimized process for separating a variety of gadolinium metallofullerenes is provided in Figure 2.17. A strong motivation for developing methods to isolate $Gd@C_{2n}$ endohedrals is their use as magnetic resonance imaging (MRI) contrast agents.[80–87]

2.3.1.4   Diels-Alder Approaches

It is well known that fullerenes and metallofullerenes react with dienes. Their reactivity differences can be the basis of a pre-cleanup method. In 2008, Angeli et al.[88] used the higher reactivity of empty-cage fullerenes and many endohedrals toward 9-methylanthracene to form functionalized species, which could then be separated from unreactive endohedrals by solubility differences. The recovered, unreacted species obtained after this Diels-Alder step were subject to column chromatography (i.e., step 2) to afford a sample enriched in $Sc_3N@C_{80}$ (60% purity).[88] A one-step HPLC pass of this enriched $M_3N@C_{2n}$ sample leads to purified $Sc_3N@C_{80}$. Figure 2.18 summarizes this enrichment step with 9-methylanthracene (stage 1) and subsequent HPLC injection (stage 2). The method was also applied to Lu-based extracts, from which an enriched sample of $Lu_3N@C_{80}$ and $Lu_2@C_{82}$ isomers was

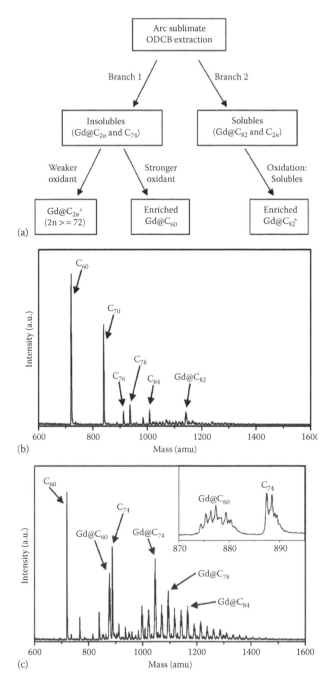

FIGURE 2.17   (a) Sublimation and oxidation flowchart for the separation of different classes (i.e., branches of metallofullerenes), and MALDI mass spectra of (b) branch 1 and (c) branch 2. (Reprinted with permission from Raebiger, J. W. and Bolskar, R. D., *J. Phys. Chem. C*, 112, 6605–6612, 2008. Copyright 2008 American Chemical Society.)

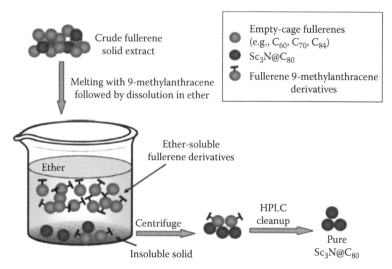

FIGURE 2.18 Separation scheme based on selective reactivity with 9-methylanthracene to provide an enriched $M_3N@C_{2n}$ sample (stage 1), further separated by HPLC for pure $Sc_3N@C_{80}$ (stage 2). (Reprinted with permission from Angeli, C. D. et al., *Chem. Mater.*, 20, 4993–4997, 2008. Copyright 2008 American Chemical Society.)

obtained. This mixture of metallofullerenes could then be further separated for subsequent purification.[88]

2.3.1.5 Uptake and Release from Aminosilica

Other pre-cleanup approaches as well have been explored. Selectively immobilizing fullerenes and endohedrals on to a reactive support, such as aminosilica or diaminosilica,[89,90] can be achieved by manipulating reactivity differences between empty-cage fullerenes, classical metallofullerenes (i.e., $M@C_{2n}$, $M_2@C_{2n}$), metallic nitride ($M_3N@C_{2n}$), and metallic oxide fullerenes ($M_xO_y@C_{2n}$). The kinetics of uptake is different among these families of endohedrals and empty-cage fullerenes.

Anhydrous solvents and vacuum oven-dried[91] aminosilica should immediately be added to solutions of metallofullerene extract when using the *stir and filter approach* (SAFA). Otherwise, *wet* conditions from the aminosilica and/or solvent can prevent the effective uptake of contaminant fullerenes, and the SAFA purification process would thus be affected.[91]

In the SAFA method, empty-cage fullerenes and classical metallofullerenes exhibit a higher reactivity toward amines and become immobilized onto aminosilica.[89–92] The more inert endohedrals, such as $M_3N@C_{2n}$, $M_xO_y@C_{2n}$, remain in solution. Many of the mono-metal, di-metal, and di-metal carbide species are reactive and bind to aminosilica. The reaction slurry is filtered at a strategic time point (i.e., monitor the desired endohedral peak area versus the loss of peak area from fullerene contaminants). In this way, enriched samples of desired metallofullerenes are readily obtained and saved for subsequent purification. A summary of the SAFA process is shown in Figure 2.19.[90]

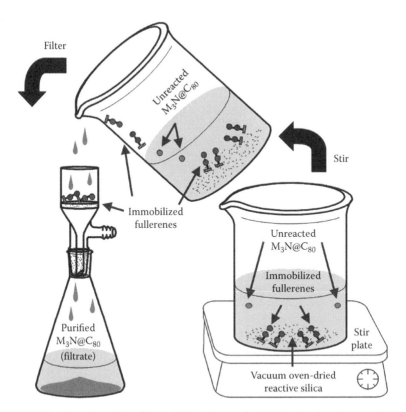

FIGURE 2.19 Overview of the *Stir and Filter Approach* (SAFA) showing the use of vacuum oven-dried reactive silica (e.g., diamino silica). (Reprinted with permission from Stevenson, S. et al., *J. Am. Chem. Soc.*, 128, 8829–8835, 2006. Copyright 2006 American Chemical Society.)

Stevenson et al. in 2008[55] manipulated the uptake differences onto aminosilica as a pre-cleanup step to obtain a simplified mixture of fewer endohedrals (five species) and empty-cage fullerenes (two species). A second stage of HPLC led to the isolation of the first metallic oxide endohedral, $Sc_4O_2@C_{80}$, as shown in Figure 2.20.[55]

A key question, however, lingered with the SAFA method. Can the fullerenes and endohedrals that were immobilized onto aminosilica be recovered? From the discovery of SAFA in 2006 until 2013, the spent aminosilica was chemical waste. This disposal was unfortunate because there were gram quantities of metallofullerenes and empty-cage fullerenes being discarded as waste. In 2014, Stevenson et al.[92] reported a chemical method to release these bound metallofullerenes and empty-cage fullerenes. An exciting discovery was the ability to manipulate the kinetics of release to obtain an array of fractions, with each sample containing a unique product distribution. Fractionated samples of Er- and Gd-based metallofullerenes were released and recovered from the spent aminosilica. In the Er endohedral experiments, about 70% of the immobilized Er metallofullerenes and empty-cage fullerenes were recovered from the spent aminosilica.[92] An overview of the selective uptake and release approach is described in Figure 2.21.

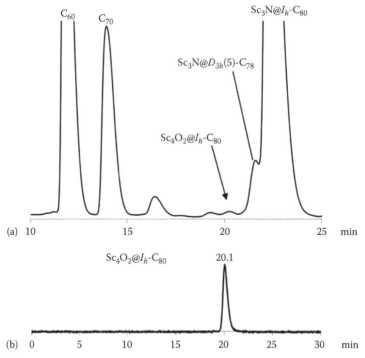

FIGURE 2.20 HPLC chromatograms of (a) SAFA filtrate after 2 h of uptake (stage 1) and (b) purified $Sc_4O_2@I_h\text{-}C_{80}$ by HPLC (stage 2). (Reprinted with permission from Stevenson, S. et al., *J. Am. Chem. Soc.*, 128, 8829–8835, 2006. Copyright 2008 American Chemical Society.)

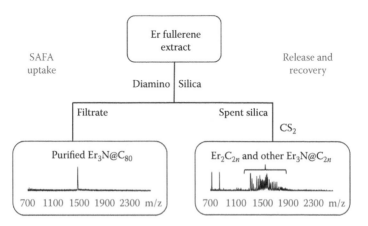

FIGURE 2.21 Overview of the fractionation method that uses selective fullerene uptake to purify $Er_3N@C_{80}$ (left) and subsequent release and recovery of immobilized endohedrals from spent waste aminosilica (right). (From Stevenson, S. et al., *Dalton Trans.*, 2014, 43, 7435–7441. Copyright 2014 The Royal Society of Chemistry, reproduced with permission.)

2.3.1.6 Lewis Acid Complexation–Decomplexation Enrichment

Another recent method for obtaining enriched samples of endohedrals is based on reactivity differences with Lewis acids. In this non-chromatographic approach, a Lewis acid is added to an extract solution. The reactive species are typically those endohedrals having lower first oxidation potentials. These metallofullerenes with the lowest first oxidation potentials are reactive toward Lewis acids, readily complex, and are more easily removed from solution via precipitation.[46,47,93,94] Other endohedrals and low molecular weight, empty-cage fullerenes with higher first oxidation potentials are more inert and can remain in solution (i.e., filtrate). In this manner, a crude separation and enrichment of endohedrals is quickly achieved.[46,47,93,94]

Manipulating the selective precipitation of reactive endohedrals with Lewis acids was recently published by Stevenson et al.,[47] who in 2009 used $AlBr_3$, $FeCl_3$, and $AlCl_3$ in a new method to separate metallofullerene soot extracts into two samples. The recovered precipitates contained endohedral metallofullerenes, whereas the filtrate consisted of unreacted, empty-cage fullerenes. This pre-cleanup method of complexation–decomplexation with $AlCl_3$ (step 1) followed by HPLC fraction collection (step 2) was used by Stevenson et al. to isolate the second and third members of the metallic oxide endohedrals, $Sc_4O_3@I_h\text{-}C_{80}$[57] and $Sc_2O@C_s\text{-}C_{82}$,[56] respectively. Figure 2.22 shows the metallofullerene-enriched sample from Lewis acid treatment (step 1) and subsequent purification of $Sc_4O_3@I_h\text{-}C_{80}$ by HPLC fraction collection (step 2).

In 2012, Shinohara's research group[93,94] used a different Lewis acid, $TiCl_4$, to quickly separate endohedral extract into two fractions. Akiyama et al.[94] obtained,

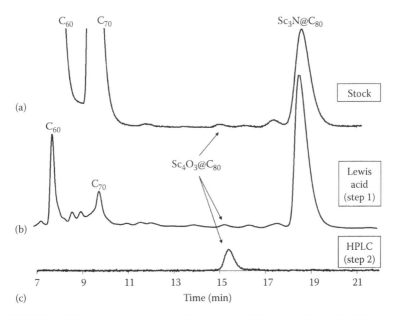

FIGURE 2.22 HPLC of (a) soot extract, (b) recovered fullerenes and metallofullerene after Lewis acid treatment (step 1), and (c) purified $Sc_4O_3@C_{80}$ after HPLC fraction collection (step 2). (From Mercado, B. Q. et al., *Chem. Commun.*, 2010, 46: pp. 279–281. Copyright 2010 The Royal Society of Chemistry, reproduced with permission.)

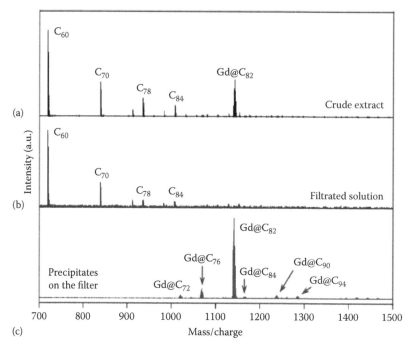

FIGURE 2.23 LDI mass spectra of (a) extract before reaction with TiCl$_4$ compared with the subsequent, (b) filtrate, and (c) recovered precipitate. (Reprinted with permission from Akiyama, K. et al., *J. Am. Chem. Soc.*, 134, 9762–9767, 2012. Copyright 2012 American Chemical Society.)

within only a few minutes of reaction, a sample of unreacted fullerenes (filtrate). A second sample contained decomplexed metallofullerenes, which had precipitated with TiCl$_4$ and had subsequently been recovered by addition of CS$_2$ and water. Figure 2.23 demonstrates that enriched samples of metallofullerenes can be obtained by leaving unreactive empty-cage fullerenes in solution.

In 2012, the origin and mechanism of endohedral reaction with TiCl$_4$ was reported by Wang et al.[93] They determined a threshold of reactivity for endohedrals with TiCl$_4$ to be between 0.62 and 0.72 V (vs Fc/Fc$^+$).[93] Endohedrals and higher empty-cage fullerenes possessing first oxidation potentials below this threshold would react and precipitate. Fullerene species having first oxidation potentials above this threshold would tend to remain in solution (i.e., filtrate).[93,94]

Whether it was TiCl$_4$,[93,94] or AlCl$_3$,[47] FeCl$_3$,[47] or AlBr$_3$,[47] the good news was the ability of Lewis acids to precipitate the entirety of endohedrals from soot extracts. The drawback was a limited selectivity among fellow metallofullerenes. Therefore, a weaker Lewis acid was sought that would have a precipitation threshold lower than these stronger Lewis acids. In 2013, Stevenson et al.[46] reported the Lewis acid, CuCl$_2$, which decreased the precipitation threshold from 0.62 to 0.72 V to an estimated first oxidation potential of 0.19 V (vs Fc/Fc$^+$).

The ability of CuCl$_2$ to react with much fewer endohedrals permitted fractionation among the different types of metallofullerenes. The use of CuCl$_2$ in a pre-cleanup

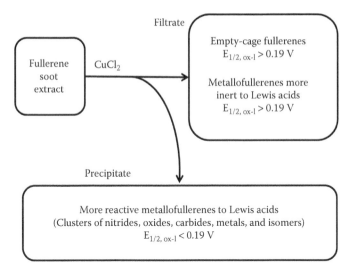

FIGURE 2.24 Overview of the use of $CuCl_2$ to selectively complex metallofullerenes with lower first oxidation potentials. (Reprinted with permission from Stevenson, S. and Rottinger, K. A., *Inorg. Chem.*, 52, 9606–9612, 2013. Copyright 2013 American Chemical Society.)

stage (step 1), followed by HPLC (step 2), led to the separation of $Er_2@C_{82}$ structural isomers. One $Er_2@C_{82}$ isomer was much more reactive to $CuCl_2$ and easily complexed and precipitated.[46] $CuCl_2$ selectively precipitates only the most reactive endohedrals from broad families of metallofullerenes, including metallic, metallic carbide, metallic oxide, and metallic nitride endohedrals (Figure 2.24).[46]

2.3.1.7 Electrochemical, Chemical Oxidation, and Chemical Reduction Methods

Empty-cage fullerenes, metallofullerenes, and their structural isomers have unique electrochemical properties. Knowledge of their oxidation and reduction potentials aids in the design of separation schemes, which are based on the ease or difficulty to reduce or oxidize a species. In 1998, Diener and Alford,[95] starting with sublimed extract, developed a scalable electrochemical process that was used to isolate small-band gap fullerenes (e.g., C_{74}, C_{80}-I_h) and endohedrals (e.g., $Gd@C_{74}$). These small-band gap, insoluble species may be put into solution as their anions by electrochemical reduction. These solubilized endohedral and fullerene anions may then be chemically reoxidized via addition of $FcPF_6$ to form precipitates.[95] This approach demonstrates a novel technique for obtaining enriched samples of endohedrals.

The 2000s represented a decade of advancing the use of selective reductions or selective oxidations of endohedrals. Sun et al.[96] reported in 2002 a way to manipulate solvent polarity and shift the reduction potentials to create a novel enrichment method for $Gd@C_{82}$ and $Gd_2@C_{80}$. By performing a reduction experiment directly with soot and using a strategic solvent mixture of THF/toluene (5:7), they successfully created an enriched sample of $Gd@C_{82}$ and $Gd@C_{80}$ by selectively producing anions of these species while not reducing C_{60}, C_{70}, and other fullerene contaminants.[96]

Bolskar and Alford[97] reported in 2003 a chemical oxidation method to selectively fractionate Gd endohedrals from arc-sublimed soot. The insoluble portion of the ODCB-extracted sublimate contained a rich array of lower cage $Gd@C_{2n}$ species (e.g., $Gd@C_{60}$, $Gd@C_{72}$, $Gd@C_{74}$). This selective chemical oxidation approach was also demonstrated with sublimed extracts of Tm endohedrals and empty cages.[97] Therein, an enriched sample of $Tm@C_{60}$ and $Tm@C_{70}$ was obtained by also using tris(4-bromophenyl)aminium hexachloroantimonate as the chemical oxidant.[97]

Selective extraction has also been demonstrated with a mixture of solvents. In 2005, Kodama et al.[98] used a mixed solvent of triethylamine and acetone to selectively extract Ce-based metallofullerenes. The triethylamine was added to reduce the endohedral to its anion or to form a donor–acceptor complex. Acetone was necessary to solvate the resulting anion or D–A complex. By mixing triethylamine and acetone together with soot, a single-step extraction process was performed.[98]

Another approach was reported shortly thereafter in 2004 by Tsuchiya et al.,[99] who extracted La-endohedral soot with 1,2,4-trichlorobenzene. Through selective electrochemical reduction, anions of La-metallofullerenes were separated from unreactive, neutral fullerenes and other endohedrals. Upon isolating the endohedral anions and with subsequent oxidation, highly enriched samples of $La@C_{82}$ and $La_2@C_{80}$ were obtained.[99] A comparison of this approach (10 mg of purified $La@C_{82}$, 1 day) versus the traditional method of HPLC purification (10 mg, 7 days) demonstrates the usefulness of this selective, electrochemical reduction method.[99]

A similar approach was used a year later for the selective reduction and extraction of $Gd@C_{82}$ and $Gd_2@C_{80}$ in 2005 by Lu et al.,[100] who selectively reduced and extracted $Gd@C_{82}$ and $Gd_2@C_{80}$ directly from soot as a separation and enrichment method (Section 2.3.1).

Given the desire to use $Gd@C_{2n}$ as MRI contrast agents, an optimized production and separation process was developed by Raebiger and Bolskar.[79] In 2008, they reported an improved approach to manipulate solubility and redox reactivity to obtain access to 90% of the Gd metallofullerenes generated by the arc process. In comparison with traditional methods, about 10 times more Gd endohedrals in soot could be obtained with their method. Advantages of their approach are (1) avoiding nucleophilic, reactive solvents (e.g., amines) and (2) being a non-chromatographic method.[79]

Selective chemical oxidation has recently been used for separating $Sc_3N@C_{78}$ and the structural isomers of $Sc_3N@D_{5h}$-C_{80} and $Sc_3N@I_h$-C_{80}. In 2013, Echegoyen's group[101] used an acetylferrocenium salt for the selective chemical oxidation of $Sc_3N@C_{78}$ and $Sc_3N@D_{5h}$-C_{80}, followed by subsequent column chromatographic separation and reduction with CH_3SNa.[101] Using the approach shown in Figure 2.25, Ceron et al.[101] reported the separation and isolation of $Sc_3N@I_h$-C_{80}, $Sc_3N@C_{78}$, and $Sc_3N@D_{5h}$-C_{80}.

2.3.1.8 Host–Guest Complexation

Another technique to obtain enriched samples of endohedrals is based on their unique, host–guest chemistry with azacrown ethers. In 2006, Akasaka, Nagase, and coworkers (Tsuchiya et al.)[102] reported $La@C_{82}$ complexation with 1,4,7,10,13,16-hexaazacyclooctadecane. Because empty-cage fullerenes have no encapsulated metal, they

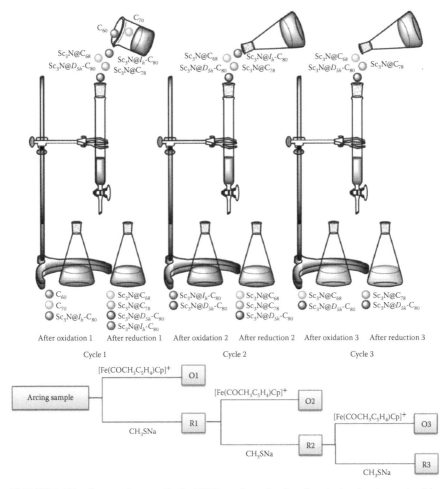

FIGURE 2.25 Approach to purify $Sc_3N@D_{5h}$ using selective chemical reduction and oxidation. (From Ceron, F. F. et al., *Chem. -A Eur. J.*, 2013. 19. 7410–7415. Copyright Wiley-VCH Verlag GmbH & Co. KGaA. Reproduced with permission.)

lack the charge transfer mechanism that metallofullerenes exhibit. Based on a lack of reactivity between azacrown ether and empty-cage fullerenes, a new method to separate and enrich metallofullerenes was developed.[102] From a solution of soot extract, La endohedrals (e.g., $La@C_{82}$ isomers, $La_2@C_{80}$) selectively precipitated upon addition of azacrown ether. Subsequent decomplexation with CS_2 and sonication permitted highly enriched, metallofullerene samples for these neutral endohedrals.[102]

2.3.2 HPLC PURIFICATION

The majority of non-chromatographic methods to purify individual metallofullerenes directly from extract arrived after 2005. In contrast, the initial separation strategies for endohedrals during 1990–1999 were based on HPLC. When metallofullerene soot extracts from the electric-arc process were first obtained in the early 1990s, there was excitement to have synthesized and extracted metallofullerenes. Enthusiasm grew as additional elements were being encapsulated. By 1992, the family of discovered metallofullerenes was expanded to several transition and rare-earth metals. But nobody knew how to isolate and purify them. At the time, metallofullerene extracts were viewed as an even more challenging and burdensome mixture to separate in comparison with extracts containing only empty-cage fullerenes.

HPLC efforts in the early 1990s focused on finding a suitable stationary phase and appropriate mobile phase for metallofullerenes. Separation attempts were based on transferring HPLC conditions that were developed for empty-cage fullerenes and extending their use toward endohedral extracts. Elution times for metallofullerenes in relation to empty-cage fullerenes were unknown. Because early endohedral extracts often contained less than 1% metallofullerenes (i.e., >99% empty-cage fullerenes), the extremely low abundance of endohedrals made it difficult in the 1990s to find their HPLC peaks. Co-elution of endohedrals with contaminant empty-cage fullerenes and their structural isomers (e.g., C_{84}) represented a significant technical hurdle. Note that during these early years, the specialty stationary phases (e.g., Buckyprep-M, Buckyprep, PYE, PBB) in use today were not available.

2.3.2.1 Automated HPLC

In the early 1990s, multiple research groups were simultaneously trying to isolate the first purified endohedrals. Two research groups (Shinohara and Dorn) in particular were each working on a method to purify metallofullerenes from their complex soot extracts. Because co-elution was a problem with HPLC-injected extracts, it was necessary during those years to develop a method with multiple types of columns, with each one having a different stationary phase or mode of separation.

In 1993, the first endohedrals to be isolated included La@C_{82}, as reported by Kikuchi et al.,[103] who used a two-stage HPLC approach. A polystyrene column and CS_2 eluent were used in a first step to obtain an enriched metallofullerene sample. A second stage used a Buckyclutcher column, comprised of a stationary phase developed by Welch and Pirkle.[104] This two-stage method demonstrated that endohedrals, such as La@C_{82}, were separable by HPLC.[103] Also in 1993, Shinohara et al.[34] purified the first scandium endohedrals using a similar two-stage approach. The first step used a preparative, recycling system with polystyrene polymer columns to obtain several fractions enriched in scandium endohedrals. Injection of collected samples into a Buckyclutcher column (i.e., stage 2) permitted purified samples corresponding to molecular formulas Sc_2C_{74}, Sc_2C_{82}, and Sc_2C_{84}.[34] Figure 2.26 shows the chromatograms from this two-stage separation.

To provide further insight into the debate of whether the metal atoms were exohedral or endohedral, Beyers et al.,[105] in 1994, purified via HPLC a sample corresponding to Sc_2C_{84}. This isolated sample was analyzed by electron diffraction and high-resolution TEM. An endohedral placement of the two Sc metal atoms was concluded.[105]

Meanwhile, Dorn's group in the early 1990s was also exploring a two-stage separation approach to isolate endohedrals. They were developing an initial step to separate the empty-cage fullerenes (99% of extract) from the endohedrals (1%), of which a

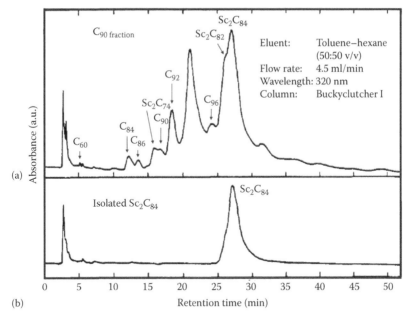

FIGURE 2.26 HPLC of the (a) C_{90} fraction after a first stage polystyrene pass to obtain a sample enriched in Sc metallofullerenes and (b) purified Sc_2C_{84} with a Buckyclutcher column, second stage. (Reprinted with permission from Shinohara, H. et al., *J. Phys. Chem.*, 97, 4259–4261, 1993. Copyright 1993 American Chemical Society.)

few species were electron paramagnetic resonance (EPR) active. In 1994, Stevenson et al.[106] reported a method to use EPR as an online HPLC detector to track the elution of EPR-active metallofullerenes by direct observation of their spectra in real time. As a pre-cleanup stage, metallofullerene extract was injected into two polystyrene columns connected in series. The resulting chromatogram, however, showed only a single, broad peak with a tail, which was the elution region for the EPR-active endohedrals (Figure 2.27).[106] By tracking the online spectra as a function of retention time, they knew when to initiate fraction collection. Continual collection and reinjection of this EPR-active fraction ultimately led to an endohedral sample enriched in a paramagnetic endohedral (e.g., $Y@C_{82}$) and ready for stage 2 (HPLC purification with a different stationary phase). This online HPLC-UV-EPR process has been useful in separating $Sc@C_{82}$,[106,107] $Y@C_{82}$,[106,107] and $La@C_{82}$[108] extracts.

Online EPR detection was advantageous for several reasons. It eliminated the need for solvent removal and avoided the time-consuming, offline EPR analysis of collected fractions, which were dilute and often required solvent removal. Due to instabilities of paramagnetic endohedrals (i.e., $M@C_{2n}$) in air,[39,109] the *in situ*, online EPR monitoring was especially amenable to anaerobic conditions.

For yttrium endohedral extracts, a convenient metallofullerene to monitor by online HPLC-EPR[106,107] was YC_{82}, which had been previously shown to be EPR active.[35,110] For scandium extracts, both ScC_{82} (8 lines) and Sc_3C_{82} (22 lines) were EPR active.[32,33,111] In the second separation stage, a Trident-Tri-DNP (Buckyclutcher) column was used, also with online EPR detection, in the purification of an endohedral

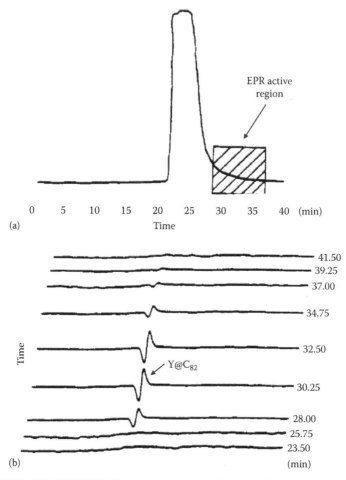

FIGURE 2.27 HPLC-UV-EPR (a) chromatogram and (b) spectra demonstrating the benefit of using online EPR detection to monitor the elution of paramagnetic endohedrals. Fraction collection begins when the EPR signal emerges. (Reprinted with permission from Stevenson, S. et al., *Anal. Chem.*, 66, 2680–2685, 1994. Copyright 1994 American Chemical Society.)

sample with a chemical formula, Sc_3C_{82}.[106] Likewise, the online HPLC-UV-EPR approach has been used for second stage Buckyclutcher injections for EPR-active yttrium metallofullerenes. Based on the online EPR spectra shown in Figure 2.28, an assignment of Y@C82, the EPR-active peak, can be made to HPLC fraction 5.[106]

Because the first separation stage involved multiple injections, collections, and repeated reinjections into the polystyrene column to obtain a sample enriched in metallofullerenes, Stevenson et al.[107] developed an automated HPLC separation for stage 1 (i.e., polystyrene pre-cleanup) for yttrium and scandium metallofullerene extracts. This automation[107] of polystyrene injections, done without use of a computer, permitted unattended separations and a more rapid isolation of individual metallofullerenes. This approach was also used for purifying an endohedral with a formula corresponding to Sc_2C_{84}.[105]

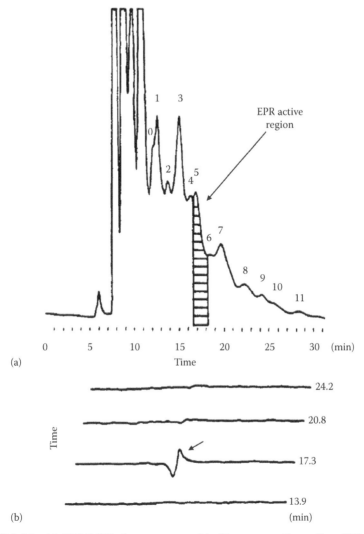

FIGURE 2.28 (a) HPLC-UV chromatogram with (b) corresponding online, HPLC-EPR spectra for the second stage, separation step with the Buckyclutcher column. The injected sample in (a) was the enriched EPR active fraction from the first stage polystyrene separations. (Reprinted with permission from Stevenson, S. et al., *Anal. Chem.*, 66, 2680–2685, 1994. Copyright 1994 American Chemical Society.)

2.3.2.2 Multicolumn HPLC

By the mid-1990s, a clear need was emerging for stationary phases that were designed for fullerenes and metallofullerenes. A goal was to develop a stationary phase which would strongly interact with fullerenes and endohedrals to permit the use of mobile phases possessing a high solubility for these species. Kimata et al.[112] reported in 1995 an array of candidate stationary phases. Of particular interest was the pentabromo-benzyl (PBB) column, which almost 20 years later is still used in endohedral and

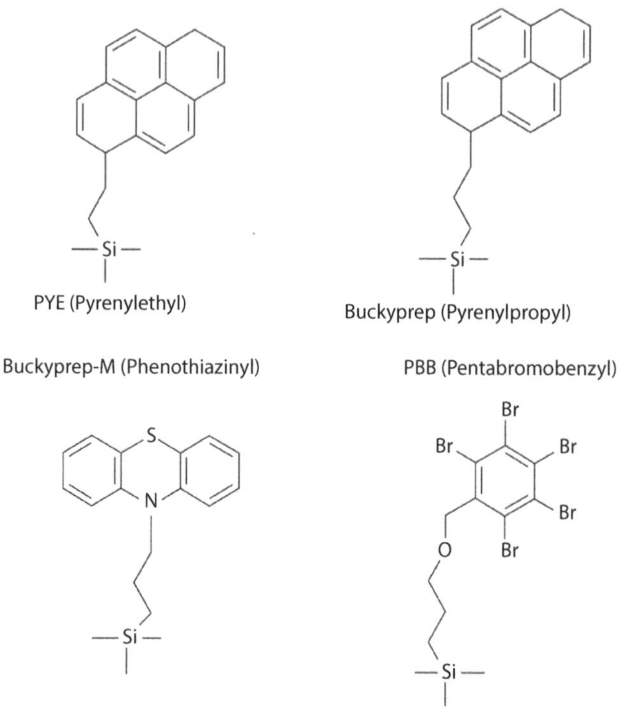

FIGURE 2.29 Popular stationary phases developed for HPLC separations of metallofullerenes.

fullerene separations. The structures for several key stationary phases, developed for HPLC separations of empty-cage fullerenes and metallofullerenes, are provided in Figure 2.29. Two companies that made early discoveries and contributions to developing stationary phases were Regis and Nacalai Tesque.

An advantage of the PBB column is the high sample loading that can be injected and separated with this column. Tanaka's group[112] also described in their seminal 1995 research a 2-[(1-pyrenyl)ethyl]silyl (PYE) stationary phase, which can separate structural isomers. Isomers of empty-cage fullerenes (e.g., C_{84}),[113] classical metallofullerenes (e.g. $Er_2@C_{82}$),[46,114] and metallic nitride endohedrals (e.g., $Sc_3N@I_h\text{-}C_{80}$ and $Sc_3N@D_{5h}\text{-}C_{80}$)[89] may be separated and purified with the PYE column.

2.3.2.3 Recycling HPLC

From 2000 till date, recycling HPLC and the strategic use of specialty columns (e.g., Buckyprep, Buckyprep-M, PBB, PYE) has led to numerous papers. The recycling HPLC approach has permitted the isolation of new endohedrals that have retention times too similar for conventional fraction collection. In addition to separating M_3N clusters in interesting cages,[115] recycling HPLC with specialty columns for metallofullerene purification includes metallofullerenes with the following endohedral clusters: MC_2,[21,116–126] MCH,[64] MNC,[127–130] M_2S,[65,66,131–133] M_3,[134–136] M_2,[134,137–141] and M.[134,142,143] For extracts containing mixed-metal endohedral clusters, a separation method using recycling

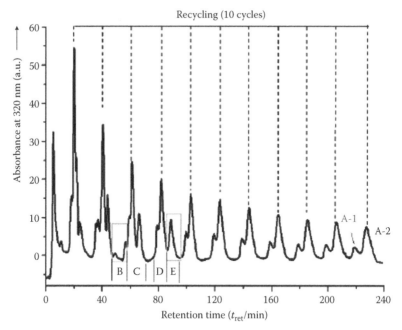

FIGURE 2.30 Use of recycling HPLC to separate $Sc_2S@C_{82}$ (dominant species in A-1) from contaminants (A-2, B, C, D, and E). (Reprinted with permission from Dunsch, L. et al., *J. Am. Chem. Soc.*, 132, 5413–5421, 2010. Copyright 2010 American Chemical Society.)

HPLC has been especially useful.[144–148] As shown in Figure 2.30, recycling HPLC was beneficial in the separation and purification of the newly discovered $Sc_2S@C_{82}$, whose retention time was very similar to that of several contaminant endohedrals.[65]

2.3.3 NON-CHROMATOGRAPHIC PURIFICATION FROM SOOT EXTRACT

Only non-chromatographic methods (i.e., no HPLC) that begin with soot extract and result in isolated samples with >95% purity are discussed in this section. Other approaches are described in the Section 2.3.1

2.3.3.1 Chemical Reduction

Fifteen years after the first endohedrals were produced and found in soot, a non-chromatographic method was reported in 2005 by Lu et al.,[100] who developed an approach whereby soot was extracted with DMF to increase the Gd endohedral content and minimize the empty-cage fullerenes. They performed a selective reduction directly with the mixture of fullerenes and endohedrals in the soot. $Gd@C_{82}$ and $Gd_2@C_{80}$ were selectively reduced and extracted with toluene/THF. As demonstrated in Figure 2.31, strategically varying the ratio of toluene versus THF permits an isolation of $Gd_2@C_{80}$ anion with a purity of >95% without using a HPLC method.[100]

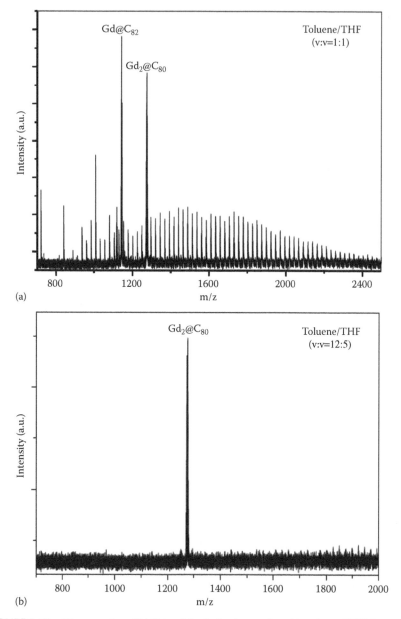

FIGURE 2.31 Mass spectra of (a) Gd endohedral anions reduced in toluene/THF (v:v = 1:1) compared to (b) the strategic use of a different ratio of toluene/THF (v:v = 12:5) for the isolation of $Gd_2@C_{80}$ anion. (Reprinted from *Carbon*, 43, Lu, X. et al., Selective reduction and extraction of $Gd@C_{82}$ and $Gd_2@C_{80}$ from soot and the chemical reaction of their anions, 1546–1549, Copyright 2005, with permission from Elsevier.)

2.3.3.2   Chemical Oxidation

In 2013, Echegoyen's group demonstrated a purification method for isolating $Sc_3N@$ I_h-C_{80} from soot extract, as reported by Ceron et al.[101] Their isolation approach is based on differences among the first oxidation potentials for Sc metallofullerenes. In addition to purifying the D_{5h} isomer of $Sc_3N@C_{80}$, the I_h isomer is readily obtained by selective chemical oxidation using an acetylferrocenium salt (i.e., $[Fe(COCH_3C_5H_4)Cp]^+$), as shown in Figure 2.32.[101]

2.3.3.3   Reactive Solid Supports

In 2005, a non-chromatographic method to purify endohedrals was developed by Dorn's group,[149] who took advantage of the high reactivity of empty cages and many metallofullerenes with cyclopentadiene. By placing the diene onto a solid support (i.e., Merrifield resin), empty-cage fullerenes and reactive endohedrals could leave solution and become immobilized.[149] The most inert species would remain unreacted. This separation approach was demonstrated by Ge et al.[149] in the isolation of metallic nitride fullerenes. Examples of purified endohedrals, isolated from their soot extracts, included $Sc_3N@C_{80}$ (Figure 2.33), $Lu_3N@C_{80}$, and $Gd_3N@C_{80}$.[149] The significance of non-chromatographic isolation strategies for $Lu_3N@C_{80}$ and $Gd_3N@C_{80}$ is their use in medical applications.[150–155]

In 2006, Stevenson et al.[90] synthesized an array of functionalized aminosilica gels and used reactive silica as a solid support. In addition to its use as an endohedral enrichment method (Section 2.3.1.5), SAFA could also isolate $Sc_3N@I_h$-C_{80} directly from soot extract.[90] The $Sc_3N@I_h$-C_{80} sample obtained from SAFA was isomerically purified because its other isomer, $Sc_3N@D_{5h}$-C_{80}, is removed from solution by its immobilization onto aminosilica.[89] More recently, in 2014, using vacuum oven-dried, diamino silica, $Er_3N@I_h$-C_{80} (Figure 2.34) has also been purified directly from soot extract by the SAFA method.[92] See Section 2.3.1.5 for the importance[91] of anhydrous solvents and adding vacuum oven-dried, amino-based silica gels immediately prior to use for purifying endohedrals with SAFA.

2.3.3.4   Lewis Acids

Another non-chromatographic method which can isolate metallofullerenes from soot extract is selective complexation with Lewis acids. In 2009, Stevenson et al.[47] reported a multistage separation approach with Lewis acids. By manipulating reaction parameters, such as time, amounts, and strength of Lewis acid (i.e., $AlCl_3$ versus $FeCl_3$), isomerically purified $Sc_3N@I_h$-C_{80} from extract was obtained (Figure 2.35).[47]

Shortly thereafter, another Lewis acid was found to purify endohedrals from soot extract and was reported in 2012 by Akiyama et al.[94] Addition of $TiCl_4$ to a solution of soot extract resulted in the selective precipitation of $Gd@C_{82}$ after only 10 minutes of reaction time. Decomplexation of the collected precipitate led to an isolated sample of $Gd@C_{82}$ with >99% purity, as shown in Figure 2.36. Key features of this work were direct isolation of $Gd@C_{82}$ from soot extract and lesser time for purification.[94]

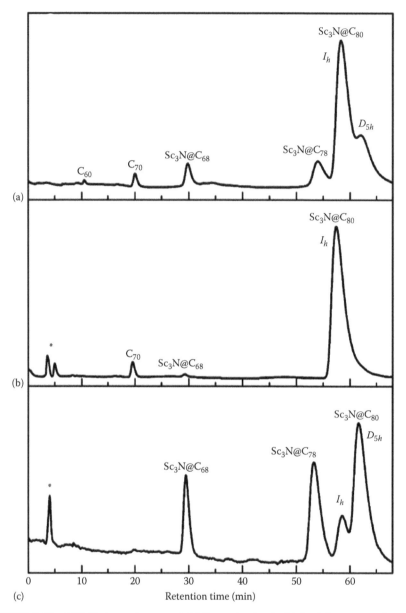

FIGURE 2.32 HPLC chromatogram of (a) Sc soot extract obtained from a reactive gas atmosphere, NH_3, (b) eluted, purified $Sc_3N@I_h$-C_{80} after selective chemical oxidation, and (c) endohedrals recovered and eluted with the reductant CH_3SNa. (Ceron, M. R. et al., *Chem. —A Eur. J.*, 2013. 19. 7410–7415. Copyright Wiley-VCH Verlag GmbH & Co. KGaA. Reproduced with permission.)

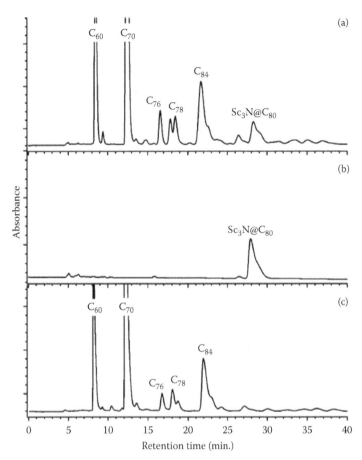

FIGURE 2.33 HPLC traces of (a) Sc metallofullerene extract, (b) toluene eluent from a column of cyclopentadiene-functionalized resin, and (c) fullerene released and recovered from the spent resin after heat treatment with maleic anhydride. (Reprinted with permission from Ge, Z., Duchamp, J. C., Cai, T., Gibson, H. W., and Dorn, H. C., *J. Am. Chem. Soc.*, 127, 16292–16298, 2005. Copyright 2005 American Chemical Society.)

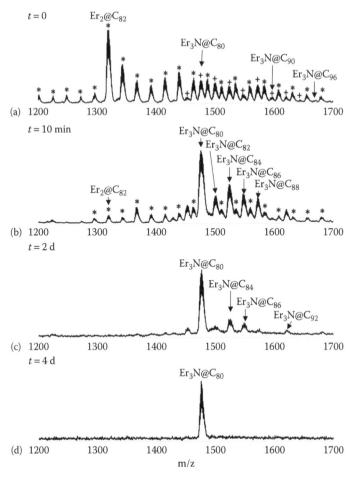

FIGURE 2.34   Aliquots of the SAFA uptake were monitored by HPLC for (a) Er metallofullerene extract before SAFA, (b) after 10 min, (c) after 2 days, and (d) purified Er$_3$N@I_h-C$_{80}$ after 4 days of SAFA uptake of contaminants. (Reproduced from Stevenson, S., Rottinger, K. A., and Field, J. S., *Dalton Trans*, 2014, 43: pp. 7435–7441. Copyright 2014 The Royal Society of Chemistry, reproduced with permission.)

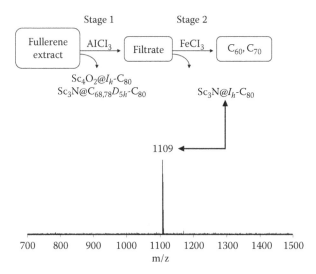

FIGURE 2.35 Overview of the nonchromatographic isolation of $Sc_3N@I_h\text{-}C_{80}$ from Sc metallofullerene soot extract using $AlCl_3$ (round 1) and $FeCl_3$ (round 2). (Reproduced and adapted with permission from Stevenson, S., Mackey, M. A., Pickens, J. E., Stuart, M. A., Confait, B. S., and Phillips, J. P., *Inorg. Chem.*, 48, 11685–11690, 2009. Copyright 2009 American Chemical Society.)

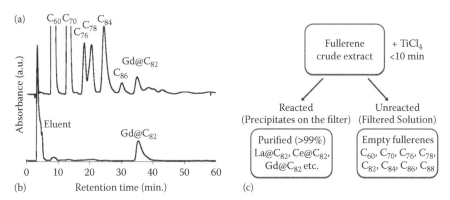

FIGURE 2.36 HPLC chromatograms of (a) Gd soot extract before, (b) after 10 min of reaction with TiCl4 using (c) their purification scheme. (Reprinted with permission from Akiyama, K., Hamano, T., Nakanishi, Y., Takeuchi, E., Noda, S., Wang, Z. Y., Kubuki, S., and Shinohara, H., *J. Am. Chem. Soc.*, 134, 9762–9767, 2012. Copyright 2012 American Chemical Society.)

ACKNOWLEDGMENT

The author thanks the National Science Foundation (RUI Grant 1151668) for financial support.

REFERENCES

1. L. Dunsch and S. F. Yang, *Physical Chemistry Chemical Physics*, 2007, **9**, 3067–3081.
2. L. Dunsch and S. Yang, *Small*, 2007, **3**, 1298–1320.
3. R. Kitaura and H. Shinohara, *Japanese Journal of Applied Physics, Part 1: Regular Papers Brief Communications & Review Papers*, 2007, **46**, 881–891.
4. M. N. Chaur, A. J. Athans, and L. Echegoyen, *Tetrahedron*, 2008, **64**, 11387–11393.
5. M. N. Chaur, F. Melin, A. L. Ortiz, and L. Echegoyen, *Angewandte Chemie-International Edition*, 2009, **48**, 7514–7538.
6. H. Shinohara, *Reports on Progress in Physics*, 2000, **63**, 843–892.
7. R. Valencia, A. Rodriguez-Fortea, A. Clotet, C. de Graaf, M. N. Chaur, L. Echegoyen, and J. M. Poblet, *Chemistry—A European Journal*, 2009, **15**, 10997–11009.
8. A. A. Popov, *Journal of Computational and Theoretical Nanoscience*, 2009, **6**, 292–317.
9. M. Yamada, T. Akasaka, and S. Nagase, *Accounts of Chemical Research*, 2010, **43**, 92–102.
10. X. Lu, T. Akasaka, and S. Nagase, *Chemical Communications*, 2011, **47**, 5942–5957.
11. A. Rodriguez-Fortea, S. Irle, and J. M. Poblet, *Wiley Interdisciplinary Reviews—Computational Molecular Science*, 2011, **1**, 350–367.
12. G. Q. Liu, Y. M. Wu, and K. Porfyrakis, *Current Organic Chemistry*, 2011, **15**, 1197–1207.
13. S. F. Yang, F. P. Liu, C. B. Chen, M. Z. Jiao, and T. Wei, *Chemical Communications*, 2011, **47**, 11822–11839.
14. A. Rodriguez-Fortea, A. L. Balch, and J. M. Poblet, *Chemical Society Reviews*, 2011, **40**, 3551–3563.
15. S. F. Yang, *Current Organic Chemistry*, 2012, **16**, 1079–1094.
16. X. Lu, L. Feng, T. Akasaka, and S. Nagase, *Chemical Society Reviews*, 2012, **41**, 7723–7760.
17. A. A. Popov, S. M. Avdoshenko, A. M. Pendas, and L. Dunsch, *Chemical Communications*, 2012, **48**, 8031–8050.
18. D. M. Rivera-Nazario, J. R. Pinzon, S. Stevenson, and L. A. Echegoyen, *Journal of Physical Organic Chemistry*, 2013, **26**, 194–205.
19. J. Y. Zhang, S. Stevenson, and H. C. Dorn, *Accounts of Chemical Research*, 2013, **46**, 1548–1557.
20. H. L. Cong, B. Yu, T. Akasaka, and X. Lu, *Coordination Chemistry Reviews*, 2013, **257**, 2880–2898.
21. X. Lu, T. Akasaka, and S. Nagase, *Accounts of Chemical Research*, 2013, **46**, 1627–1635.
22. A. A. Popov, S. F. Yang, and L. Dunsch, *Chemical Reviews*, 2013, **113**, 5989–6113.
23. T. Akasaka and X. Lu, *Chemical Record*, 2012, **12**, 256–269.
24. H. W. Kroto, J. R. Heath, S. C. Obrien, R. F. Curl, and R. E. Smalley, *Nature*, 1985, **318**, 162–163.
25. J. R. Heath, S. C. Obrien, Q. Zhang, Y. Liu, R. F. Curl, H. W. Kroto, F. K. Tittel, and R. E. Smalley, *Journal of the American Chemical Society*, 1985, **107**, 7779–7780.
26. Y. Chai, T. Guo, C. M. Jin, R. E. Haufler, L. P. F. Chibante, J. Fure, L. H. Wang, J. M. Alford, and R. E. Smalley, *Journal of Physical Chemistry*, 1991, **95**, 7564–7568.
27. S. W. McElvany, *Journal of Physical Chemistry*, 1992, **96**, 4935–4937.
28. W. Kratschmer, K. Fostiropoulos, and D. R. Huffman, *Chemical Physics Letters*, 1990, **170**, 167–170.

29. W. Kratschmer, L. D. Lamb, K. Fostiropoulos, and D. R. Huffman, *Nature*, 1990, **347**, 354–358.
30. D. W. Cagle, T. P. Thrash, M. Alford, L. P. F. Chibante, G. J. Ehrhardt, and L. J. Wilson, *Journal of the American Chemical Society*, 1996, **118**, 8043–8047.
31. S. Bandow, H. Shinohara, Y. Saito, M. Ohkohchi, and Y. Ando, *Journal of Physical Chemistry*, 1993, **97**, 6101–6103.
32. C. S. Yannoni, M. Hoinkis, M. S. Devries, D. S. Bethune, J. R. Salem, M. S. Crowder, and R. D. Johnson, *Science*, 1992, **256**, 1191–1192.
33. H. Shinohara, H. Sato, M. Ohkohchi, Y. Ando, T. Kodama, T. Shida, T. Kato, and Y. Saito, *Nature*, 1992, **357**, 52–54.
34. H. Shinohara, H. Yamaguchi, N. Hayashi, H. Sato, M. Ohkohchi, Y. Ando, and Y. Saito, *Journal of Physical Chemistry*, 1993, **97**, 4259–4261.
35. H. Shinohara, H. Sato, Y. Saito, M. Ohkohchi, and Y. Ando, *Journal of Physical Chemistry*, 1992, **96**, 3571–3573.
36. M. M. Ross, H. H. Nelson, J. H. Callahan, and S. W. McElvany, *Journal of Physical Chemistry*, 1992, **96**, 5231–5234.
37. M. M. Alvarez, E. G. Gillan, K. Holczer, R. B. Kaner, K. S. Min, and R. L. Whetten, *Journal of Physical Chemistry*, 1991, **95**, 10561–10563.
38. J. H. Weaver, Y. Chai, G. H. Kroll, C. Jin, T. R. Ohno, R. E. Haufler, T. Guo et al. *Chemical Physics Letters*, 1992, **190**, 460–464.
39. S. Bandow, H. Kitagawa, T. Mitani, H. Inokuchi, Y. Saito, H. Yamaguchi, N. Hayashi, H. Sato, and H. Shinohara, *Journal of Physical Chemistry*, 1992, **96**, 9609–9612.
40. M. M. Olmstead, A. de Bettencourt-Dias, J. C. Duchamp, S. Stevenson, H. C. Dorn, and A. L. Balch, *Journal of the American Chemical Society*, 2000, **122**, 12220–12226.
41. M. H. Olmstead, A. de Bettencourt-Dias, J. C. Duchamp, S. Stevenson, D. Marciu, H. C. Dorn, and A. L. Balch, *Angewandte Chemie-International Edition*, 2001, **40**, 1223–1225.
42. S. Stevenson, H. M. Lee, M. M. Olmstead, C. Kozikowski, P. Stevenson, and A. L. Balch, *Chemistry—A European Journal*, 2002, **8**, 4528–4535.
43. Y. F. Lian, Z. J. Shi, X. H. Zhou, X. R. He, and Z. N. Gu, *Carbon*, 2000, **38**, 2117–2121.
44. Y. F. Lian, S. F. Yang, and S. H. Yang, *Journal of Physical Chemistry B*, 2002, **106**, 3112–3117.
45. S. Stevenson, M. A. Mackey, M. C. Thompson, H. L. Coumbe, P. K. Madasu, C. E. Coumbe, and J. P. Phillips, *Chemical Communications*, 2007, **41**, 4263–4265.
46. S. Stevenson and K. A. Rottinger, *Inorganic Chemistry*, 2013, **52**, 9606–9612.
47. S. Stevenson, M. A. Mackey, J. E. Pickens, M. A. Stuart, B. S. Confait, and J. P. Phillips, *Inorganic Chemistry*, 2009, **48**, 11685–11690.
48. S. Stevenson, M. C. Thompson, H. L. Coumbe, M. A. Mackey, C. E. Coumbe, and J. P. Phillips, *Journal of the American Chemical Society*, 2007, **129**, 16257–16262.
49. H. Takikawa, M. Imamura, M. Kouchi, and T. Sakakibara, *Fullerene Science and Technology*, 1998, **6**, 339–349.
50. L. Dunsch, M. Krause, J. Noack, and P. Georgi, *Journal of Physics and Chemistry of Solids*, 2004, **65**, 309–315.
51. S. Stevenson, G. Rice, T. Glass, K. Harich, F. Cromer, M. R. Jordan, J. Craft et al. *Nature*, 1999, **401**, 55–57.
52. S. F. Yang, L. Zhang, W. F. Zhang, and L. Dunsch, *Chemistry—A European Journal*, 2010, **16**, 12398–12405.
53. M. Z. Jiao, W. F. Zhang, Y. Xu, T. Wei, C. B. Chen, F. P. Liu, and S. F. Yang, *Chemistry—A European Journal*, 2012, **18**, 2666–2673.
54. F. P. Liu, J. Guan, T. Wei, S. Wang, M. Z. Jiao, and S. F. Yang, *Inorganic Chemistry*, 2013, **52**, 3814–3822.
55. S. Stevenson, M. A. Mackey, M. A. Stuart, J. P. Phillips, M. L. Easterling, C. J. Chancellor, M. M. Olmstead, and A. L. Balch, *Journal of the American Chemical Society*, 2008, **130**, 11844–11845.

56. B. Q. Mercado, M. A. Stuart, M. A. Mackey, J. E. Pickens, B. S. Confait, S. Stevenson, M. L. Easterling et al. *Journal of the American Chemical Society*, 2010, **132**, 12098–12105.
57. B. Q. Mercado, M. M. Olmstead, C. M. Beavers, M. L. Easterling, S. Stevenson, M. A. Mackey, C. E. Coumbe et al. *Chemical Communications*, 2010, **46**, 279–281.
58. R. Valencia, A. Rodriguez-Fortea, S. Stevenson, A. L. Balch, and J. M. Poblet, *Inorganic Chemistry*, 2009, **48**, 5957–5961.
59. S. Stevenson, Y. Ling, C. E. Coumbe, M. A. Mackey, B. S. Confait, J. P. Phillips, H. C. Dorn, and Y. Zhang, *Journal of the American Chemical Society*, 2009, **131**, 17780–17782.
60. Y. Ling, S. Stevenson, and Y. Zhang, *Chemical Physics Letters*, 2011, **508**, 121–124.
61. S. Stevenson, C. B. Rose, J. S. Maslenikova, J. R. Villarreal, M. A. Mackey, B. Q. Mercado, K. Chen, M. M. Olmstead, and A. L. Balch, *Inorganic Chemistry*, 2012, **51**, 13096–13102.
62. S. Stevenson, in *Handbook of Carbon Nano Materials*, eds. F. D. D'Souza and K. M. Kadish, World Scientific, Singapore, 2011, vol. 1, pp. 185–205.
63. S. Stevenson, in *Endohedral Fullerenes: From Fundamentals to Applications*, eds. S. Yang and C. R. Wang, World Scientific Publishing Company, Singapore, 2014, 179–210.
64. M. Krause, F. Ziegs, A. A. Popov, and L. Dunsch, *Chemphyschem*, 2007, **8**, 537–540.
65. L. Dunsch, S. F. Yang, L. Zhang, A. Svitova, S. Oswald, and A. A. Popov, *Journal of the American Chemical Society*, 2010, **132**, 5413–5421.
66. N. Chen, M. N. Chaur, C. Moore, J. R. Pinzon, R. Valencia, A. Rodriguez-Fortea, J. M. Poblet, and L. Echegoyen, *Chemical Communications*, 2010, **46**, 4818–4820.
67. E. Krokos, *Journal of Physical Chemistry C*, 2010, **114**, 7626–7630.
68. T. S. Wang, N. Chen, J. F. Xiang, B. Li, J. Y. Wu, W. Xu, L. Jiang et al. *Journal of the American Chemical Society*, 2009, **131**, 16646–16647.
69. T. A. Murphy, T. Pawlik, A. Weidinger, M. Hohne, R. Alcala, and J. M. Spaeth, *Physical Review Letters*, 1996, **77**, 1075–1078.
70. S. Aoyagi, E. Nishibori, H. Sawa, K. Sugimoto, M. Takata, Y. Miyata, R. Kitaura et al. *Nature Chemistry*, 2010, **2**, 678–683.
71. R. S. Ruoff, D. S. Tse, R. Malhotra, and D. C. Lorents, *Journal of Physical Chemistry*, 1993, **97**, 3379–3383.
72. X. H. Zhou, Z. N. Gu, Y. Q. Wu, Y. L. Sun, Z. X. Jin, Y. Xiong, B. Y. Sun, Y. Wu, H. Fu, and J. Z. Wang, *Carbon*, 1994, **32**, 935–937.
73. B. Y. Sun, L. Feng, Z. J. Shi, and Z. N. Gu, *Carbon*, 2002, **40**, 1591–1595.
74. J. Q. Ding and S. H. Yang, *Chemistry of Materials*, 1996, **8**, 2824–2827.
75. Y. F. Lian, Z. J. Shi, X. H. Zhou, and Z. N. Gu, *Chemistry of Materials*, 2004, **16**, 1704–1714.
76. W. R. Creasy, J. A. Zimmerman, and R. S. Ruoff, *Journal of Physical Chemistry*, 1993, **97**, 973–979.
77. T. Tsuchiya, T. Wakahara, Y. F. Lian, Y. Maeda, T. Akasaka, T. Kato, N. Mizorogi, and S. Nagase, *Journal of Physical Chemistry B*, 2006, **110**, 22517–22520.
78. C. Yeretzian, J. B. Wiley, K. Holczer, T. Su, S. Nguyen, R. B. Kaner, and R. L. Whetten, *Journal of Physical Chemistry*, 1993, **97**, 10097–10101.
79. J. W. Raebiger and R. D. Bolskar, *Journal of Physical Chemistry C*, 2008, **112**, 6605–6612.
80. E. Toth, R. D. Bolskar, A. Borel, G. Gonzalez, L. Helm, A. E. Merbach, B. Sitharaman, and L. J. Wilson, *Journal of the American Chemical Society*, 2005, **127**, 799–805.
81. S. Laus, B. Sitharaman, V. Toth, R. D. Bolskar, L. Helm, S. Asokan, M. S. Wong, L. J. Wilson, and A. E. Merbach, *Journal of the American Chemical Society*, 2005, **127**, 9368–9369.
82. B. Sitharaman, R. D. Bolskar, I. Rusakova, and L. J. Wilson, *Nano Letters*, 2004, **4**, 2373–2378.

83. X. Lu, J. X. Xu, Z. J. Shi, B. Y. Sun, Z. N. Gu, H. D. Liu, and H. B. Han, *Chemical Journal of Chinese Universities-Chinese*, 2004, **25**, 697–700.
84. H. Kato, Y. Kanazawa, M. Okumura, A. Taninaka, T. Yokawa, and H. Shinohara, *Journal of the American Chemical Society*, 2003, **125**, 4391–4397.
85. R. D. Bolskar, A. F. Benedetto, L. O. Husebo, R. E. Price, E. F. Jackson, S. Wallace, L. J. Wilson, and J. M. Alford, *Journal of the American Chemical Society*, 2003, **125**, 5471–5478.
86. M. Okumura, M. Mikawa, T. Yokawa, Y. Kanazawa, H. Kato, and H. Shinohara, *Academic Radiology*, 2002, **9**, S495–S497.
87. M. Mikawa, H. Kato, M. Okumura, M. Narazaki, Y. Kanazawa, N. Miwa, and H. Shinohara, *Bioconjugate Chemistry*, 2001, **12**, 510–514.
88. C. D. Angeli, T. Cai, J. C. Duchamp, J. E. Reid, E. S. Singer, H. W. Gibson, and H. C. Dorn, *Chemistry of Materials*, 2008, **20**, 4993–4997.
89. S. Stevenson, M. A. Mackey, C. E. Coumbe, J. P. Phillips, B. Elliott, and L. Echegoyen, *Journal of the American Chemical Society*, 2007, **129**, 6072–6073.
90. S. Stevenson, K. Harich, H. Yu, R. R. Stephen, D. Heaps, C. Coumbe, and J. P. Phillips, *Journal of the American Chemical Society*, 2006, **128**, 8829–8835.
91. S. Stevenson, C. B. Rose, A. A. Robson, D. T. Heaps, and J. P. Buchanan, *Fullerenes Nanotubes and Carbon Nanostructures*, 2013, **22**, 182–189, DOI:10.1080/15363 83X.1532013.1798725.
92. S. Stevenson, K. A. Rottinger, and J. S. Field, *Dalton Transactions*, 2014, **43**, 7435–7441, Under Review.
93. Z. Y. Wang, Y. Nakanishi, S. Noda, K. Akiyama, and H. Shinohara, *Journal of Physical Chemistry C*, 2012, **116**, 25563–25567.
94. K. Akiyama, T. Hamano, Y. Nakanishi, E. Takeuchi, S. Noda, Z. Y. Wang, S. Kubuki, and H. Shinohara, *Journal of the American Chemical Society*, 2012, **134**, 9762–9767.
95. M. D. Diener and J. M. Alford, *Nature*, 1998, **393**, 668–671.
96. B. Y. Sun and Z. N. Gu, *Chemistry Letters*, 2002, 1164–1165.
97. R. D. Bolskar and J. M. Alford, *Chemical Communications*, 2003, 1292–1293.
98. T. Kodama, K. Higashi, T. Ichikawa, S. Suzuki, H. Nishikawa, I. Ikemoto, K. Kikuchi, and Y. Achiba, *Chemistry Letters*, 2005, **34**, 464–465.
99. T. Tsuchiya, T. Wakahara, S. Shirakura, Y. Maeda, T. Akasaka, K. Kobayashi, S. Nagase, T. Kato, and K. M. Kadish, *Chemistry of Materials*, 2004, **16**, 4343–4346.
100. X. Lu, H. J. Li, B. Y. Sun, Z. J. Shi, and Z. N. Gu, *Carbon*, 2005, **43**, 1546–1549.
101. M. R. Ceron, F. F. Li, and L. Echegoyen, *Chemistry—A European Journal*, 2013, **19**, 7410–7415.
102. T. Tsuchiya, K. Sato, H. Kurihara, T. Wakahara, T. Nakahodo, Y. Maeda, T. Akasaka et al. *Journal of the American Chemical Society*, 2006, **128**, 6699–6703.
103. K. Kikuchi, S. Suzuki, Y. Nakao, N. Nakahara, T. Wakabayashi, H. Shiromaru, K. Saito, I. Ikemoto, and Y. Achiba, *Chemical Physics Letters*, 1993, **216**, 67–71.
104. C. J. Welch and W. H. Pirkle, *Journal of Chromatography*, 1992, **609**, 89–101.
105. R. Beyers, C. H. Kiang, R. D. Johnson, J. R. Salem, M. S. Devries, C. S. Yannoni, D. S. Bethune et al. *Nature*, 1994, **370**, 196–199.
106. S. Stevenson, H. C. Dorn, P. Burbank, K. Harich, Z. Sun, C. H. Kiang, J. R. Salem et al. *Analytical Chemistry*, 1994, **66**, 2680–2685.
107. S. Stevenson, H. C. Dorn, P. Burbank, K. Harich, J. Haynes, C. H. Kiang, J. R. Salem et al. *Analytical Chemistry*, 1994, **66**, 2675–2679.
108. S. Stevenson, P. Burbank, K. Harich, Z. Sun, H. C. Dorn, P. H. M. van Loosdrecht, M. S. deVries et al. *Journal of Physical Chemistry A*, 1998, **102**, 2833–2837.
109. M. D. Diener, C. A. Smith, and D. K. Veirs, *Chemistry of Materials*, 1997, **9**, 1773–1777.
110. K. Kikuchi, Y. Nakao, S. Suzuki, Y. Achiba, T. Suzuki, and Y. Maruyama, *Journal of the American Chemical Society*, 1994, **116**, 9367–9368.

111. P. H. M. Vanloosdrecht, R. D. Johnson, M. S. Devries, C. H. Kiang, D. S. Bethune, H. C. Dorn, P. Burbank, and S. Stevenson, *Physical Review Letters*, 1994, **73**, 3415–3418.

112. K. Kimata, T. Hirose, K. Moriuchi, K. Hosoya, T. Araki, and N. Tanaka, *Analytical Chemistry*, 1995, **67**, 2556–2561.

113. M. R. Anderson, H. C. Dorn, S. A. Stevenson, and S. M. Dana, *Journal of Electroanalytical Chemistry*, 1998, **444**, 151–154.

114. N. Tagmatarchis, E. Aslanis, H. Shinohara, and K. Prassides, *Journal of Physical Chemistry B*, 2000, **104**, 11010–11012.

115. C. M. Beavers, M. N. Chaur, M. M. Olmstead, L. Echegoyen, and A. L. Balch, *Journal of the American Chemical Society*, 2009, **131**, 11519–11524.

116. T. Inoue, T. Tomiyama, T. Sugai, T. Okazaki, T. Suematsu, N. Fujii, H. Utsumi, K. Nojima, and H. Shinohara, *Journal of Physical Chemistry B*, 2004, **108**, 7573–7579.

117. T. Inoue, T. Tomiyama, T. Sugai, and H. Shinohara, *Chemical Physics Letters*, 2003, **382**, 226–231.

118. J. Y. Zhang, T. Fuhrer, W. J. Fu, J. C. Ge, D. W. Bearden, J. Dallas, J. Duchamp et al. *Journal of the American Chemical Society*, 2012, **134**, 8487–8493.

119. K. Tan and X. Lu, *Journal of Physical Chemistry A*, 2006, **110**, 1171–1176.

120. Y. Yamazaki, K. Nakajima, T. Wakahara, T. Tsuchiya, M. O. Ishitsuka, Y. Maeda, T. Akasaka, M. Waelchli, N. Mizorogi, and S. Nagase, *Angewandte Chemie-International Edition*, 2008, **47**, 7905–7908.

121. H. Kurihara, X. Lu, Y. Iiduka, N. Mizorogi, Z. Slanina, T. Tsuchiya, T. Akasaka, and S. Nagase, *Journal of the American Chemical Society*, 2011, **133**, 2382–2385.

122. H. Kurihara, X. Lu, Y. Iiduka, N. Mizorogi, Z. Slanina, T. Tsuchiya, S. Nagase, and T. Akasaka, *Chemical Communications*, 2012, **48**, 1290–1292.

123. Y. H. Ma, T. S. Wang, J. Y. Wu, Y. Q. Feng, H. Li, L. Jiang, C. Y. Shu, and C. R. Wang, *Journal of Physical Chemistry Letters*, 2013, **4**, 464–467.

124. H. Kurihara, X. Lu, Y. Iiduka, H. Nikawa, M. Hachiya, N. Mizorogi, Z. Slanina, T. Tsuchiya, S. Nagase, and T. Akasaka, *Inorganic Chemistry*, 2012, **51**, 746–750.

125. H. Kurihara, Y. Iiduka, Y. Rubin, M. Waelchli, N. Mizorogi, Z. Slanina, T. Tsuchiya, S. Nagase, and T. Akasaka, *Journal of the American Chemical Society*, 2012, **134**, 4092–4095.

126. H. Kurihara, X. Lu, Y. Iiduka, H. Nikawa, N. Mizorogi, Z. Slanina, T. Tsuchiya, S. Nagase, and T. Akasaka, *Journal of the American Chemical Society*, 2012, **134**, 3139–3144.

127. T. S. Wang, L. Feng, J. Y. Wu, W. Xu, J. F. Xiang, K. Tan, Y. H. Ma et al. *Journal of the American Chemical Society*, 2010, **132**, 16362–16364.

128. P. Jin, Z. Zhou, C. Hao, Z. X. Gao, K. Tan, X. Lu, and Z. F. Chen, *Physical Chemistry Chemical Physics*, 2010, **12**, 12442–12449.

129. S. F. Yang, C. B. Chen, F. P. Liu, Y. P. Xie, F. Y. Li, M. Z. Jiao, M. Suzuki et al. *Scientific Reports*, 2013, **1487**, 1–5.

130. Y. Q. Feng, T. S. Wang, J. Y. Wu, Y. H. Ma, Z. X. Zhang, L. Jiang, C. H. Ge, C. Y. Shu, and C. R. Wang, *Chemical Communications*, 2013, **49**, 2148–2150.

131. N. Chen, M. Mulet-Gas, Y. Y. Li, R. E. Stene, C. W. Atherton, A. Rodriguez-Fortea, J. M. Poblet, and L. Echegoyen, *Chemical Science*, 2013, **4**, 180–186.

132. N. Chen, C. M. Beavers, M. Mulet-Gas, A. Rodriguez-Fortea, E. J. Munoz, Y. Y. Li, M. M. Olmstead, A. L. Balch, J. M. Poblet, and L. Echegoyen, *Journal of the American Chemical Society*, 2012, **134**, 7851–7860.

133. B. Q. Mercado, N. Chen, A. Rodriguez-Fortea, M. A. Mackey, S. Stevenson, L. Echegoyen, J. M. Poblet, M. M. Olmstead, and A. L. Balch, *Journal of the American Chemical Society*, 2011, **133**, 6752–6760.

134. N. Tagmatarchis, E. Aslanis, K. Prassides, and H. Shinohara, *Chemistry of Materials*, 2001, **13**, 2374–2379.

135. W. Xu, L. Feng, M. Calvaresi, J. Liu, Y. Liu, B. Niu, Z. J. Shi, Y. F. Lian, and F. Zerbetto, *Journal of the American Chemical Society*, 2013, **135**, 4187–4190.

136. A. A. Popov, L. Zhang, and L. Dunsch, *Acs Nano*, 2010, **4**, 795–802.

137. Y. Maeda, J. Miyashita, T. Hasegawa, T. Wakahara, T. Tsuchiya, T. Nakahodo, T. Akasaka et al. *Journal of the American Chemical Society*, 2005, **127**, 12190–12191.

138. H. Yang, H. X. Jin, B. Hong, Z. Y. Liu, C. M. Beavers, H. Y. Zhen, Z. M. Wang, B. Q. Mercado, M. M. Olmstead, and A. L. Balch, *Journal of the American Chemical Society*, 2011, **133**, 16911–16919.

139. M. Yamada, M. Minowa, S. Sato, M. Kako, Z. Slanina, N. Mizorogi, T. Tsuchiya, Y. Maeda, S. Nagase, and T. Akasaka, *Journal of the American Chemical Society*, 2010, **132**, 17953–17960.

140. H. Yang, C. X. Lu, Z. Y. Liu, H. X. Jin, Y. L. Che, M. M. Olmstead, and A. L. Balch, *Journal of the American Chemical Society*, 2008, **130**, 17296–17300.

141. C. M. Beavers, H. X. Jin, H. Yang, Z. M. Wang, X. Q. Wang, H. L. Ge, Z. Y. Liu, B. Q. Mercado, M. M. Olmstead, and A. L. Balch, *Journal of the American Chemical Society*, 2011, **133**, 15338–15341.

142. H. Nikawa, T. Yamada, B. P. Cao, N. Mizorogi, Z. Slanina, T. Tsuchiya, T. Akasaka, K. Yoza, and S. Nagase, *Journal of the American Chemical Society*, 2009, **131**, 10950–10954.

143. H. Yang, H. X. Jin, X. Q. Wang, Z. Y. Liu, M. L. Yu, F. K. Zhao, B. Q. Mercado, M. M. Olmstead, and A. L. Balch, *Journal of the American Chemical Society*, 2012, **134**, 14127–14136.

144. A. L. Svitova, A. A. Popov, and L. Dunsch, *Inorganic Chemistry*, 2013, **52**, 3368–3380.

145. S. R. Plant, T. C. Ng, J. H. Warner, G. Dantelle, A. Ardavan, G. A. D. Briggs, and K. Porfyrakis, *Chemical Communications*, 2009, 4082–4084.

146. H. Okimoto, R. Kitaura, T. Nakamura, Y. Ito, Y. Kitamura, T. Akachi, D. Ogawa et al. *Journal of Physical Chemistry C*, 2008, **112**, 6103–6109.

147. A. A. Popov, C. B. Chen, S. F. Yang, F. Lipps, and L. Dunsch, *Acs Nano*, 2010, **4**, 4857–4871.

148. S. F. Yang, C. B. Chen, A. A. Popov, W. F. Zhang, F. P. Liu, and L. Dunsch, *Chemical Communications*, 2009, **45**, 6391–6393.

149. Z. X. Ge, J. C. Duchamp, T. Cai, H. W. Gibson, and H. C. Dorn, *Journal of the American Chemical Society*, 2005, **127**, 16292–16298.

150. P. P. Fatouros and M. D. Shultz, *Nanomedicine*, 2013, **8**, 1853–1864.

151. J. F. Zhang, P. P. Fatouros, C. Y. Shu, J. Reid, L. S. Owens, T. Cai, H. W. Gibson et al. *Bioconjugate Chemistry*, 2010, **21**, 610–615.

152. M. D. Shultz, J. C. Duchamp, J. D. Wilson, C. Y. Shu, J. C. Ge, J. Y. Zhang, H. W. Gibson et al. *Journal of the American Chemical Society*, 2010, **132**, 4980–4981.

153. H. C. Dorn and P. P. Fatouros, *Nanoscience and Nanotechnology Letters*, 2010, **2**, 65–72.

154. C. Y. Shu, X. Y. Ma, J. F. Zhang, F. D. Corwin, J. H. Sim, E. Y. Zhang, H. C. Dorn et al. *Bioconjugate Chemistry*, 2008, **19**, 651–655.

155. P. P. Fatouros, F. D. Corwin, Z. J. Chen, W. C. Broaddus, J. L. Tatum, B. Kettenmann, Z. Ge et al. *Radiology*, 2006, **240**, 756–764.

3 Computational Studies of Endohedral Metallofullerenes

Alexey A. Popov

CONTENTS

3.1 BONDING INTERACTIONS IN ENDOHEDRAL METALLOFULLERENES

3.1.1 METAL–CAGE ELECTRON TRANSFER AND BONDING INTERACTIONS

Since the first indication of the existence of endohedral metallofullerenes (EMFs) reported in 1985, the nature of the interaction between endohedral metal atoms and the host carbon cage has been recognized as one of the crucial points of EMF research. With the isolation of the first bulk amounts of EMFs in early 1990s, electron spin resonance (ESR) spectroscopic studies of paramagnetic mono-metallofullerenes $Y@C_{82}$ and $La@C_{82}$ showed that metal atoms transfer their valence electrons to the carbon cage.[1,2] X-ray photoelectron spectroscopy (XPS) studies also showed that the spectra in the range of metal-based transition levels correspond well with the spectra of metal oxides in the corresponding oxidation states.[3] Hence, the electronic structure of EMFs can be formally described as $M^{3+}@C_{82}^{3-}$ and metal–cage interactions are understood as ionic bonding. Ionic conjecture is also supported by frontier molecular orbital (MO) analysis, which shows that the MOs of the EMFs can be described as MOs of the fullerene cores with additional electrons borrowed from metal atoms.[4,5] Furthermore, the electrostatic potential (ESP) has large negative values inside negatively charged carbon cages,[6–8] which results in the strong stabilization of the metal cations if they are placed inside fullerene anions, thus favoring the formation of EMFs.

Developed first for mono-metallofullerenes, the ionic model is now used to describe the bonding situation in all other EMFs with complex endohedral species, such as dimetallofullerenes [e.g., $(La^{3+})_2@C_{80}^{6-}$], nitride clusterfullerenes [e.g., $(Sc^{3+})_3N^{3-}@C_{80}^{6-}$], carbide clusterfullerenes [e.g., $(Sc^{3+})_2C_2^{2-}@C_{82}^{4-}$], and sulfide clusterfullerenes [e.g., $(Sc^{3+})_2S^{2-}@C_{82}^{4-}$].[9–11] The ionic model is also indirectly supported by absorption and vibrational spectra as well as electrochemical behavior of EMFs: spectra and redox potentials of isostructural EMFs with different metal atoms are usually very similar if these metal atoms are in the same oxidation state, which points to the lack of metal-specific interactions.[9,10,12] Computational studies have also shown that the stability order of the isomers of EMFs corresponds well with the stability order of the isomers of empty fullerenes in the appropriate charge states (see below).[13–15]

The ionic model of bonding in EMFs is thus widely accepted and appears useful to explain many spectroscopic and structural properties of EMFs. However, there are also evidences that the ionic model is over-simplified. For instance, vibrational and UV-Vis-NIR absorption spectra of EMFs with the same isomeric cage but different metals are indeed very similar and in many cases are well reproduced by calculations (e.g., vibrational spectra of all $M_3N@C_{80}$ NCFs are almost identical except for the cluster-based modes), but analogous calculations for the charged *empty* cages do not match experimental spectra of EMFs for the fullerene-based vibrations or excitations.[16–20] Thus, the transfer of a large integer number of electrons from an endohedral metal atom to the fullerene cage is rather formal and should not be understood literally. Metal–cage bonding in EMFs cannot be considered to be purely ionic; covalent contribution must be also taken into account. Although analysis of the frontier MO energies and populations in EMFs and corresponding empty fullerenes can be interpreted as complete electron transfer of metal valence electrons, many experimental[21,22] and theoretical[6,23–26] studies clearly show that endohedral metal atoms have a substantial population of nd-levels. This formal inconsistency results from small contributions of metal atoms to individual π-MO orbitals, which therefore often remain hidden in a frontier MO analysis. To reveal significant degree of covalency in metal–cage interactions, it is necessary to analyze net populations, atomic charges (which are considerably smaller than formal ones), bond orders, or total electron density.[19,26–29] The mechanism of the metal–cage interactions in EMFs can be described as the formal transfer of an appropriate number of electrons from the endohedral species to the carbon cage with subsequent π-nd coordination of the metal cations by the fullerene π-system and reoccupation of the metal nd orbitals.

Because covalent contribution to the metal–cage bonding is built up by the overlap of many fullerene-based π-orbitals with nd orbitals of the metal atoms, it appears more convenient to analyze these phenomena using approaches based on total density such as Bader's quantum theory of atoms in molecules (QTAIM).[30] The detailed study of chemical bonding in EMFs with the use of topological QTAIM analysis was reported in reference 31. In this chapter, we focus mainly on the QTAIM atomic charges and delocalization indices $\delta(A, B)$, which are defined as the number of electron pairs shared between the atoms A and B. Sharing of electrons is the essence of covalent bonding, and for nonpolar bonds, $\delta(A, B)$ values have the physical meaning of bond orders in Lewis definition. The sum of all $\delta(M, C)$ indices for a given metal atom M and carbon atoms from the carbon cage defines a *metal–cage bond order*, $\delta(M, cage)$.[31]

In the majority of EMFs, individual $\delta(M, C)$ values were found to be a smooth function of M—C distances and did not exceed 0.25; however, when summed over all metal–cage interactions, $\delta(M, cage)$ values close to 2–3 were obtained.[31] Thus, whereas QTAIM atomic charges of metal atoms were found to be approximately two times smaller than their formal oxidation states, the total number of the electron pairs shared by metal atoms with the EMF molecule was found to be close to the typical valence of the given element (≈ 3 for Sc, Y, and La and ≈ 4 for Ti).[31] An analysis of these values circumvents an apparent contradiction between the ionic model and substantial degree of covalency in metal–cage interactions revealed by more comprehensive analyses. The surplus electrons (e.g., six electrons in $M_3N@C_{2n}$ compounds) are indeed available to the carbon cage in EMFs, but these electrons are not completely transferred to the cage; instead, they are *shared* between the endohedral cluster and a fullerene.

The complex nature of metal–cage bonding raises a question on the locality of these interactions. That is, when the fullerene is charged, negative charge is rather homogeneously distributed over the cage. However, metal–cage interactions have limited spatial extension. As an illustration, Figure 3.1 visualizes distribution of $\delta(M, C)$ delocalization indices and atomic charges of carbon atoms, $q(C)$, in two EMFs: $La_2@C_{78}$ and $Sc_3N@C_{78}$. Carbon atoms are shown as spheres, whose radii scale as the values of $\delta(M, C)$ or negative of $q(C)$. It can be well seen that both parameters decay very fast with the increase in the metal–carbon distance. Thus, metal–cage interactions adjust local fragments of a fullerene π-system, each comprising roughly a dozen of carbon atoms.

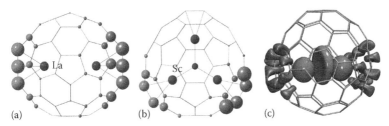

(a)   (b)   (c)

FIGURE 3.1   (a, b) Visualization of atomic charges of carbon atoms and $\delta(M, C)$ indices in $La_2@C_{78}$ (a) and $Sc_3N@C_{78}$ (b) in the vicinity of metal atoms: carbon atoms are shown as spheres whose radii scale as absolute values of atomic charges (green) or $\delta(M, C)$ values (pink). (From Popov, A. A. and Dunsch, L.: The bonding situation in endohedral metallofullerenes as studied by quantum theory of atoms in molecules (QTAIM). *Chem. Eur. J.* 2009. 15. 9707–9729. Copyright Wiley-VCH Verlag GmbH & Co. KGaA. Reproduced with permission.) (c) Color-coded representation of electron localization function (ELF) in $Y_2@C_{82}$-$C_{3v}(8)$: trisynaptic (Y, C, C) basins are blue, disynaptic (Y, Y) basin is green, yttrium core basins are dark orange. (From Popov, A. A. et al., *Chem. Commun.*, 48, 8031–8050, 2012. Reproduced by permission of The Royal Society of Chemistry.)

Convenient visualization of metal–cage bonding is obtained with the use of electron localization function (ELF). The ELF was introduced to visualize electron localization in atoms, molecules, and crystals and is in some respect similar to the Laplacian of the electronic density.[32–34] Topological analysis of ELF yield basins, which are characterized by a synaptic order (usually a number of atoms, whose bonding is described by the basin; for example, monosynaptic basins are typical for core electrons or lone electron pairs, disynaptic basins—for bonds between two atoms, trisynaptic basins—for three-center bonding). In empty fullerenes, only disynaptic valence carbon–carbon basins V(C, C) are found, whereas in EMFs, formation of the metal–cage bonds is seen in ELF representation as a spatial extension of some V(C, C) basins in the vicinity of the metal atoms and their transformation into the *trisynaptic* V(M, C, C) basins (Figure 3.1).[35] Figure 3.1 shows that both QTAIM and ELF approaches yield similar spatial extension of the metal–cage bonding in EMFs.

3.1.2 METAL–METAL BONDING

Despite the considerable degree of covalency in metal–cage bonding, metal atoms in EMFs still have large positive charges. Therefore, encapsulation of more than one metal atom inside the fullerene cage results in a strong Coulomb repulsion, which forces metal ions to be as far as possible from each other. However, analysis of the formal oxidation states in a significant portion of metallofullerenes with nominally trivalent metals (e.g., $Y_2@C_{82}$,[36] $Lu_2@C_{76}$,[37] $Sc_2@C_{82}$,[38] $Er_2@C_{82}$[39–41]) shows that here metal atoms have lower oxidation state. This means that their valence electrons are not completely transferred to the carbon cage, which can lead to bonding interactions between metal atoms. Computational studies show that in such di-EMFs, the HOMO or one of the highest occupied MOs has an M—M bonding character (Figure 3.2a).[36,38,42] Occupied metal–metal bonding MOs are also found in $M_2@C_{79}N$,[43,44] $Sc_4O_2@C_{80}$ (Figure 3.2b–c),[45] or $Y_3@C_{80}$ (Figure 3.2d).[35] Furthermore, even in di-EMFs with trivalent metal atoms (e.g., in $La_2@C_{2n}$ and $Ce_2@C_{2n}$), the lowest unoccupied molecular orbital (LUMO) is usually an M—M bonding orbital (Figure 3.2e), and therefore, this MO can be populated (i.e., metal–metal bond can be formed) by chemical or electrochemical reduction.[46] Detailed analysis of such bonding situations and related phenomena with the use of MO, QTAIM, and ELF approaches is found in reference 46. Numerical estimations show that the energy of repulsive M···M interactions in di-EMFs is 2–3 times higher than the energy of covalent bonding. However, the carbon cage acts as a frame that prevents complete *dissociation* of M_2 dimers, thus creating an "oxymoron": bonding between strongly repulsive metal atoms.[46]

A valence state of metal atoms in a neutral di-EMF molecule depends on the energy matching between the carbon-cage orbitals and the lowest-energy valence MO of the free metal dimer, which usually has $(ns)\sigma_g^2$ character.[46] In due turn, the energy of the $(ns)\sigma_g^2$ orbital in M_2 dimers correlates with the $ns^2(n-1)d^1 \rightarrow ns^1(n-1)d^2$ excitation energy of the free metal atom. Therefore, the valence state of metal atoms in di-EMFs is determined by this excitation energy, rather than by the third ionization potential (which determines the valence state of metals in

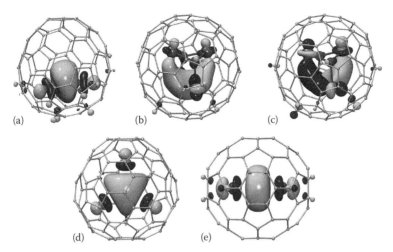

FIGURE 3.2 Representative molecular orbitals of EMFs with metal–metal bonding character: (a) HOMO of $Sc_2@C_{82}$-C_{3v}(8), (b) HOMO of $Sc_4O_2@C_{80}$-I_h(7), (c) LUMO of $Sc_4O_2@$ C_{80}-I_h(7), (d) HOMO-1 of $Y_3@C_{80}$-I_h(7), (e) LUMO of $La_2@C_{80}$-I_h(7).

mono-metallofullerenes).[47,48] Namely, the $ns^2(n-1)d^1 \rightarrow ns^1(n-1)d^2$ excitation energies are increasing from La (0.33 eV) to Sc (1.43 eV)/Y (1.36 eV) to Lu (2.34 eV),[49] which parallels a systematic stabilization of the $(ns)\sigma_g^2$ MO from La_2 to Sc_2/Y_2 to Lu_2 (Figure 3.3).[46] As a result, La is always trivalent in di-EMFs, Lu prefers a divalent state, whereas a valence state of Sc and Y can be either di- or trivalent depending on the energy of carbon-cage MOs.[46]

Metal–metal bonding can be found not only in di-EMFs, but also in more complex clusterfullerenes. A special type of M—M bonding is found in trimetallofullerenes such as $Y_3@C_{80}$. MO analysis showed that the highest twofold occupied orbital of $Y_3@C_{80}$ has metal–metal bonding character involving all three Y atoms, that is, the bonding situation can be described as a three center–two electron bond.[35] QTAIM analysis revealed the presence of a *pseudoatom* (also known as *non-nuclear attractor*, the maximum of the electron density not associated with the nuclei) in the center of the molecule. Each Y atom forms a bond path to the pseudoatom, that is, bonding in the Y_3 cluster in $Y_3@C_{80}$ mimics bonding in the nitride cluster in $Y_3N@C_{80}$ with pseudoatom replacing nitrogen.

Another special example of metal–metal interactions in EMFs is $Sc_4O_2@C_{80}$, whose Sc atoms form a tetrahedral cluster, whereas two μ^3-coordinated oxygen atoms are located above the centers of two faces of the tetrahedron. A formal charge distribution in the molecule can be described as $(Sc^{2+})_2(Sc^{3+})_2(\mu^3\text{-}O^{2-})_2@C_{80}^{6-}$. Sc atoms with one oxygen neighbor are in a formal divalent state, and HOMO of the molecule is largely a Sc—Sc bonding orbital reminiscent of the Sc—Sc bonding MO in $Sc_2@C_{82}$ (Figure 3.2). LUMO of $Sc_4O_2@C_{80}$ is also localized on the Sc_4O_2 cluster, which means that reduction in the molecule forms a different type of metal–metal bonding. According to the shape of the LUMO as well as results of ELF analysis, additional bonds formed in mono- and dianions of $Sc_4O_2@C_{80}$ can be described as

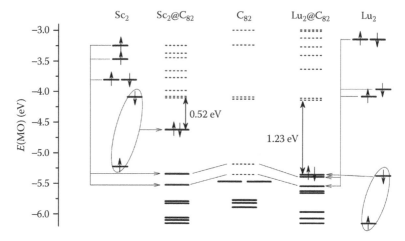

FIGURE 3.3 Correlation between molecular orbitals of metal dimers (Sc_2, Lu_2), MOs of empty C_{82}-$C_{3v}(8)$, and MOs of $M_2@C_{82}$ (M = Sc, Lu). Energy levels of twofold occupied MOs are shown in thick black lines and those of vacant MOs in dashed lines. Spin-polarized solutions are shown for metal dimers. $(ns)\sigma_g^2$ MOs of Sc_2 and Lu_2 are encircled; corresponding MOs in $M_2@C_{82}$ molecules are HOMOs. Four electrons of metal dimers are transferred to the two lowest energy unoccupied MOs of C_{82}.

a three-center Sc—Sc—Sc bonding spatially localized over the oxygen-free faces of the Sc_4 tetrahedron. The HOMO and LUMO of $Sc_4O_2@C_{80}$ have substantially different compositions in terms of Sc atomic orbitals. In particular, HOMO has approximately twofold higher 4s-Sc contribution than LUMO, which results in dramatically larger [45]Sc hyperfine coupling constants (hfcc) in ESR spectrum of the cation-radical $Sc_4O_2@C_{80}^+$ than those in the spectrum of the anion-radical $Sc_4O_2@C_{80}^-$.

3.1.3 Bonding between Metal and Nonmetal Atoms

Strong Coulomb repulsion between encapsulated metal atoms is an obvious destabilizing factor and presumably a reason why encapsulation of more than two metal atoms in conventional EMFs is very rare. However, this destabilizing factor can be balanced if endohedral species also include nonmetal atoms. QTAIM analysis shows that the nonmetals in clusterfullerenes localize a substantial excess of electron density: typical charges are -1.7 to -1.9 for nitrogen in $M_3N@C_{2n}$, -1.3 to -1.5 for a C_2 unit in $M_2C_2@C_{2n}$, approximately -1.2 for sulfur in $M_2S@C_{82}$, and -1.3 to -1.5 for oxygen in $Sc_4O_{2,3}@C_{80}$.[31,50–52] In terms of the ionic model, formal charges of these species can be written down as N^{3-}, C_2^{2-}, S^{2-}, O^{2-}, and so on. Repulsive metal–metal interactions in clusterfullerenes are compensated by attractive interactions with nonmetals, and hence, clusters comprising up to four metal atoms can be stabilized (e.g., $Sc_4O_{2,3}@C_{80}$ or $Sc_4C_2@C_{80}$).

Numerical contributions of different types of intra-cluster interactions in $Sc_4O_2@C_{80}$ were estimated in reference 46 by means of interaction quantum atom model, which allows partitioning of the interatomic interaction energies into physically sound ionic and covalent terms.[53–55] Net Sc—Sc and O—O interactions were found to be strongly destabilizing (68.8 and 7.8 eV, respectively) because of the dominating role of Coulomb repulsion (70.6 and 8.5 eV, respectively). However, Sc—O interactions are strongly stabilizing with the net interaction energy of −126.7 eV, which includes the Coulomb attraction of −109.4 eV and a covalent term of −17.2 eV. As a result, the whole Sc_4O_2 cluster is stabilized by −50.2 eV. Importantly, although ionic Sc—O interaction energy is 6.4 times higher than that of the covalent term, the former is to a large extent balanced by Coulomb Sc—Sc and O—O repulsion. As a result, intra-cluster interaction energy has comparable covalent and Coulomb contributions of −19.8 and −30.4 eV, respectively.[46]

Analysis of the intra-cluster interaction in $Sc_4O_2@C_{80}$ clearly shows that covalent contribution to the bonding in clusterfullerenes should not be underestimated. According to the QTAIM analysis, significant degree of covalency is also manifested in considerable metal/nonmetal delocalization indices.[31,52,56] For instance, δ(M, N), δ(M, C_2), and δ(M, S) values in respective clusterfullerenes span the range of 0.7–0.8, whereas δ(Sc, O) indices in oxide clusterfullerenes are in the range 0.5–0.6.[31,50] By all QTAIM indicators, metal/nonmetal bonds can be described as ordinary covalent bonds with high degree of polarity.

An unprecedented intra-cluster bonding situation was recently discovered in $TiLu_2C@C_{80}$, which is the first structurally characterized member of a new type of Ti-based clusterfullerenes. The $TiLu_2C$ cluster has central μ^3-coordinated carbon atom and is isoelectronic to $ScLu_2N$ (a *shift* from scandium to the next element in the periodic table, titanium, is *compensated* by a shift from nitrogen to the previous element, carbon). The formal charge distribution in $TiLu_2C$ might be described as $[(Lu^{3+})_2Ti^{4+}C^{4-}]^{6+}$ with fourfold charged titanium and carbon atoms; however, structural and computational studies showed that a more appropriate description of the cluster structure is $[(Lu^{3+})_2Ti^{3+} = C^{3-}]^{6+}$ with a double bond between the titanium and the central carbon atoms. X-ray-determined Ti—C distance in the cluster is short, at 1.874(6) Å, and the QTAIM-derived δ(Ti, C) delocalization index is 1.49, which is close to the value in $(C_5H_5)_2Ti = CH_2$ (1.34) and is twice that of the value computed for $Ti_2C_2@C_{78}$ (0.71). On the other hand, Lu—C bonding in $TiLu_2C@C_{80}$ is similar to Lu—N bonding in $Lu_2ScN@C_{80}$ nitride clusterfullerenes: δ(Lu, C) and δ(Lu, N) indices are both equal 0.73.

3.2 MOLECULAR STRUCTURES OF ENDOHEDRAL METALLOFULLERENES

3.2.1 CHARGE TRANSFER AND CARBON-CAGE ISOMERISM

The electron transfer from encapsulated species to the fullerene cage dramatically affects molecular structures of EMFs. In particular, carbon-cage isomers in structurally characterized EMFs are usually different from those of empty fullerenes. The fact that charging can affect relative stabilities of fullerene isomers was recognized in early 1990s.[57,58] The milestone in the computational studies of EMFs was the finding that the $I_h(7)$ isomer of C_{80}, being the least stable structure for neutral C_{80} and C_{80}^{2+}, has the lowest energy structure for multiply charged C_{80}^{4-} and C_{80}^{6-} states.[58,59] The preference of the C_{80}-I_h(7) cage in highly charged state was a clear explanation why the most stable isomer of $La_2@C_{80}$ has I_h(7) cage, as opposed to the D_2-symmetric hollow C_{80}.[60] In another work, variation of the relative energies of negatively charged isomers of C_{82} was analyzed to explain isomerism of $M@C_{82}$.[7,61] The C_2(3) cage was shown to be the lowest energy isomer of the neutral C_{82}, whereas C_{2v}(9) cage was proved to be the most stable isomer of C_{82}^{2-} and C_{82}^{3-}. Likewise, the lowest energy isomers of $M@C_{82}$ (M = Ca, Sc, Y, La, etc.) also had the C_{2v}(9) cage isomer.[7,61] These computational studies were performed when structural characterization of EMFs was not well developed yet, and hence, their results were an important step toward understanding that EMFs have different cage isomers compared to empty fullerenes.

The influence of charge on the stability of isolated pentagon rule (IPR) isomers has further consequences. In 1995, Fowler and Zerbetto showed that the non-IPR C_{2v}(1809) isomer of C_{60} with two pairs of adjacent pentagons is stabilized when the fullerene is negatively charged.[58] In the C_{60}^{4-} state, its relative energy equals that of the IPR C_{60}-I_h, whereas in the hexa-anionic state, the non-IPR isomer is already more stable. Pentalene (a unit of two fused pentagons) is a 8π anti-aromatic system in the neutral state, but its dianion with 10π electrons is aromatic.[62] Therefore, electron transfer from metal atoms to the carbon cage can stabilize non-IPR isomers. The first computational proof was reported by Kobayashi et al. in 1997 in the studies of $Ca@C_{72}$ isomers,[63] whereas the first experimental evidence of the violation of IPR in EMFs was provided in 2000, with the isolation of $Sc_2@C_{66}$-C_{2v}(4348)[64] and $Sc_3N@$ C_{68}-D_3(6140).[65] Since that time, a lot of non-IPR EMFs have been reported, including $Sc_3N@C_{70}$-C_{2v}(7854),[14] $Sc_2S@C_{70}$-C_2(7892),[66] $Sc_2S@C_{72}$-C_s(10528),[67] $La@C_{72}$-C_2(10612),[68] $La_2@C_{72}$-D_2(10611),[69,70] $DySc_2N@C_{76}$-C_s(17490),[71] $M_3N@C_{78}$-C_2(22010) (M = Y, Dy, Tm, Gd),[13,72,73] $Sc_3NC@C_{78}$-C_2(22010),[74] $Gd_3N@C_{82}$-C_s(39663),[75] $M_3N@$ C_{84}-C_s(51365) (M = Tb, Gd, Tm),[76,77] and $Y_2C_2@C_{84}$-C_1(51383) [78,79]. Based on computational studies, non-IPR isomers have also been proposed for $Sc_2@C_{70}$-C_{2v}(7854),[80] $Ca@C_{72}$-C_2(10612), or C_{2v}(11188),[63,81] the minor isomer of $Yb@C_{74}$ [C_1(13393) or C_1(14049)],[82] $M_2@C_{74}$ [C_2(13295) and C_2(13333)] (M = Sc, La),[83] $M@C_{76}$ [C_{2v}(19138) and C_1(17459)] (M = Yb, Ca, Sr, Ba),[84,85] $Gd_2@C_{98}$-C_1(168785).[86] These results show that the IPR is not valid for EMFs.

Violation of the IPR makes prediction of the most stable EMF isomers more complex than for empty fullerenes. Besides, in addition to a dramatically increased number of isomers, one should also consider that metal atoms and clusters can have different bonding sites inside the fullerene, and hence, several such conformers have to be considered for each cage isomer. Fortuitously, correlation between stability of EMF and appropriately charged empty fullerene isomers can substantially facilitate a search of the most stable isomers of EMFs because computations of empty fullerenes are less demanding and more reliable than those of metal-containing EMFs. Typically, thousands of isomers of empty fullerenes in their charged form are screened at the semiempirical level (e.g., AM1), whereas DFT calculations of EMFs are then performed for a limited number of the most stable isomers (10–20 isomers is usually sufficient).[15] This procedure has become a routine now and allows one to locate the most stable isomers within 1–2 weeks. For instance, screening through a large number of C_{70}^{6-}, C_{76}^{6-}, and C_{78}^{6-} isomers helped in the elucidation of molecular structure of non-IPR $Sc_3N@C_{70}$-C_{2v}(7854),[14] $DySc_2N@C_{76}$-C_s(17490),[71] and $M_3N@$

C_{78}-C_2(22010) (M = Dy, Tm).[13] An exhaustive search of the most favorable cage isomers of $Sc_3N@C_{2n}$ and $Y_3N@C_{2n}$ ($2n = 68$–98) at the PBE/TZ2P level was reported in 2007.[15] Computational studies aimed at localization of the most stable isomers were also published for $Sc_2@C_{70}$,[80] $Sc_2S@C_{70}$,[66] $Ti_2S@C_{78}$,[87] $Sc_2S@C_{82}$,[50] $Y_2C_2@$ C_{84},[78] $M_3N@C_{88}$ (M = La, Gd, Lu),[88] $La_3N@C_{92}$,[89] $Gd_2C_2@C_{92}/Gd_2@C_{94}$,[90] $M_2C_2@$ $C_{96}/M_2@C_{98}$ (M = Sc, Y, La, Gd, Lu),[86,91] and so on (see also the list of non-IPR EMFs above).

Experimental studies of EMFs with endohedral species donating six electrons to the carbon cage, such as M_2, M_3N, Sc_3NC, $Sc_4O_{2,3}$, Sc_4C_2, and $TiLu_2C$, show that EMFs with C_{80}-I_h(7) isomer are always dominant in the produced mixtures. Exhaustive computations of $C_{2n}{}^{6-}$ isomers show that this phenomenon is caused by the enhanced stability of the $C_{80}{}^{6-}$-I_h(7) hexa-anion.[15] Figure 3.4 plots normalized absolute energies of the most stable isomers of $C_{2n}{}^{6-}$ hexa-anions versus fullerene size. Decrease in the cage curvature and on-site Coulomb repulsions of six surplus electrons result in a smooth decay of the normalized energy with the increase in fullerene size. However, the point corresponding to the $C_{80}{}^{6-}$-I_h(7) isomer is found well below the decay curve, which corresponds to the additional stabilization energy of approximately 185 kJ/mol. Stability of the second most stable isomer of $C_{80}{}^{6-}$ with D_{5h}(6) cage is also enhanced by approximately 100 kJ/mol.

In references 15 and 19, enhanced stability of $C_{80}{}^{6-}$-I_h(7) isomer has been related to the most uniform distribution of pentagons possible in a fullerene cage, as quantified in terms of hexagon indices established by Raghavachari.[92] These indices were introduced in early 1990s to measure distribution of pentagon-induced curvature in

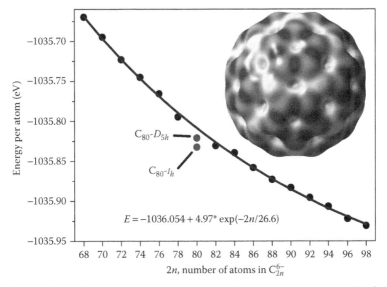

FIGURE 3.4 Normalized energy per carbon atom in lowest energy isomers of $C_{2n}{}^{6-}$ hexaanions as a function of the cage size. (Reprinted with permission from Popov, A. A. and Dunsch, L., *J. Am. Chem. Soc.* 129, 11835–11849, 2007. Copyright 2007 American Chemical Society.) The inset show electrostatic potential in $[C_{80}$-$I_h]^{6-}$ hexa-anion mapped on the isosurface of the electron density. Darker fragments on 5/6 edges correspond to enhanced negative charge.

fullerenes. However, correlation between pentagon distribution and relative energies of non-charged empty fullerenes was not sufficiently good, presumably because the steric strain factor is often outweighed by other factors such as stability of the π-system. Surprisingly, a correlation between relative energies of the IPR C_{2n}^{6-} isomers and pentagon distribution was much better than for non-charged fullerenes. Poblet and coworkers showed that this phenomenon can be explained by enhanced role of on-site Coulomb repulsion in negatively charged fullerenes.[93] Because surplus charge in fullerene anions is mainly accumulated in pentagons (Figure 3.4), the lowest-energy isomers should have the maximal separation (and hence the most uniform distribution) of pentagons. To quantify distribution of pentagons, Poblet and coworkers introduced an inverse pentagon separation index and showed that its correlation with the relative stability of isomers improves with the increase in charge. Thus, minimization of the on-site Coulomb repulsion and minimization of the steric strain are achieved by fulfillment of the same condition, the most uniform distribution of the pentagons.

3.2.2 Cage Form Factor and Strain in the Endohedral Cluster

As discussed in the previous section, the charge transfer from the encased species to the fullerene cage is the dominant factor determining stability of the EMFs. The role of charge transfer to the carbon cage in stabilizing a particular cage isomer is well established now,[15,19,60,93] and the most preferable isomers can be routinely predicted through the computations of the empty fullerene isomers in the appropriate charge states.[13–15,66] However, this is not the sole stability factor which should be taken into account. The size and the shape of the carbon cage should fit the size and the shape of the encapsulated cluster. When this condition is violated, distortions of the cluster may increase its energy and even destabilize the whole clusterfullerene molecule. As a result, the preferable cage isomer can be switched or a given fullerene may not be formed at all. In short, this factor has been defined as the *cage form factor*.[19,94]

An illustrative example of the crucial effect of the cage form factor is isomerism of EMFs with C_{78} carbon cage. The lowest energy isomer of C_{78}^{6-} is $D_{3h}(5)$, and EMFs with a sixfold charged endohedral cluster usually have this cage isomer, for example, $M_2@C_{78}$ (M = La, Ce),[95] $Ti_2C_2@C_{78}$,[96] $Ti_2S@C_{78}$,[87] and $Sc_3N@C_{78}$.[97] However, in $M_3N@C_{78}$ EMFs with larger metal atoms (such as Y or Lu), the size of the M_3N cluster turns to be so large that the C_{78}-$D_{3h}(5)$ cage does not provide sufficient space to accommodate it. As a result, the most stable isomers of nitride clusterfullerenes $M_3N@C_{78}$ and even mixed-metal $MSc_2N@C_{78}$ (M = Y or lanthanides) have a non-IPR C_{78}-$C_2(22010)$ cage, whose flattened shape provides more space for the large cluster.[13] Although C_{78}^{6-}-$C_2(22010)$ is less stable than C_{78}^{6-}-$D_{3h}(5)$ by 59 kJ/mol, $Y_3N@C_{78}$-$C_2(22010)$ is more stable than $Y_3N@C_{78}$-$D_{3h}(5)$ by 84 kJ/mol.

The unfavorable cage form factor also explains why nitride clusterfullerenes $M_3N@C_{72}$ have never been synthesized, whereas dimetallofullerenes $M_2@C_{72}$ (M = Ce, La) are well known. The most stable C_{72}^{6-} isomer with $D_2(10611)$ cage has an elongated shape with two pentagon pairs located on the opposite poles of the molecule. This shape perfectly suits di-EMFs (metal atoms coordinate adjacent pentagon pairs and have a relatively large M⋯M distance) but is not compatible with the triangular shape of the M_3N cluster (Figure 3.5). The latter cannot be coordinated to two pentagon pairs of C_{72}-D_2(10611) without its regular triangular shape being preserved, and hence, $M_3N@C_{72}$ with the D_2(10611) cage is very unstable.[15] A similar unfavorable shape of the low energy $C_{74}{}^{6-}$ isomers explains why $M_3N@C_{74}$ NCFs have never been observed.[15] Likewise, the yields of $Sc_3N@C_{70}$-C_{2v}(7854), in which the cluster has to be significantly deformed from the equilateral shape, are much lower than those of other Sc-based NCFs.[14] In nitride $M_3N@C_{76}$ clusterfullerenes with the C_{76}-C_s(17490) cage, neither $Sc_3N@C_{76}$ nor $Dy_3N@C_{76}$ has a good match of the size

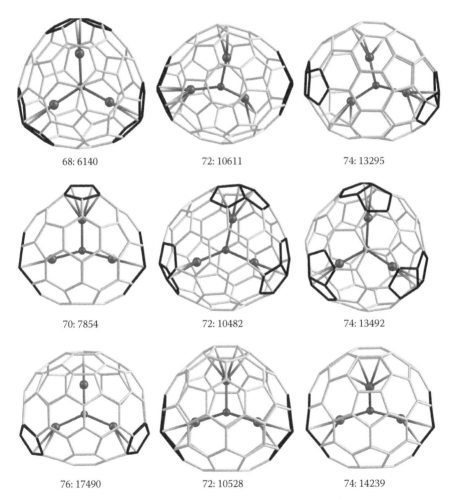

68: 6140   72: 10611   74: 13295

70: 7854   72: 10482   74: 13492

76: 17490   72: 10528   74: 14239

FIGURE 3.5 DFT-optimized molecular structures of selected $Sc_3N@C_{2n}$ nitride cluster-fullerenes ($2n = 68$–76). Adjacent pentagon pairs are highlighted in black. (Reprinted with permission from Popov, A. A. and Dunsch, L., *J. Am. Chem. Soc.*, 129, 11835–11849, 2007. Copyright 2007 American Chemical Society.) Note that the shape of the lowest energy isomers $C_{72}{}^{6-}$-D_2(10611) and $C_{74}{}^{6-}$-C_2(13295) is not suitable to coordinate all adjacent pentagon pairs by Sc atoms. As a result, the lowest energy isomers $Sc_3N@C_{72}$-C_s(10528) and $Sc_3N@C_{74}$-C_{2v}(14239) are based on unstable carbon cages.

and shape of the cluster and the cage. However, the asymmetric shape of the mixed-metal cluster $DySc_2N$ fits the cage shape much better, and hence, $DySc_2N@C_{76}$ is the only C_{76}-based NCF obtained in isolable amounts.[71] Finally, even archetypical $M_3N@C_{80}$-$I_h(7)$ NCFs are not free from the strain when the endohedral cluster is larger than Sc_3N, as will be discussed below.

The decisive role of the cage form factor in carbon-cage isomerism has also been reported in studies on some carbonitride and sulfide clusterfullerenes. Similar to $Y_3N@C_{78}$, $Sc_3NC@C_{78}$ prefers the $C_2(22010)$ cage, whereas $Sc_3NC@C_{78}$-$D_{3h}(5)$ is 19 kJ/mol less stable.[52,74] $Sc_2S@C_{70}$ has a non-IPR $C_2(10528)$ cage,[66] although $C_{70}{}^{4-}$-$C_2(10528)$ is 86 kJ/mol less stable than IPR $C_{70}{}^{4-}$-D_{5h}. In the former, two adjacent pentagon pairs are located suitably well to be coordinated by Sc atoms of the sulfide cluster, and the non-IPR isomer is additionally stabilized by these interactions.

The size of the fullerene cage can also dramatically influence the bond lengths or even the shape of the internal clusters. For example, a systematic experimental and computational NMR study of a series of $Y_2C_2@C_{2n}$ ($2n$ = 82, 84, 92, 100) by Dorn et al. showed that the internal yttrium carbide cluster prefers to adopt a stretched linear shape when the cage is sufficiently large (e.g., in C_{100}), whereas it bends to a compressed *butterfly shape* in relatively small cages (e.g., in C_{82}).[98]

From the examples discussed above, it should be obvious that the cage form factor is much more important for stability of clusterfullerene than that of conventional EMFs. Although for the former the shape and size of the cluster play a limiting role in the cage size distribution, for mono- and dimetallofullerenes these factors are not so crucial and a much broader range of structures can be obtained.[99–103]

In spite of the obvious vital role played by the cage form factor in stability of clusterfullerenes, numerical estimation of the strain energy and optimal geometrical characteristics of encapsulated clusters are not known because in all real systems they are always templated by the carbon cage. That is, it is hard to predict which cluster geometry is strained and which is not, and how large the strain is if it is considered to be present. A computational study of the *isolated* cluster (e.g., Sc_3N) taken outside of the carbon cage is not relevant because of the different electronic structures of the free cluster and the same cluster inside the carbon cage. At the same time, the study of the isolated charged cluster with the formal charge taken as a real one (e.g., Sc_3N^{6+}) also does not give a reliable answer. The real charges of the endohedral atoms in EMFs are much smaller than the formal ones because of the significant degree of covalency in metal–cage interactions,[31] and hence, computations for highly charged *naked* clusters will severely overestimate Coulomb repulsion of metal atoms.

Deng and Popov[104] circumvented this problem by using a simple scheme in which metal atoms are coordinated with small organic π-systems mimicking the real electronic situation in EMFs and yet bestowing an endohedral cluster sufficient freedom to adjust its geometrical parameters in the most optimal way. In the majority of clusterfullerenes, metal atoms formally donate two electrons to the carbon cage and one electron to the nonmetal central atom or more diatomic unit (N, S, C_2, CN, etc.). Although endohedral clusters interact with the whole cage, the nature of metal–cage interactions remains rather local, as discussed above. Metal atoms strongly interact with *islands* of the fullerene cage comprising approximately 8–12 carbon atoms.[31] Thus, pentalene C_8H_6 (two fused pentagons; see Figure 3.6) is a good model of the

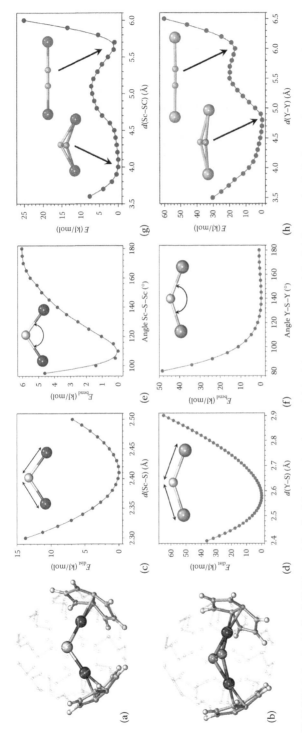

FIGURE 3.6 (a, b) $Sc_2S(C_8H_6)_2$ and $Sc_2C_2(C_8H_6)_2$ molecules overlaid with carbon cages of $Sc_2S@C_{82}$ and $Sc_2C_2@C_{82}$; (c, d) energy profiles along metal–sulfur bond length in $Sc_2S(C_8H_6)_2$ (c) and $Y_2S(C_8H_6)_2$ (d); (e, f) energy profiles along metal–sulfur–metal angle in $Sc_2S(C_8H_6)_2$ (e) and $Y_2S(C_8H_6)_2$ (f); (g, h) energy profiles along metal–metal distance in $Sc_2C_2(C_8H_6)_2$ (g) and $Y_2C_2(C_8H_6)_2$ (h). (Reprinted with permission from Deng, Q. and Popov, A. A., *J. Am. Chem. Soc.*, 2014, *136*, 4257–4264, DOI:10.1021/ja4122582, 2014. Copyright 2014 American Chemical Society.)

metal-coordinated cage fragment because pentalene units are two-electron acceptors and are small enough to avoid steric hindrances when the cluster geometry is varied.

The PBE/TZ2P calculations of the $Sc_3N(C_8H_6)_3$ molecule showed that the Sc_3N cluster is planar with the optimal Sc—N bond length of 2.020 Å. This value is very close to the Sc—N bond length in $Sc_3N@C_{80}$-I_h (2.020 Å). The cluster strain energies [estimated as the energy difference between optimized $Sc_3N(C_8H_6)_3$ molecule and the one whose cluster geometry is identical to that in a given nitride clusterfullerene] are all found to be below 2 kJ/mol in $Sc_3N@C_{68}$, $Sc_3N@C_{78}$, and $Sc_3N@C_{80}$; however, in $Sc_3N@C_{70}$ the cluster strain is as high as 50 kJ/mol, which agrees with the low yield of this compound. Overall, the Sc_3N cluster in NCFs is rather flexible: the cluster distortion energy remains below 10 kJ/mol in the 2.02 ± 0.06 Å range of the Sc—N bond lengths and in the 110–135°Å range of the Sc—N—Sc bending angles.

In the free $Y_3N(C_8H_6)_3$ molecule, the Y—N bond length is elongated to 2.190 Å. This value is close to bond lengths in $Y_3N@C_{86}$-D_3(19) and $Y_3N@C_{88}$-D_2(35), which therefore have negligible cluster strain energies (1–2 kJ/mol). The cluster strain energy in $Y_3N@C_{78}$-C_2(22010) is 21 kJ/mol, whereas in the hypothetical $Y_3N@C_{78}$-D_{3h}(5), the strain is increased to 72 kJ/mol. In $Y_3N@C_{80}$-I_h(7), the Y—N bond length is 2.060 Å, and the cluster strain energy is increased to 42 kJ/mol. In fact, this is the highest value for all experimentally available $Y_3N@C_{2n}$ molecules. Thus, a extraordinarily high stability of the $[C_{80}$-$I_h]^{6-}$ cage results in the higher yield of $Y_3N@C_{80}$-I_h in the synthesis in comparison with $Y_3N@C_{2n}$ ($2n$ = 82–88), but a further increase in the metal ionic radius (Gd and beyond) leads to drastic decrease in the relative yield of $M_3N@C_{80}$. Besides, the yield of $Y_3N@C_{80}$ is much lower than that of $Sc_3N@C_{80}$ (the cluster is not strained in the latter); high strain of the nitride cluster is presumably one of the reasons.

Analogues computations for $M_2S(C_8H_6)_2$ molecules (M = Sc, Y) also showed high flexibility of the M_2S clusters both in terms of the M—S bond lengths and M—S—M angles (Figure 3.6). In $Sc_2S(C_8H_6)_2$, the optimized Sc—S bond length is 2.411 Å and the strain energy remains below 10 kJ/mol when the Sc—S distance is varied in the range 2.4 ± 0.09 Å. The optimized Sc—S—Sc angle is 110°, but the energy is below 6 kJ/mol in the whole studied range of 95–180°. The highest cluster strain energy in experimentally available sulfide clusterfullerene is found in $Sc_2S@C_{70}$-C_2(7892), whereas in $Sc_2S@C_{72}$-C_s(10528) and both isomers of $Sc_2S@C_{82}$ with C_s(6) and C_{3v}(8) cages, the energy is below 10 kJ/mol. Optimized Y—S bond length and Y—S—Y angle in $Y_2S(C_8H_6)_2$ are 2.578 Å and 140°, respectively. The cluster strain energy in the only experimentally available sulfide, $Y_2S@C_{82}$-C_{3v}(8), is 26 kJ/mol (DFT-predicted Y—S distance is rather short, at 2.486 Å, and the Y—S—Y angle is 97°).

A specific situation is found in carbide clusterfullerenes. Figure 3.6 shows the energy profile in $Sc_2C_2(C_8H_6)_2$ and $Y_2C_2(C_8H_6)_2$ molecules along the metal–metal distance. Calculations reveal the presence of two energy minima corresponding to a "butterfly" and linear forms of the M_2C_2 cluster at shorter and longer metal–metal distances, respectively. For comparison, $[M_2C_2]^{4+}$ energy profiles have only one minimum corresponding to the linear cluster, whereas strong repulsion is predominant at shorter distances.

In $Sc_2C_2(C_8H_6)_2$, both minima are almost isoenergetic (linear form is 1 kJ/mol less stable). All experimentally characterized $Sc_2C_2@C_{2n}$ clusterfullerenes have bent cluster structure; the cluster strain energies in $Sc_2C_2@C_{2n}$ ($2n$ = 72, 82, and 84) are below 10, whereas in $Sc_2C_2@C_{70}$ and $Sc_2C_2@C_{80}$, they reach 21 and 15 kJ/mol, respectively. In $Y_2C_2(C_8H_6)_2$, the butterfly configuration at the Y—Y distance of 4.809 Å is 16 kJ/mol more stable than the linear one with the Y—Y distance of 5.991 Å. The Y—Y distances in $Y_2C_2@C_{84}$-$C_1(51383)$, $Y_2C_2@C_{88}$-$D_2(35)$, and $Y_2C_2@$ C_{92}-$D_3(85)$ span a range of 4.26–4.88 Å, the cluster has the butterfly shape, and the cluster strain energies are below 10 kJ/mol. At shorter distances (3.793 Å in $Y_2C_2@C_{82}$), the strain energy is increased to 25 kJ/mol. In the hypothetical $Y_2C_2@$ C_{100}-$D_5(450)$ with the Y—Y distance of 5.516 Å, the cluster is closer to the linear shape and the strain energy reaches 27 kJ/mol. These results show that the basis for the *nanoscale fullerene compression*, which implies that the least strained structure of Y_2C_2 is linear and that the butterfly shape is forced by cage-induced compression, should be reconsidered. In fact, the butterfly shape of the M_2C_2 cluster is more energetically favorable and it is realized when the metal–metal distance is approximately 4–5 Å. At shorter distances, the energy increases and *nanoscale compression* is really an appropriate term. However, at longer distances, the energy also increases and the linear form is higher in energy, especially for Y_2C_2. Thus, *nanoscale stretching* would be an appropriate term for longer metal–metal distances. In fact, the shape and the size of the carbon cage determine the position of metal atoms and the M···M distance (M-cage distances are more or less constant), whereas the C_2 unit then finds its best configuration for a given position of metal atoms.

3.3 DYNAMICS OF ENCAPSULATED SPECIES IN ENDOHEDRAL FULLERENES

3.3.1 DYNAMICS FROM STATIC CALCULATIONS

The discussion of metal–cage interactions and cage form factor in the previous sections implied that the endohedral species are fixed inside the EMF molecule. However, the interior of a fullerene can have different metal–cage bonding sites. The endohedral species can therefore occupy different positions inside a carbon cage (such configurations are denoted as *conformers* in this chapter), and relative energies of such conformers can vary in a rather broad range depending on the nature of endohedral species and topology of the carbon cage. If relative energies of different conformers are similar and barriers of interconversion between them are low, an endohedral cluster can rotate inside the cage at ambient conditions (e.g., room temperature). Highly symmetric C_{80}-$I_h(7)$ is a typical example of the carbon cage which forms EMFs with free rotation of endohedral species. On the contrary, rotation of clusters in non-IPR fullerenes is usually hindered. Thus, the dynamic behavior of endohedral species can vary from the tight fixed position inside the carbon cage to nearly free rotation.

Dynamics of endohedral cluster can seriously affect the results of structural or spectroscopic studies of EMFs. For instance, disorder of endohedral species in single-crystal X-ray diffraction studies is obviously related to the internal dynamics. Likewise, results of cage symmetry determination by ^{13}C NMR spectroscopy depend on the rotational behavior of the endohedral cluster. An illustrative example is $Sc_2C_2@C_{80}$-$C_{2v}(5)$, which has C_s-symmetric C_{80} cage at room temperature due to frozen cluster position, whereas at 413 K, the cluster can rotate at NMR timescales resulting in apparent C_{2v}-symmetry of the cage.[105]

From a computational viewpoint, the problem of internal dynamics in EMFs is usually addressed via static calculations of relative energies of different conformers and, more rarely, transition states and trajectories (i.e., intrinsic reaction coordinate). The most detailed computational studies were performed for $M_2@C_{80}$ (M = La, Ce) or $M_3N@C_{80}$ (M = Sc, Y, Lu, etc.) with the $I_h(7)$-symmetric carbon cage.

Computational studies of $La_2@C_{80}$ have shown the presence of several unique conformers with relative energies of only a few kilojoules per mole and equally low interconversion barriers.[60,106–108] Each conformer is multiplied by symmetry operations of the I_h point group, yielding equivalent structures with La atoms distributed all over the carbon cage, and hence, free rotation of La_2 unit may be expected. The same conclusion can be drawn from the analysis of the spatial distribution of the ESP, which has no well-distinguished minima inside C_{80}^{6-}-$I_h(7)$.[8] Experimental NMR[109] and powder diffraction[110] studies of $La_2@C_{80}$ prove that La atoms circulate freely inside the carbon cage.

Several groups have reported relative energies of conformers of $M_3N@C_{80}$-$I_h(7)$ (M = Sc,[25,29,111–114] Y,[29,112,114] Lu,[29] Gd[114,115]). Similar to $La_2@C_{80}$, the energies of the majority of structures found lie within the range of 10–15 kJ/mol. The lowest energy is predicted for a C_3-symmetric conformer with metal atoms coordinated to hexagons but somewhat displaced toward the pentagon/hexagon edges (Figure 3.7).[29] An exhaustive search for transition states accompanied by intrinsic reaction coordinate (IRC) computations has showed that the barrier to the cluster rotation in $Sc_3N@C_{80}$ does not exceed 11.4 kJ/mol at the PBE/TZ2P level.[29] Computational data agree well with the results of NMR spectroscopic studies, which confirmed free rotation of the cluster.

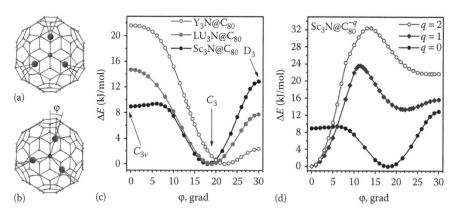

FIGURE 3.7 (a) Molecular structure of $Sc_3N@C_{80}$-$I_h(7)$ in C_{3v}-symmetric conformer; (b) the same for the lowest energy C_3-symmetric conformer; here definition of the cluster rotation angle φ is also shown (rotation axis is perpendicular to the paper); (c, d) section of the potential energy surface along rotation of the M_3N cluster in: (c) $Sc_3N@C_{80}$, $Lu_3N@C_{80}$, and $Y_3N@C_{80}$; (d) $Sc_3N@C_{80}$, $Sc_3N@C_{80}^-$, and $Sc_3N@C_{80}^{2-}$. (Reprinted with permission from Popov, A. A. and Dunsch, L., *J. Am. Chem. Soc.*, 130, 17726–17742, 2008. Copyright 2008 American Chemical Society.)

Extended search for possible conformers has also been performed for NCFs with other carbon cages,[56,114,116,117] as well as for $Ca@C_{74}$,[118] $Sc_2C_2@C_{82}$,[50,119] $Sc_2S@C_{82}$,[50] $Y_2C_2@C_{92}$,[120] and $Sc_4O_2@C_{80}$,[121,122] and so on. Importantly, the structure of frontier MOs in some clusterfullerenes was found to be significantly dependent on the orientation of the cluster.[29,50,116] Therefore, the search for the most stable conformers is a necessary prerequisite for the computational analysis of the electronic structure of EMFs, because calculations performed for an arbitrarily chosen high energy conformer can give erroneous results.

Exohedral derivatization of EMFs (e.g., cycloaddition) changes the topology of the fullerene π-system and hence can significantly alter the dynamic behavior of endohedral species. In the first report on the possible manipulation of internal dynamics by cycloaddition, Kobayashi et al.[123] showed that three-dimensional random circulation of La atoms in $La_2@C_{80}$ turns into in-plane motion in mono- and bis-adducts $La_2@C_{80}(Si_2H_4CH_2)_{1,2}$. Significant changes in the dynamics of metal atoms were indeed found in experimental studies of $M_2@C_{80}$ (M = La, Ce) cycloadducts.[124–128]

Computational studies of $Sc_3N@C_{80}(CF_3)_2$[129] and [5,6]-pyrrolidino-$Sc_3N@C_{80}$[130] adducts and their anions revealed more than 40 different conformers for each derivative. Their relative energies span the range of 60–70 kJ/mol, which is several times larger than in the non-derivatized $Sc_3N@C_{80}$. In both derivatives, only two to three closely related conformers have their energies below 10 kJ/mol with small barriers of interconversion; a few more structures are found in the 10–20 kJ/mol range, whereas the majority of conformers have higher energy. In the anions, the energy gap between a few stable and all other conformers is further increased.[129,130] Thus, exohedral derivatization can significantly hinder rotation of the endohedral cluster, and multiple functionalization can completely block it, as found in $Sc_3N@C_{80}(CF_3)_x$ and proved by ESR spectroscopy of corresponding anion radicals.[131–133]

Several studies showed that changing the charge state of the EMF molecule can affect the internal dynamics of the endohedral species. For instance, the lowest energy conformer of $Sc_3N@C_{80}^-$ has C_{3v} symmetry and corresponds to the higher-energy conformer of the neutral molecule, whereas the cluster rotation barrier around the C_3 axis is increased to 25 kJ/mol (Figure 3.7).[29] Significant changes in the relative energies of conformers and increase in the internal barriers in the anionic state have also been reported for $La_2@C_{80}$,[123] $Sc_3N@C_{80}(CF_3)_2$,[129] [5,6]-pyrrolidino-$Sc_3N@C_{80}$,[130] and $Sc_4O_2@C_{80}$.[121] At the same time, computational studies show that the rotation of the cluster is facilitated in the cations of $TiSc_2N@C_{80}$,[56] $Sc_3N@C_{80}(CF_3)_2$,[129] and $Sc_4O_2@C_{80}$.[121]

3.3.2 MOLECULAR DYNAMICS SIMULATIONS

Another computational approach to address the rotational behavior of endohedral clusters in EMFs is molecular dynamics (MD). MD simulation is a perfect tool to address dynamic problems because it provides on-time description of the process. However, computational costs of MD simulations are very high. For instance, a 1-ps-long MD trajectory propagated with a time step of 1 fs requires 1000 evaluations of energy and energy gradients. For this reason, first principle MD studies of endohedral fullerenes are still very rare.[35,50,56,121,129,134–137] Much longer dynamics (reaching the nanosecond scale) can be studied using the density-functional-based tight-binding (DFTB) approach, but it is less reliable and requires careful parameterization of interatomic interaction potentials.[137–141]

The first *ab initio* MD study of EMFs was reported by Andreoni and Curioni in 1996.[136] Car-Parrinello (CP) MD study of La@C_{60} and La@C_{82} on a picosecond timescale showed that La atoms in C_{60} exhibit fluxional motion. On the contrary, in La@C_{82}, the metal atom was trapped in two different configurations [correct isomeric structure of La@C_{82}-C_{2v}(9) was not known yet in 1996, and calculations were performed for a C_2 cage isomer].

Several groups have reported computational studies of internal dynamics of the metal atom in M@C_{74} EMFs. Due to the high symmetry of the C_{74}-D_{3h} cage, the metal atom in M@C_{74} can be in one of three equivalent positions. Experimental NMR studies of M@C_{74} (M = Ca,[142] Yb[143]) showed that although metal atoms circulate inside the cage at room temperature, the fast-exchange limit is not reached. Static DFT calculation at the B3LYP/3-21G~dz level showed that the barrier of such circulation in Ca@C_{74} is in the range of 30–40 kJ/mol.[118] Several MD studies were performed for M@C_{74} to analyze the dynamic behavior of metal atoms. In the CPMD study of Ba@C_{74}-D_{3h}[134] at room temperature, the metal atom accomplished one hopping event between two minima over the 15-ps trajectory. Longer timescales were achieved in the DFTB-BOMD study of Eu@C_{74}.[141] Simulations showed that the metal atom oscillates along the C_3 axis with large amplitude already at low temperature (approximately 0.17 Å at 140 K and 0.22 Å at 200 K). At 600 K, the amplitude of oscillations became larger than the harmonic region of the potential (0.9 Å along the axis and 0.5 Å in the transverse direction), and at 900 K, the Eu atom started to circulate randomly.

The dynamics of carbide cluster and in particular acetylide unit in carbide clusterfullerenes have been addressed in several MD studies. DFTB-MD simulations of Sc_2C_2@C_{84} at different temperatures showed that positions of Sc atoms are almost fixed, whereas the C_2 unit freely rotates around the Sc—Sc axis.[139] The dynamics of Sc_3C_2@C_{80} on the timescale of 3.5–6.0 ps was studied by Taubert et al. at the BP86/def2-SVP level.[135] Sc atoms were shown to form a relatively rigid triangle, which exhibited ratchet-like rotational hopping inside the carbon cage, whereas the C_2 unit flipped rapidly through the Sc_3 plane (but did not rotate as in Sc_2C_2@C_{84}). Comparative MD simulation of Sc_2S@C_{82}-C_{3v}(8) and Sc_2C_2@C_{82}-C_{3v}(8) have been performed at the PBE/DZV(P) level.[50] Both Sc_2C_2 and Sc_2S tend to rotate around the C_3-axis of the carbon cage. On the 10-ps timescale followed in that work, the sulfide cluster exhibited faster rotation (approximately 180° over 10 ps), whereas the carbide cluster showed large-amplitude librations. The intra-cluster dynamics of the C_2 unit in Sc_2C_2@C_{82} was rather chaotic and could not be described as in-plane rotation or through-plane flipping.[50]

Several MD studies have also been reported for nitride clusterfullerenes. Vietze and Seifert studied nanosecond-long dynamics of Sc_3N@C_{78} and Sc_3N@C_{80} at the DFTB level.[137,139,140] Complex fluxional behavior of the Sc_3N cluster was found in Sc_3N@C_{80} with activation barrier of 0.144 eV, which agrees well with static DFT as well as experimental studies.[140] On the contrary, the position of the cluster in Sc_3N@C_{78} was locked inside. Sc_3N exhibited slow in-plane rotational vibrations and in-plane deformations of the cluster.[139] A comparative DFT-MD study of Y_3@C_{80} and Y_3N@C_{80}

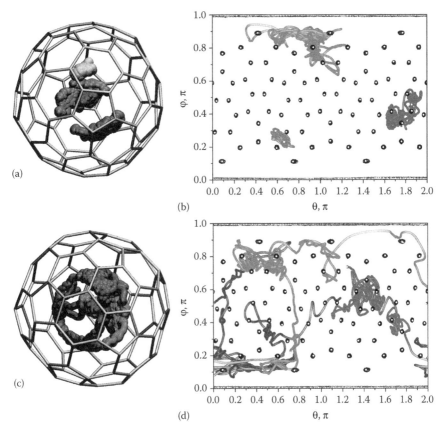

FIGURE 3.8 MD trajectories of $TiSc_2N@C_{80}$ (a, b) and isoelectronic $Sc_3N@C_{80}^-$ (c, d): Ti atom is cyan, Sc atoms are in different shades of magenta and violet, nitrogen atom is blue, and carbon atoms are gray. In (a) and (c), molecular presentations of trajectories are given (displacement of the carbon atoms are not shown); in (b) and (d), the trajectories are shown in polar coordinates (only angular part) emphasizing rotational component of the cluster dynamics (carbon atoms are shown as black dots with relatively small displacements). (Reprinted with permission from Popov, A. A. et al., *ACS Nano*, 4, 4857–4871, 2010. Copyright 2010 American Chemical Society.)

revealed similarity of rotational motion of Y_3 and Y_3N clusters, both remaining rather rigid.[35] Specific dynamic behavior was found in $TiSc_2N@C_{80}$.[56] In contrast to the freely rotating Sc_3N cluster, the $TiSc_2N$ cluster exhibits only restricted rotations around the Ti–N bond, at the 10-ps timescale, and Ti atoms remain locked near its bonding site (Figure 3.8). MD study of $Sc_3N@C_{80}(CF_3)_2$ confirmed that exohedral derivatization restricts rotation of the endohedral cluster and that cluster dynamics strongly depends on the charge of the molecule.[129] Recently, 50-ps-long DFT-MD trajectories have been studied for $Sc_4O_2@C_{80}$, its cation, and anion at 300 K.[121]

In addition to the description of dynamics of endohedral species, MD simulations can be useful in the prediction and interpretation of spectroscopic properties of EMFs with rotating clusters. For such molecules, results of static calculations can give significant errors, and more reliable modeling requires averaging of the desired spectroscopic property over MD trajectories.

For example, averaging of cage carbon signals in ^{13}C NMR spectra of $Sc_3N@C_{80}$ was studied by Heine et al.[137] DFTB-based MD simulations were used to follow motions of the cluster on the nanosecond timescale, and chemical shifts were then computed for selected snapshots at the DFT level. This procedure resulted in better agreement of experimental and computed chemical shifts than static DFT calculations.[137]

Dynamical averaging is also important for prediction of hfcc in ESR spectra of EMF radicals, because hfcc values are very sensitive to the cluster orientation inside the cage. This factor is especially important for those EMFs whose spin density has a large contribution from endohedral clusters such as $Sc_3N@C_{80}^-$, and its $a(^{45}Sc)$ value averaged over the PBE-MD trajectory was found to be significantly larger than the constant computed at the same level of theory in the lowest energy static configuration.[129] MD-based computations of ESR parameters were also reported for mono- and trianion of $Sc_3N@C_{80}(CF_3)_2$[129] and for cation and anion radicals of $Sc_4O_2@C_{80}$.[121] *Ab initio* MD studies of $TiSc_2N@C_{80}$[56] and charged states of $Sc_3N@C_{80}$,[129] $Sc_3N@C_{80}(CF_3)_2$,[129] and $Sc_4O_2@C_{80}$[121] revealed important aspects of the spin density dynamics in these radicals, defined as *spin flow*.[56] Spin density distribution in these EMFs is very flexible and even small displacements of the metal atoms result in drastic changes of spin populations.

Vibrational spectra can be obtained from Fourier transformation of the autocorrelation function of kinetic energy calculated along the MD trajectory. Kemner and Zerbetto computed vibrational density of states for $M@C_{82}$ and $Eu@C_{74}$ molecules using classical MD simulation and showed that metal motions are chaotic at temperatures above 50 K.[144] Vietze and Seifert reported vibrational spectra of several EMF molecules computed based on the DFTB-BOMD trajectories.[139–141,145] They also showed that MD-based spectra can be noticeably different from those computed in the static approximation.

3.4   CONCLUDING REMARKS

Computational studies of endohedral metallofullerenes have developed dramatically over last 20 years and now can successfully address many problems raised in experimental studies. The factors which determine molecular structures of EMFs are well understood, and structures of newly isolated EMFs can be predicted with high degree of certainty. Likewise, analysis of the electronic structure of EMFs and interpretation of their spectroscopic properties benefit a lot from DFT computations. Study of the dynamical properties of EMFs, however, is a more complex problem. Static DFT calculations can give a general idea of the possible dynamic behavior of endohedral clusters, but more detailed description requires molecular dynamics simulations. At the *ab initio* level, such simulations are still limited to tens of picoseconds, but the situation is likely to improve in the near future with the continuous development of computational hardware and software. Finally, understanding of the formation mechanism of EMFs still remains a challenging problem. Its solution is hardly possible by means of static calculations and requires reliable description of complex dynamic processes, which take place during the arc-discharge synthesis.

REFERENCES

1. Johnson, R. D.; Devries, M. S.; Salem, J.; Bethune, D. S.; Yannoni, C. S., Electron-paramagnetic resonance studies of Lanthanum-containing C_{82}. *Nature* **1992**, *355*, 239–240.
2. Shinohara, H.; Sato, H.; Saito, Y.; Ohkohchi, M.; Ando, Y., Mass spectroscopic and ESR characterization of soluble yttrium-containing metallofullerenes YC_{82} and Y_2C_{82}. *J. Phys. Chem.* **1992**, *96*, 3571–3573.
3. Weaver, J. H.; Chai, Y.; Kroll, G. H.; Jin, C.; Ohno, T. R.; Haufler, R. E.; Guo, T. et al. Xps probes of Carbon-caged metals. *Chem. Phys. Lett.* **1992**, *190*, 460–464.
4. Rosen, A.; Wastberg, B., 1st-Principle calculations of the ionization-potentials and electron-affinities of the spheroidal molecules C_{60} and LaC_{60}. *J. Am. Chem. Soc.* **1988**, *110*, 8701–8703.
5. Rosen, A.; Wastberg, B., Electronic-structure of spheroidal metal containing Carbon shells—study of the LaC_{60} and C_{60} clusters and their ions within the local density approximation. *Z. Phys. D Atoms Mol. Clusters* **1989**, *12*, 387–390.
6. Nagase, S.; Kobayashi, K.; Akasaka, T., Recent progress in endohedral dimetallofullerenes. *J. Mol. Struct. (Theochem)* **1997**, *398*, 221–227.
7. Kobayashi, K.; Nagase, S., Structures and electronic states of $M@C_{82}$ (M = Sc, Y, La and lanthanides). *Chem. Phys. Lett.* **1998**, *282*, 325–329.
8. Kobayashi, K.; Nagase, S.; Akasaka, T., Endohedral dimetallofullerenes $Sc_2@C_{84}$ and $La_2@C_{80}$. Are the metal atoms still inside the fullerene cages? *Chem. Phys. Lett.* **1996**, *261*, 502–506.
9. Shinohara, H., Endohedral metallofullerenes. *Rep. Prog. Phys.* **2000**, *63*, 843–892.
10. Chaur, M. N.; Melin, F.; Ortiz, A. L.; Echegoyen, L., Chemical, electrochemical, and structural properties of endohedral metallofullerenes. *Angew. Chem. Int. Ed.* **2009**, *48*, 7514–7538.
11. Dunsch, L.; Yang, S., Metal Nitride clusterfullerenes: Their current state and future prospects. *Small* **2007**, *3*, 1298–1320.
12. Suzuki, T.; Kikuchi, K.; Oguri, F.; Nakao, Y.; Suzuki, S.; Achiba, Y.; Yamamoto, K.; Funasaka, H.; Takahashi, T., Electrochemical properties of fullerenolanthanides. *Tetrahedron* **1996**, *52*, 4973–4982.
13. Popov, A. A.; Krause, M.; Yang, S. F.; Wong, J.; Dunsch, L., C_{78} cage isomerism defined by trimetallic nitride cluster size: A computational and vibrational spectroscopic study. *J. Phys. Chem. B* **2007**, *111*, 3363–3369.
14. Yang, S. F.; Popov, A. A.; Dunsch, L., Violating the isolated pentagon rule (IPR): The endohedral non-IPR cage of $Sc_3N@C_{70}$. *Angew. Chem. Int. Ed.* **2007**, *46*, 1256–1259.
15. Popov, A. A.; Dunsch, L., Structure, stability, and cluster-cage interactions in Nitride clusterfullerenes $M_3N@C_{2n}$ (M = Sc, Y; 2n = 68–98): A density functional theory study. *J. Am. Chem. Soc.* **2007**, *129*, 11835–11849.
16. Krause, M.; Popov, A.; Dunsch, L., Vibrational structure of endohedral fullerene $Sc_3N@C_{78}(D_{3h}')$: Evidence for a strong coupling between the Sc_3N cluster and C_{78} cage. *ChemPhysChem* **2006**, *7*, 1734–1740.
17. Yang, S. F.; Kalbac, M.; Popov, A.; Dunsch, L., A facile route to the non-IPR fullerene $Sc_3N@C_{68}$: Synthesis, spectroscopic characterization, and density functional theory computations (IPR = isolated pentagon rule). *Chem. Eur. J.* **2006**, *12*, 7856–7863.
18. Popov, A. A.; Yang, S. F.; Kalbac, M.; Rapta, P.; Dunsch, L., Electronic structure of $Sc_3N@C_{68}$ in neutral and charged states: An experimental and TD-DFT study. In *Computational Methods in Science and Engineering, Vol 2—Advances in Computational Science*, Simos, T. E.; Maroulis, G., Eds. Amer Institute of Physics: Melville, NY, 2009; Vol. 1148, pp. 712–715.
19. Popov, A. A., Metal-cage bonding, molecular structures and vibrational spectra of endohedral fullerenes: Bridging experiment and theory *J. Comput. Theor. Nanosci.* **2009**, *6*, 292–317.

20. Popov, A. A.; Kästner, C.; Krause, M.; Dunsch, L., Carbon cage vibrations of M@C$_{82}$ and M$_2$@C$_{2n}$ (M = La, Ce; 2n = 72, 78, 80): The role of the metal atoms. *Fuller. Nanotub. Carbon Nanostruct.* **2014**, *22*, 202–214.

21. Kessler, B.; Bringer, A.; Cramm, S.; Schlebusch, C.; Eberhardt, W.; Suzuki, S.; Achiba, Y.; Esch, F.; Barnaba, M.; Cocco, D., Evidence for incomplete charge transfer and La-derived states in the valence bands of endohedrally doped La@C$_{82}$. *Phys. Rev. Lett.* **1997**, *79*, 2289–2292.

22. Alvarez, L.; Pichler, T.; Georgi, P.; Schwieger, T.; Peisert, H.; Dunsch, L.; Hu, Z. et al. Electronic structure of pristine and intercalated Sc$_3$N@C$_{80}$ metallofullerene. *Phys. Rev. B* **2002**, *66*, 035107.

23. Lu, J.; Zhang, X. W.; Zhao, X. G.; Nagase, S.; Kobayashi, K., Strong metal-cage hybridization in endohedral La@C$_{82}$, Y@C$_{82}$ and Sc@C$_{82}$. *Chem. Phys. Lett.* **2000**, *332*, 219–224.

24. Muthukumar, K.; Larsson, J. A., A density functional study of Ce@C$_{82}$: Explanation of the Ce preferential bonding site. *J. Phys. Chem. A* **2008**, *112*, 1071–1075.

25. Campanera, J. M.; Bo, C.; Olmstead, M. M.; Balch, A. L.; Poblet, J. M., Bonding within the endohedral fullerenes Sc$_3$N@C$_{78}$ and Sc$_3$N@C$_{80}$ as determined by density functional calculations and reexamination of the crystal structure of Sc$_3$N@C$_{78}$•Co(OEP)•1.5(C$_6$H$_6$)•0.3(CHCl$_3$). *J. Phys. Chem. A* **2002**, *106*, 12356–12364.

26. Yang, S.; Yoon, M.; Hicke, C.; Zhang, Z.; Wang, E., Electron transfer and localization in endohedral metallofullerenes: Ab initio density functional theory calculations. *Phys. Rev. B* **2008**, *78*, 115435.

27. Liu, D.; Hagelberg, F.; Park, S. S., Charge transfer and electron backdonation in metallofullerenes encapsulating NSc$_3$. *Chem. Phys.* **2006**, *330*, 380–386.

28. Wu, J.; Hagelberg, F., Computational study on C$_{80}$ enclosing mixed Trimetallic Nitride clusters of the form Gd$_x$M$_{3-x}$N (M = Sc, Sm, Lu) *J. Phys. Chem. C* **2008**, *112*, 5770–5777.

29. Popov, A. A.; Dunsch, L., Hindered cluster rotation and ^{45}Sc hyperfine splitting constant in distonoid anion radical Sc$_3$N@C$_{80}$, and spatial spin charge separation as a general principle for anions of endohedral fullerenes with metal-localized lowest unoccupied molecular orbitals. *J. Am. Chem. Soc.* **2008**, *130*, 17726–17742.

30. Bader, R. F. W., *Atoms in Molecules—A Quantum Theory*. Oxford University Press: Oxford, 1990.

31. Popov, A. A.; Dunsch, L., The bonding situation in endohedral metallofullerenes as studied by quantum theory of atoms in molecules (QTAIM). *Chem. Eur. J.* **2009**, *15*, 9707–9729.

32. Becke, A. D.; Edgecombe, K. E., A simple measure of electron localization in atomic and molecular systems. *J. Chem. Phys.* **1990**, *92*, 5397–5403.

33. Savin, A.; Jepsen, O.; Flad, J.; Andersen, O. K.; Preuss, H.; von Schnering, H. G., Electron localization in solid-state structures of the elements: The diamond structure. *Angew. Chem. Int. Ed. Engl.* **1992**, *31*, 187–188.

34. Bader, R. F. W.; Johnson, S.; Tang, T. H.; Popelier, P. L. A., The electron pair. *J. Phys. Chem.* **1996**, *100*, 15398–15415.

35. Popov, A. A.; Zhang, L.; Dunsch, L., A pseudoatom in a cage: Trimetallofullerene Y$_3$@C$_{80}$ mimics Y$_3$N@C$_{80}$ with Nitrogen substituted by a pseudoatom. *ACS Nano* **2010**, *4*, 795–802.

36. Inoue, T.; Tomiyama, T.; Sugai, T.; Okazaki, T.; Suematsu, T.; Fujii, N.; Utsumi, H.; Nojima, K.; Shinohara, H., Trapping a C$_2$ radical in endohedral metallofullerenes: Synthesis and structures of (Y$_2$C$_2$)@C$_{82}$ (Isomers I, II, and III). *J. Phys. Chem. B* **2004**, *108*, 7573–7579.

37. Umemoto, H.; Ohashi, K.; Inoue, T.; Fukui, N.; Sugai, T.; Shinohara, H., Synthesis and UHV-STM observation of the T_d-symmetric Lu metallofullerene: Lu$_2$@C$_{76}$(T_d). *Chem. Commun.* **2010**, *46*, 5653–5655.

38. Kurihara, H.; Lu, X.; Iiduka, Y.; Mizorogi, N.; Slanina, Z.; Tsuchiya, T.; Nagase, S.; Akasaka, T., $Sc_2@C_{3v}(8)$-C_{82} vs. $Sc_2C_2@C_{3v}(8)$-C_{82}: Drastic effect of C_2 capture on the redox properties of scandium metallofullerenes. *Chem. Commun.* **2012**, *48*, 1290–1292.

39. Olmstead, M. M.; de Bettencourt-Dias, A.; Stevenson, S.; Dorn, H. C.; Balch, A. L., Crystallographic characterization of the structure of the endohedral fullerene {$Er_2@$ C_{82} Isomer I} with C_s cage symmetry and multiple sites for erbium along a band of ten contiguous hexagons. *J. Am. Chem. Soc.* **2002**, *124*, 4172–4173.

40. Olmstead, M. M.; Lee, H. M.; Stevenson, S.; Dorn, H. C.; Balch, A. L., Crystallographic characterization of Isomer 2 of $Er_2@C_{82}$ and comparison with Isomer 1 of $Er_2@C_{82}$. *Chem. Commun.* **2002**, 2688–2689.

41. Ito, Y.; Okazaki, T.; Okubo, S.; Akachi, M.; Ohno, Y.; Mizutani, T.; Nakamura, T.; Kitaura, R.; Sugai, T.; Shinohara, H., Enhanced 1520 nm photoluminescence from Er^{3+} ions in di-erbium-carbide metallofullerenes (Er_2C_2)$@C_{82}$ (isomers I, II, and III). *ACS Nano* **2007**, *1*, 456–462.

42. Yang, T.; Zhao, X.; Osawa, E., Can a metal–metal bond hop in the fullerene cage? *Chem. Eur. J.* **2011**, *17*, 10230–10234.

43. Zuo, T.; Xu, L.; Beavers, C. M.; Olmstead, M. M.; Fu, W.; Crawford, T. D.; Balch, A. L.; Dorn, H. C., $M_2@C_{79}N$ (M = Y, Tb): Isolation and characterization of stable endohedral metallofullerenes exhibiting M⋯M bonding interactions inside Aza[80]fullerene cages. *J. Am. Chem. Soc.* **2008**, *130*, 12992–12997.

44. Fu, W.; Zhang, J.; Fuhrer, T.; Champion, H.; Furukawa, K.; Kato, T.; Mahaney, J. E. et al. $Gd_2@C_{79}N$: Isolation, characterization, and monoadduct formation of a very stable heterofullerene with a magnetic spin state of S = 15/2. *J. Am. Chem. Soc.* **2011**, *133*, 9741–9750.

45. Stevenson, S.; Mackey, M. A.; Stuart, M. A.; Phillips, J. P.; Easterling, M. L.; Chancellor, C. J.; Olmstead, M. M.; Balch, A. L., A distorted tetrahedral metal oxide cluster inside an Icosahedral Carbon cage. Synthesis, isolation, and structural characterization of $Sc_4(\mu_3\text{-}O)_2@I_h$-$C_{80}$. *J. Am. Chem. Soc.* **2008**, *130*, 11844–11845.

46. Popov, A. A.; Avdoshenko, S. M.; Pendás, A. M.; Dunsch, L., Bonding between strongly repulsive metal atoms: An oxymoron made real in a confined space of endohedral metallofullerenes. *Chem. Commun.* **2012**, *48*, 8031–8050.

47. Okazaki, T.; Suenaga, K.; Lian, Y. F.; Gu, Z. N.; Shinohara, H., Intrafullerene electron transfers in Sm-containing metallofullerenes: $Sm@C_{2n}$ ($74 \leq 2n \leq 84$). *J. Mol. Graph.* **2001**, *19*, 244–251.

48. Okazaki, T.; Suenaga, K.; Lian, Y. F.; Gu, Z. N.; Shinohara, H., Direct EELS observation of the oxidation states of Sm atoms in $Sm@C_{2n}$ metallofullerenes ($74 \leq 2n \leq 84$). *J. Chem. Phys.* **2000**, *113*, 9593–9597.

49. Ralchenko, Y.; Kramida, A. E.; Reader, J., NIST Atomic Spectra Database (ver. 4.1.0), [Online]. National Institute of Standards and Technology, Gaithersburg, MD: 2011.

50. Dunsch, L.; Yang, S.; Zhang, L.; Svitova, A.; Oswald, S.; Popov, A. A., Metal sulfide in a C_{82} fullerene cage: A new form of endohedral clusterfullerenes. *J. Am. Chem. Soc.* **2010**, *132*, 5413–5421.

51. Mercado, B. Q.; Sruart, M. A.; Mackey, M. A.; Pickens, J. E.; Confait, B. S.; Stevenson, S.; Easterling, M. L. et al. $Sc_2(\mu_2\text{-}O)$ trapped in a fullerene cage: The isolation and structural characterization of $Sc_2(\mu_2\text{-}O)@C_s(6)$-$C_{82}$ and the relevance of the thermal and entropic effects in fullerene isomer selection. *J. Am. Chem. Soc.* **2010**, *132*, 12098–12105.

52. Jin, P.; Zhou, Z.; Hao, C.; Gao, Z.; Tan, K.; Lu, X.; Chen, Z., NC unit trapped by fullerenes: A density functional theory study on $Sc_3NC@C_{2n}$ (2n = 68, 78 and 80). *Phys. Chem. Chem. Phys.* **2010**, *12*, 12442–12449.

53. Martín Pendás, A.; Francisco, E.; Blanco, M. A., Binding energies of first row diatomics in the light of the interacting quantum atoms approach. *J. Phys. Chem. A* **2006**, *110*, 12864–12869.

54. Francisco, E.; Martín Pendás, A.; Blanco, M. A., A molecular energy decomposition scheme for atoms in molecules. *J. Chem. Theory Comput.* **2005**, *2*, 90–102.

55. Blanco, M. A.; Martín Pendás, A.; Francisco, E., Interacting quantum atoms: A correlated energy decomposition scheme based on the quantum theory of atoms in molecules. *J. Chem. Theory Comput.* **2005**, *1*, 1096–1109.

56. Popov, A. A.; Chen, C.; Yang, S.; Lipps, F.; Dunsch, L., Spin-flow vibrational spectroscopy of molecules with flexible spin density: Electrochemistry, ESR, cluster and spin dynamics, and bonding in $TiSc_2N@C_{80}$. *ACS Nano* **2010**, *4*, 4857–4871.

57. Fowler, P. W.; Manolopoulos, D. E., Magic numbers and stable structures for fullerenes, fullerides and fullerenium ions. *Nature* **1992**, *355*, 428–430.

58. Fowler, P. W.; Zerbetto, F., Charging and equilibration of fullerene isomers. *Chem. Phys. Lett.* **1995**, *243*, 36–41.

59. Nakao, K.; Kurita, N.; Fujita, M., Ab-Initio molecular-orbital calculation for C_{70} and seven isomers of C_{80}. *Phys. Rev. B* **1994**, *49*, 11415–11420.

60. Kobayashi, K.; Nagase, S.; Akasaka, T., A theoretical study of C_{80} and $La_2@C_{80}$. *Chem. Phys. Lett.* **1995**, *245*, 230–236.

61. Kobayashi, K.; Nagase, S., Structures of the $Ca@C_{82}$ isomers: A theoretical prediction. *Chem. Phys. Lett.* **1997**, *274*, 226–230.

62. Zywietz, T. K.; Jiao, H.; Schleyer, R.; de Meijere, A., Aromaticity and antiaromaticity in oligocyclic annelated five-membered ring systems. *J. Org. Chem.* **1998**, *63*, 3417–3422.

63. Kobayashi, K.; Nagase, S.; Yoshida, M.; Osawa, E., Endohedral metallofullerenes. Are the isolated pentagon rule and fullerene structures always satisfied? *J. Am. Chem. Soc.* **1997**, *119*, 12693–12694.

64. Wang, C. R.; Kai, T.; Tomiyama, T.; Yoshida, T.; Kobayashi, Y.; Nishibori, E.; Takata, M.; Sakata, M.; Shinohara, H., C_{66} fullerene encaging a scandium dimer. *Nature* **2000**, *408*, 426–427.

65. Stevenson, S.; Fowler, P. W.; Heine, T.; Duchamp, J. C.; Rice, G.; Glass, T.; Harich, K.; Hajdu, E.; Bible, R.; Dorn, H. C., Materials science—a stable non-classical metallofullerene family. *Nature* **2000**, *408*, 427–428.

66. Chen, N.; Mulet-Gas, M.; Li, Y.-Y.; Stene, R. E.; Atherton, C. W.; Rodriguez-Fortea, A.; Poblet, J. M.; Echegoyen, L., $Sc_2S@C_2(7892)$-C_{70}: A metallic sulfide cluster inside a non-IPR C_{70} cage. *Chem. Sci.* **2013**, *4*, 180–186.

67. Chen, N.; Beavers, C. M.; Mulet-Gas, M.; Rodriguez-Fortea, A.; Munoz, E. J.; Li, Y.-Y.; Olmstead, M. M.; Balch, A. L.; Poblet, J. M.; Echegoyen, L., $Sc_2S@C_s(10528)$-C_{72}: A dimetallic sulfide endohedral fullerene with a non-IPR cage. *J. Am. Chem. Soc.* **2012**, *134*, 7851–7860.

68. Wakahara, T.; Nikawa, H.; Kikuchi, T.; Nakahodo, T.; Rahman, G. M. A.; Tsuchiya, T.; Maeda, Y. et al. $La@C_{72}$ having a non-IPR carbon cage. *J. Am. Chem. Soc.* **2006**, *128*, 14228–14229.

69. Kato, H.; Taninaka, A.; Sugai, T.; Shinohara, H., Structure of a missing-caged metallofullerene: $La_2@C_{72}$. *J. Am. Chem. Soc.* **2003**, *125*, 7782–7783.

70. Lu, X.; Nikawa, H.; Nakahodo, T.; Tsuchiya, T.; Ishitsuka, M. O.; Maeda, Y.; Akasaka, T. et al. Chemical understanding of a non-IPR metallofullerene: Stabilization of encaged metals on fused-Pentagon bonds in $La_2@C_{72}$. *J. Am. Chem. Soc.* **2008**, *130*, 9129–9136.

71. Yang, S.; Popov, A. A.; Dunsch, L., The role of an asymmetric nitride cluster on a fullerene cage: The Non-IPR endohedral $DySc_2N@C_{76}$. *J. Phys. Chem. B* **2007**, *111*, 13659–13663.

72. Beavers, C. M.; Chaur, M. N.; Olmstead, M. M.; Echegoyen, L.; Balch, A. L., Large metal ions in a relatively small fullerene cage: The structure of $Gd_3N@C_2(22010)$-C_{78} departs from the isolated Pentagon rule. *J. Am. Chem. Soc.* **2009**, *131*, 11519–11524.

73. Ma, Y.; Wang, T.; Wu, J.; Feng, Y.; Xu, W.; Jiang, L.; Zheng, J.; Shu, C.; Wang, C., Size effect of endohedral cluster on fullerene cage: Preparation and structural studies of $Y_3N@C_{78}$-C_2. *Nanoscale* **2011**, *3*, 4955–4957.

74. Wu, J.; Wang, T.; Ma, Y.; Jiang, L.; Shu, C.; Wang, C., Synthesis, isolation, characterization, and theoretical studies of $Sc_3NC@C_{78}$-C_2. *J. Phys. Chem. C* **2011**, *115*, 23755–23759.

75. Mercado, B. Q.; Beavers, C. M.; Olmstead, M. M.; Chaur, M. N.; Walker, K.; Holloway, B. C.; Echegoyen, L.; Balch, A. L., Is the isolated Pentagon rule merely a suggestion for endohedral fullerenes? The structure of a second egg-shaped endohedral fullerene— $Gd_3N@C_s(39663)$-C_{82}. *J. Am. Chem. Soc.* **2008**, *130*, 7854–7855.

76. Beavers, C. M.; Zuo, T. M.; Duchamp, J. C.; Harich, K.; Dorn, H. C.; Olmstead, M. M.; Balch, A. L., $Tb_3N@C_{84}$: An improbable, egg-shaped endohedral fullerene that violates the isolated pentagon rule. *J. Am. Chem. Soc.* **2006**, *128*, 11352–11353.

77. Zuo, T.; Walker, K.; Olmstead, M. M.; Melin, F.; Holloway, B. C.; Echegoyen, L.; Dorn, H. C. et al. New egg-shaped fullerenes: Non-isolated pentagon structures of $Tm_3N@$ $C_s(51365)$-C_{84} and $Gd_3N@C_s(51365)$-C_{84}. *Chem. Commun.* **2008**, 1067–1069.

78. Yang, T.; Zhao, X.; Li, S.-T.; Nagase, S., Is the isolated Pentagon rule always satisfied for metallic carbide endohedral fullerenes? *Inorg. Chem.* **2012**, *51*, 11223–11225.

79. Zhang, J.; Bowles, F. L.; Bearden, D. W.; Ray, W. K.; Fuhrer, T.; Ye, Y.; Dixon, C. et al. A missing link in the transformation from asymmetric to symmetric metallofullerene cages implies a top-down fullerene formation mechanism. *Nat. Chem.* **2013**, *5*, 880–885.

80. Zheng, H.; Zhao, X.; Wang, W.-W.; Yang, T.; Nagase, S., $Sc_2@C_{70}$ rather than $Sc_2C_2@$ C_{68}: Density functional theory characterization of metallofullerene Sc_2C_{70}. *J. Chem. Phys.* **2012**, *137*, 014308.

81. Nagase, S.; Kobayashi, K.; Akasaka, T., Unconventional cage structures of endohedral metallofullerenes. *Theochem. J. Mol. Struct.* **1999**, *462*, 97–104.

82. Slanina, Z.; Uhlik, F.; Nagase, S., Computed structures of two known $Yb@C_{74}$ isomers. *J. Phys. Chem. A* **2006**, *110*, 12860–12863.

83. Zheng, H.; Zhao, X.; Ren, T.; Wang, W., C_{74} Endohedral metallofullerenes violating the isolated Pentagon rule: A density functional theory study. *Nanoscale* **2012**, *4*, 4530–4536.

84. Yang, T.; Zhao, X.; Xu, Q.; Zhou, C.; He, L.; Nagase, S., Non-IPR endohedral fullerene $Yb@$ C_{76}: Density functional theory characterization. *J. Mater. Chem.* **2011**, *21*, 12206–12209.

85. Yang, T.; Zhao, X.; Xu, Q.; Zheng, H.; Wang, W.-W.; Li, S.-T., Probing the role of encapsulated alkaline earth metal atoms in endohedral metallofullerenes $M@C_{76}$ (M = Ca, Sr, and Ba) by first-principles calculations. *Dalton Trans.* **2012**, *41*, 5294–5300.

86. Zhao, X.; Gao, W.-Y.; Yang, T.; Zheng, J.-J.; Li, L.-S.; He, L.; Cao, R.-J.; Nagase, S., Violating the isolated Pentagon rule (IPR): Endohedral non-IPR C_{98} cages of $Gd_2@C_{98}$. *Inorg. Chem.* **2012**, *51*, 2039–2045.

87. Li, F.-F.; Chen, N.; Mulet-Gas, M.; Triana, V.; Murillo, J.; Rodriguez-Fortea, A.; Poblet, J. M.; Echegoyen, L., $Ti_2S@D_{3h}(24109)$-C_{78}: A sulfide cluster metallofullerene containing only transition metals inside the cage. *Chem. Sci.* **2013**, *4*, 3404–3410.

88. Xu, L.; Li, S.-F.; Gan, L.-H.; Shu, C.-Y.; Wang, C.-R., The structures of trimetallic nitride fullerenes $M_3N@C_{88}$: Theoretical evidence of corporation between electron transfer interaction and size effect. *Chem. Phys. Lett.* **2012**, *521*, 81–85.

89. Zheng, J.; Zhao, X.; Dang, J.; Chen, Y.; Xu, Q.; Wang, W., Density functional theory characterization of lanthanum nitride endohedral fullerene: $La_3N@C_{92}$. *Chem. Phys. Lett.* **2011**, *514*, 104–108.

90. Guo, Y.-J.; Yang, T.; Nagase, S.; Zhao, X., Carbide clusterfullerene $Gd_2C_2@C_{92}$ vs dimetallofullerene $Gd_2@C_{94}$: A quantum chemical survey. *Inorg. Chem.* **2014**, *53*, 2012–2021.

91. Zheng, H.; Zhao, X.; Wang, W.-W.; Dang, J.-S.; Nagase, S., Quantum chemical insight into metallofullerenes M_2C_{98}: $M_2C_2@C_{96}$ or $M_2@C_{98}$, which will survive? *J. Phys. Chem. C* **2013**, *117*, 25195–25204.

92. Raghavachari, K., Ground state of C_{84}: Two almost isoenergetic isomers. *Chem. Phys. Lett.* **1992**, *190*, 397–400.

93. Rodriguez-Fortea, A.; Alegret, N.; Balch, A. L.; Poblet, J. M., The maximum pentagon separation rule provides a guideline for the structures of endohedral metallofullerenes. *Nat. Chem.* **2010**, *2*, 955–961.

94. Popov, A. A.; Yang, S.; Dunsch, L., Endohedral fullerenes. *Chem. Rev.* **2013**, *113*, 5989–6113.

95. Cao, B. P.; Wakahara, T.; Tsuchiya, T.; Kondo, M.; Maeda, Y.; Rahman, G. M. A.; Akasaka, T.; Kobayashi, K.; Nagase, S.; Yamamoto, K., Isolation, characterization, and theoretical study of $La_2@C_{78}$. *J. Am. Chem. Soc.* **2004**, *126*, 9164–9165.

96. Tan, K.; Lu, X., Ti_2C_{80} is more likely a titanium carbide endohedral metallofullerene $(Ti_2C_2)@C_{78}$. *Chem. Commun.* **2005**, 4444–4446.

97. Olmstead, M. M.; de Bettencourt-Dias, A.; Duchamp, J. C.; Stevenson, S.; Marciu, D.; Dorn, H. C.; Balch, A. L., Isolation and structural characterization of the endohedral fullerene $Sc_3N@C_{78}$. *Angew. Chem. Int. Ed.* **2001**, *40*, 1223–1225.

98. Zhang, J.; Fuhrer, T.; Fu, W.; Ge, J.; Bearden, D. W.; Dallas, J. L.; Duchamp, J. C. et al. Nanoscale fullerene compression of a Yttrium Carbide cluster. *J. Am. Chem. Soc.* **2012**, *134*, 8487–8493.

99. Yang, H.; Jin, H.; Zhen, H.; Wang, Z.; Liu, Z.; Beavers, C. M.; Mercado, B. Q.; Olmstead, M. M.; Balch, A. L., Isolation and crystallographic identification of four isomers of $Sm@C_{90}$. *J. Am. Chem. Soc.* **2011**, *133*, 6299–6306.

100. Yang, H.; Jin, H.; Hong, B.; Liu, Z.; Beavers, C. M.; Zhen, H.; Wang, Z.; Mercado, B. Q.; Olmstead, M. M.; Balch, A. L., Large endohedral fullerenes containing two metal ions, $Sm_2@D_2(35)-C_{88}$, $Sm_2@C_1(21)-C_{90}$, and $Sm_2@D_3(85)-C_{92}$, and their relationship to endohedral fullerenes containing two Gadolinium ions. *J. Am. Chem. Soc.* **2011**, *133*, 16911–16919.

101. Mercado, B. Q.; Jiang, A.; Yang, H.; Wang, Z.; Jin, H.; Liu, Z.; Olmstead, M. M.; Balch, A. L., Isolation and structural characterization of the molecular nanocapsule $Sm_2@D_{3d}(822)-C_{104}$. *Angew. Chem. Int. Ed. Engl.* **2009**, *48*, 9114–9116.

102. Yang, H.; Lu, C.; Liu, Z.; Jin, H.; Che, Y.; Olmstead, M. M.; Balch, A. L., Detection of a family of Gadolinium-containing endohedral fullerenes and the isolation and crystallographic characterization of one member as a metal-carbide encapsulated inside a large fullerene cage. *J. Am. Chem. Soc.* **2008**, *130*, 17296–17300.

103. Che, Y.; Yang, H.; Wang, Z.; Jin, H.; Liu, Z.; Lu, C.; Zuo, T. et al. Isolation and structural characterization of two very large, and largely empty, endohedral fullerenes: $Tm@C_{3v}-C_{94}$ and $Ca@C_{3v}-C_{94}$. *Inorg. Chem.* **2009**, *48*, 6004–6010.

104. Deng, Q.; Popov, A. A., Clusters encapsulated in endohedral metallofullerenes: How strained they are? *J. Am. Chem. Soc.* **2014**, *136*, 4257–4264, DOI:10.1021/ja4122582.

105. Kurihara, H.; Lu, X.; Iiduka, Y.; Mizorogi, N.; Slanina, Z.; Tsuchiya, T.; Akasaka, T.; Nagase, S., $Sc_2C_2@C_{80}$ rather than $Sc_2@C_{82}$: Templated formation of unexpected $C_{2v}(5)-C_{80}$ and temperature-dependent dynamic motion of internal Sc_2C_2 cluster. *J. Am. Chem. Soc.* **2011**, *133*, 2382–2385.

106. Shimotani, H.; Ito, T.; Iwasa, Y.; Taninaka, A.; Shinohara, H.; Nishibori, E.; Takata, M.; Sakata, M., Quantum chemical study on the configurations of encapsulated metal ions and the molecular vibration modes in endohedral dimetallofullerene $La_2@C_{80}$. *J. Am. Chem. Soc.* **2004**, *126*, 364–369.

107. Zhang, J.; Hao, C.; Li, S. M.; Mi, W. H.; Jin, P., Which configuration is more stable for La$_2$@C$_{80}$, D_{3d} or D_{2h}? Recomputation with ZORA methods within ADF. *J. Phys. Chem. C* **2007**, *111*, 7862–7867.

108. Muthukumar, K.; Larsson, J. A., Explanation of the different preferential binding sites for Ce and La in M$_2$@C$_{80}$ (M = Ce, La)—a density functional theory prediction. *J. Mater. Chem.* **2008**, *18*, 3347–3351.

109. Akasaka, T.; Nagase, S.; Kobayashi, K.; Walchli, M.; Yamamoto, K.; Funasaka, H.; Kako, M.; Hoshino, T.; Erata, T., ^{13}C and ^{139}La NMR studies of La$_2$@C$_{80}$: First evidence for circular motion of metal atoms in endohedral dimetallofullerenes. *Angew. Chem. Int. Ed. Engl.* **1997**, *36*, 1643–1645.

110. Nishibori, E.; Takata, M.; Sakata, M.; Taninaka, A.; Shinohara, H., Pentagonal-dodecahedral La$_2$ charge density in [80-I_h]fullerene: La$_2$@C$_{80}$. *Angew. Chem. Int. Ed.* **2001**, *40*, 2998–2999.

111. Kobayashi, K.; Sano, Y.; Nagase, S., Theoretical study of endohedral metallofullerenes: Sc$_{3-n}$La$_n$N@C$_{80}$ (n = 0–3). *J. Comput. Chem.* **2001**, *22*, 1353–1358.

112. Gan, L. H.; Yuan, R., Influence of cluster size on the structures and stability of trimetallic nitride fullerenes M$_3$N@C$_{80}$. *ChemPhysChem* **2006**, *7*, 1306–1310.

113. Yanov, I.; Kholod, Y.; Simeon, T.; Kaczmarek, A.; Leszczynski, J., Local minima conformations of the Sc$_3$N@C$_{80}$ endohedral complex: Ab initio quantum chemical study and suggestions for experimental verification. *Int. J. Quantum Chem.* **2006**, *106*, 2975–2980.

114. Valencia, R.; Rodriguez-Fortea, A.; Clotet, A.; de Graaf, C.; Chaur, M. N.; Echegoyen, L.; Poblet, J. M., Electronic structure and Redox properties of metal Nitride endohedral fullerenes M$_3$N@C$_{2n}$ (M = Sc, Y, La, and Gd; 2n = 80, 84, 88, 92, 96). *Chem. Eur. J.* **2009**, *15*, 10997–11009.

115. Lu, J.; Sabirianov, R. F.; Mei, W. N.; Gao, Y.; Duan, C. G.; Zeng, X. C., Structural and magnetic properties of Gd$_3$N@C$_{80}$. *J. Phys. Chem. B* **2006**, *110*, 23637–23640.

116. Svitova, A. L.; Popov, A. A.; Dunsch, L., Gd/Sc-based mixed metal Nitride cluster-fullerenes: The mutual influence of the cage and cluster size and the role of Sc in the electronic structure. *Inorg. Chem.* **2013**, *52*, 3368–3380.

117. Park, S. S.; Liu, D.; Hagelberg, F., Comparative investigation on non-IPR C$_{68}$ and IPR C$_{78}$ fullerenes encaging Sc$_3$N molecules. *J. Phys. Chem. A* **2005**, *109*, 8865–8873.

118. Slanina, Z.; Kobayashi, K.; Nagase, S., Ca@C$_{74}$ isomers: Relative concentrations at higher temperatures. *Chem. Phys.* **2004**, *301*, 153–157.

119. Valencia, R.; Rodríguez-Fortea, A.; Poblet, J. M., Understanding the stabilization of metal carbide endohedral fullerenes M$_2$C$_2$@C$_{82}$ and related systems. *J. Phys. Chem. A* **2008**, *112*, 4550–4555.

120. Burke, B. G.; Chan, J.; Williams, K. A.; Fuhrer, T.; Fu, W.; Dorn, H. C.; Puretzky, A. A.; Geohegan, D. B., Vibrational spectrum of the endohedral Y$_2$C$_2$@C$_{92}$ fullerene by Raman spectroscopy: Evidence for tunneling of the diatomic C$_2$ molecule. *Phys. Rev. B* **2011**, *83*, 115457.

121. Popov, A. A.; Chen, N.; Pinzón, J. R.; Stevenson, S.; Echegoyen, L. A.; Dunsch, L., Redox-active Scandium Oxide cluster inside a fullerene cage: Spectroscopic, voltammetric, electron spin resonance spectroelectrochemical, and extended density functional theory study of Sc$_4$O$_2$@C$_{80}$ and its ion radicals. *J. Am. Chem. Soc.* **2012**, *134*, 19607–19618.

122. Valencia, R.; Rodriguez-Fortea, A.; Stevenson, S.; Balch, A. L.; Poblet, J. M., Electronic structures of Scandium Oxide endohedral metallofullerenes, Sc$_4$(μ_3-O)$_n$@I_h-C$_{80}$ (n = 2, 3). *Inorg. Chem.* **2009**, *48*, 5957–5961.

123. Kobayashi, K.; Nagase, S.; Maeda, Y.; Wakahara, T.; Akasaka, T., La$_2$@C$_{80}$: Is the circular motion of two La atoms controllable by exohedral addition? *Chem. Phys. Lett.* **2003**, *374*, 562–566.

124. Yamada, M.; Wakahara, T.; Nakahodo, T.; Tsuchiya, T.; Maeda, Y.; Akasaka, T.; Yoza, K.; Horn, E.; Mizorogi, N.; Nagase, S., Synthesis and structural characterization of endohedral pyrrolidinodimetallofullerene: $La_2@C_{80}(CH_2)_2NTrt$. *J. Am. Chem. Soc.* **2006**, *128*, 1402–1403.

125. Wakahara, T.; Yamada, M.; Takahashi, S.; Nakahodo, T.; Tsuchiya, T.; Maeda, Y.; Akasaka, T. et al. Two-dimensional hopping motion of encapsulated La atoms in silylated $La_2@C_{80}$. *Chem. Commun.* **2007**, 2680–2682.

126. Yamada, M.; Nakahodo, T.; Wakahara, T.; Tsuchiya, T.; Maeda, Y.; Akasaka, T.; Kako, M. et al. Positional control of encapsulated atoms inside a fullerene cage by exohedral addition. *J. Am. Chem. Soc.* **2005**, *127*, 14570–14571.

127. Ishitsuka, M. O.; Sano, S.; Enoki, H.; Sato, S.; Nikawa, H.; Tsuchiya, T.; Slanina, Z. et al. Regioselective bis-functionalization of endohedral dimetallofullerene, $La_2@C_{80}$: Extremal La-La distance. *J. Am. Chem. Soc.* **2011**, *133*, 7128–7134.

128. Yamada, M.; Akasaka, T.; Nagase, S., Endohedral metal atoms in Pristine and functionalized fullerene cages. *Acc. Chem. Res.* **2010**, *43*, 92–102.

129. Popov, A. A.; Dunsch, L., Charge controlled changes in the cluster and spin dynamics of $Sc_3N@C_{80}(CF_3)_2$: The flexible spin density distribution and its impact on ESR spectra. *Phys. Chem. Chem. Phys.* **2011**, *13*, 8977–8984.

130. Elliott, B.; Pykhova, A. D.; Rivera, J.; Cardona, C. M.; Dunsch, L.; Popov, A. A.; Echegoyen, L., Spin density and cluster dynamics in $Sc_3N@C_{80}^-$ upon [5,6] exohedral functionalization: An ESR and DFT study. *J. Phys. Chem. C* **2013**, *117*, 2344–2348.

131. Shustova, N. B.; Peryshkov, D. V.; Kuvychko, I. V.; Chen, Y.-S.; Mackey, M. A.; Coumbe, C. E.; Heaps, D. T. et al. Poly(perfluoroalkylation) of metallic Nitride fullerenes reveals addition-pattern fuidelines: Synthesis and characterization of a family of $Sc_3N@C_{80}(CF_3)_n$ (n = 2–16) and their radical anions. *J. Am. Chem. Soc.* **2011**, *133*, 2672–2690.

132. Shustova, N. B.; Chen, Y.-S.; Mackey, M. A.; Coumbe, C. E.; Phillips, J. P.; Stevenson, S.; Popov, A. A.; Boltalina, O. V.; Strauss, S. H., $Sc_3N@(C_{80}-I_h(7))(CF_3)_{14}$ and $Sc_3N@(C_{80}-I_h(7))(CF_3)_{16}$. Endohedral metallofullerene derivatives with exohedral addends on four and eight Triple-Hexagon junctions. Does the Sc_3N cluster control the addition pattern or vice versa? *J. Am. Chem. Soc.* **2009**, *131*, 17630–17637.

133. Popov, A. A.; Shustova, N. B.; Svitova, A. L.; Mackey, M. A.; Coumbe, C. E.; Phillips, J. P.; Stevenson, S.; Strauss, S. H.; Boltalina, O. V.; Dunsch, L., Redox-Tuning endohedral fullerene spin states: From the dication to the Trianion radical of $Sc_3N@C_{80}(CF_3)_2$ in five reversible single-electron steps. *Chem. Eur. J.* **2010**, *16*, 4721–4724.

134. Reich, A.; Panthofer, M.; Modrow, H.; Wedig, U.; Jansen, M., The structure of $Ba@C_{74}$. *J. Am. Chem. Soc.* **2004**, *126*, 14428–14434.

135. Taubert, S.; Straka, M.; Pennanen, T. O.; Sundholm, D.; Vaara, J., Dynamics and magnetic resonance properties of $Sc_3C_2@C_{80}$ and its monoanion. *Phys. Chem. Chem. Phys.* **2008**, *10*, 7158–7168.

136. Andreoni, W.; Curioni, A., Freedom and constraints of a metal atom encapsulated in fullerene cages. *Phys. Rev. Lett.* **1996**, *77*, 834–837.

137. Heine, T.; Vietze, K.; Seifert, G., ^{13}C NMR fingerprint characterizes long time-scale structure of $Sc_3N@C_{80}$ endohedral fullerene. *Magn. Reson. Chem.* **2004**, *42*, S199–S201.

138. Krause, M.; Hulman, M.; Kuzmany, H.; Dubay, O.; Kresse, G.; Vietze, K.; Seifert, G.; Wang, C.; Shinohara, H., Fullerene quantum gyroscope. *Phys. Rev. Lett.* **2004**, *93*, 137403.

139. Vietze, K.; Seifert, G., Intra-cage dynamics in endohedral fullerenes. *AIP Conf. Proc.* **2003**, *685*, 7–10.

140. Vietze, K.; Seifert, G., Fluxional behaviour of Sc_3N in endohedral $Sc_3N@C_{80}$. *AIP Conf. Proc.* **2002**, *633*, 39–42.

141. Vietze, K.; Seifert, G.; Fowler, P. W., Structure and dynamics of endohedral fullerenes. *AIP Conf. Proc.* **2000**, *544*, 131–134.

142. Kodama, T.; Fujii, R.; Miyake, Y.; Suzuki, S.; Nishikawa, H.; Ikemoto, I.; Kikuchi, K.; Achiba, Y., ^{13}C NMR study of Ca@C_{74}: The cage structure and the site-hopping motion of a Ca atom inside the cage. *Chem. Phys. Lett.* **2004**, *399*, 94–97.

143. Xu, J. X.; Tsuchiya, T.; Hao, C.; Shi, Z. J.; Wakahara, T.; Mi, W. H.; Gu, Z. N.; Akasaka, T., Structure determination of a missing-caged metallofullerene: Yb@C_{74} (II) and the dynamic motion of the encaged ytterbium ion. *Chem. Phys. Lett.* **2006**, *419*, 44–47.

144. Kemner, E.; Zerbetto, F., Guest dynamics in endohedrally doped fullerenes. *J. Phys. Chem. B* **2005**, *109*, 15048–15051.

145. Vietze, K.; Seifert, G.; Richter, M.; Dunsch, L.; Krause, M., Endohedral rare earth fullerenes—electronic and dynamic properties. *AIP Conf. Proc.* **1999**, *486*, 128–131.

4 NMR Spectroscopic and X-Ray Crystallographic Characterization of Endohedral Metallofullerenes

CONTENTS

4.1   NMR SPECTROSCOPY

Wenjun Zhang, Muqing Chen, Lipiao Bao, Michio Yamada, and Xing Lu

Nuclear magnetic resonance (NMR) spectroscopy is the study of molecular structure through measurement of the interaction of an oscillation radio-frequency electromagnetic field with a collection of nuclei immersed in a strong external magnetic field.

A NMR spectrum could provide detailed information about molecular structures and dynamics of fullerenes and endohedral metallofullerenes (EMFs).[1] ^{13}C NMR spectroscopy and multinuclear NMR spectroscopy have been commonly used in these characterizations.

4.1.1 ^{13}C NMR Spectroscopy

Nowadays, ^{13}C NMR spectroscopy is one common structural tool to determine the cage structures of EMFs in solution. In the beginning of the EMF research, this method was not widely engaged due to the difficulties in obtaining enough amounts of isomerically pure samples for newly isolated EMFs. Additionally, owing to the low natural abundance of ^{13}C isotope and long relaxation time in fullerenes, the extra accumulation time necessary for ^{13}C NMR spectroscopy measurements of EMFs was not available in common lab conditions. In addition, ^{13}C NMR spectroscopy gives only the symmetry of the carbon cage, and the structure could not be confirmed if the many isomers share the same symmetry.[2] In these cases, theoretical studies can be used to choose the most appropriate isomers.

4.1.1.1 NMR Characterization of Paramagnetic EMFs

Many EMFs are paramagnetic, which limits the use of ^{13}C NMR spectroscopy for the molecular structure elucidation. To solve this problem, Akasaka et al. performed bulk electrolysis on some paramagnetic EMFs, obtaining mono-anions that are diamagnetic and thus suitable for ^{13}C NMR spectroscopic research.[3] In this way, paramagnetic EMFs M@C_{82} (M = Y,[4] La,[5] Ce,[6] Pr,[7] etc.) have been structurally determined with ^{13}C NMR spectroscopy. The results showed that two cage isomers, that is, $C_{2v}(9)$-C_{82} and $C_s(6)$-C_{82}, are both suitable to encapsulate the above-mentioned lanthanoides, with the former more stable.

In principle, a strong interaction between a NMR-active nucleus and an unpaired electron gives rise to large changes in chemical shifts of nuclei located within a specific distance of the paramagnetic center. Such chemical shifts of paramagnetic molecules in solution are generally expressed as a sum of three contributions from diamagnetic (δ_{dia}), Fermi contact (δ_{fc}), and pseudocontact (δ_{pc}) shifts, in which the paramagnetic δ_{fc} and δ_{pc} are proportional to T^{-1} and T^{-2} (T = absolute temperature), respectively (see Equation 4.1).

$$\delta = \delta_{dia} + \delta_{fc} + \delta_{pc} = \delta_{dia} + \frac{c_{fc}}{T} + \frac{c_{pc}}{T^2} \qquad (4.1)$$

Constants c_{fc} and c_{pc} signify characteristic values of individual carbon signals. The Fermi contact interaction is a through-bond, scalar interaction that involves the direct transfer of unpaired electron density to the nuclear spin. The pseudocontact interaction is a through-space dipolar interaction (spin–spin interaction) and depends on the geometrical factors. This is also the case for paramagnetic EMFs. In the field of EMF chemistry, temperature-dependent ^{13}C NMR shifts were first observed in Tm@C_{82} by Kodama et al.[8] In that case, an electron transfer from the Tm atom to the carbon cage generates the divalent EMF where the electronic structure can be described as Tm^{2+}(C_{82})$^{2-}$.[9] However, it is noted that unpaired electrons exist on the Tm ion because of its 4f^{13} electronic structure. Therefore, the interaction between the ^{13}C nucleus spin on the carbon cage and the f electrons on the encaged Tm ions was observed in the variable-temperature (VT) ^{13}C NMR spectra. In contrast, Pr@C_{2v}(9)-C_{82} is a trivalent EMF and its electronic structure can be described as $Pr^{3+}(C_{82})^{3-}$. Therefore, an unpaired electron is distributed on the carbon cage. This paramagnetic nature makes the relaxation time of ^{13}C nuclear spins too short to obstruct the observation of ^{13}C NMR signals in a neutral form. Akasaka et al. prepared the Pr@C_{2v}(9)-C_{82} anion by electrochemical reduction to observe its ^{13}C NMR signals.[7] The generated Pr@C_{2v}(9)-C_{82} anion has a diamagnetic electronic structure for the carbon cage (i.e., the formal charge of the cage is even). Consequently, ^{13}C NMR signals were observed between 105 and 170 ppm accompanied with signal broadening, whereas the signals of La@C_{2v}(9)-C_{82} anion[3] appeared between 135 and 165 ppm. The difference can be ascribed to the interaction between the ^{13}C nucleus spin on the carbon cage and the $4f^2$ electrons remaining on Pr, as found in neutral Tm@C_{82}. Such paramagnetic effects on ^{13}C NMR spectra were also found in Sm@C_{2v}(3)-C_{80}[10] and PrSc@C_{80}.[11]

Cerium-encapsulated EMFs (Ce-EMFs) are suitable species for minute analyses of such paramagnetic NMR shifts, because the encaged Ce ions possess only one 4f electron and the interaction can be considered in a simple manner. The diamagnetic terms can be compared to the corresponding chemical shifts in diamagnetic analogues such as lanthanum-encaged EMFs. For instance, δ_{dia} values of Ce@C_{2v}(9)-C_{82} anion correspond to the chemical shifts of the diamagnetic La@C_{2v}(9)-C_{82} anion. Comparative analysis of their ^{13}C NMR chemical shifts disclosed that pseudocontact terms make a much larger contribution than Fermi contact terms for the chemical shifts of Ce@C_{2v}(9)-C_{82} anion, according to results of analysis of the line-fitting plots of the carbon signals.[6,12] That result is consistent with the fact that there is no significant connection between the encaged Ce atom and cage carbons, and the electronic structure of the Ce@C_{2v}(9)-C_{82} anion can be considered as an ionic model. It is noteworthy that Fermi contact interaction is isotropic, whereas pseudocontact interaction is anisotropic. Therefore, geometrical information can be provided by the analysis of pseudocontact terms in principle. In this context, encaged metal atom position in EMFs was determined by the paramagnetic NMR shift analysis. In axial symmetry, a good approximation for the δ_{pc} is given by the following equation (Equation 4.2)

$$\delta_{pc} = \frac{C(3\cos^2\theta - 1)}{r^3 T^2} \tag{4.2}$$

where:
   r is the distance between the paramagnetic center (i.e., the encaged Ce ion) and cage carbons
   θ is the angle between the r vector and the principle symmetry axis of Ce@C_{2v}(9)-C_{82} anion
   C is a common constant with a negative value for all the cage carbons (see Figure 4.1)

The Ce@C_{2v}(9)-C_{82} anion possesses C_{2v} symmetry; however, the paramagnetic circumstance can be considered to be axially symmetric because of its spherical shape of the carbon cage. In fact, the temperature-dependent NMR shift analysis of the Ce@C_{2v}(9)-C_{82} anion revealed that the encaged Ce ion is located at an off-centered

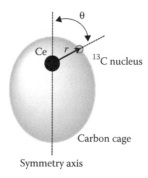

FIGURE 4.1   Definition of r and θ.

position adjacent to a hexagonal ring along the C_2 axis of the carbon cage. Based on Equation 4.2, the distance between Ce and the hexagonal ring was calculated to be approximately 2.1–2.8 Å. This metal position corresponds to the minimum of the electrostatic potential of $C_{2v}(9)$-C_{82}^{3-}.[13] This agreement is important evidence that the electrostatic metal–cage interaction plays a dominant role in determining the metal position.

EMFs containing paramagnetic lanthanide atoms were also studied with NMR studies and the results showed that paramagnetism affected the chemical shifts of the cage carbons dramatically. Utilizing this effect, binding sites of the paramagnetic metal atoms can be probed by the analysis of the temperature dependence of their ^{13}C chemical shifts in combination with theoretical calculations. The simple approximation method is also applicable for dimetallic EMFs, such as $Ce_2@D_2(10611)$-C_{72},[14] $Ce_2@D_{3h}(5)$-C_{78},[15] $Ce_2@I_h(7)$-C_{80},[16] and $Ce_2@D_{5h}(6)$-C_{80}.[17] The essential assumption in the paramagnetic NMR shift analysis of dimetallic Ce-EMFs is that the pseudocontact shifts are produced by a sum of the individual contributions from the two Ce ions. The ^{13}C NMR chemical shifts of $Ce_2@I_h(7)$-C_{80} showed only slight temperature dependence. That observation can be rationalized by the fact that the Ce ions rotate three dimensionally, causing the equalization of the paramagnetic effects for all carbon atoms. Similar week paramagnetic behaviors were also found in Ce-containing trimetallic nitride EMFs $CeSc_2N@I_h(7)$-C_{80}[18] and $Lu_2CeN@I_h(7)$-C_{80}.[19] These NMR data suggested a relatively small energy barrier for rotation of the clusters inside the icosahedral carbon cages.

Interestingly, the c_{pc} values for $Ce_2@D_{5h}(6)$-C_{80} were roughly 10 times larger than those for $Ce_2@I_h(7)$-C_{80}, despite their structural similarity. In fact, if one cuts the $D_{5h}(6)$-C_{80} along the horizontal mirror plane, rotates the top half 36°, and reattaches the top to the bottom, the $I_h(7)$-C_{80} is obtained. Therefore, the geometric factors of $Ce_2@I_h(7)$-C_{80} and $Ce_2@D_{5h}(6)$-C_{80} resemble each other very closely. The large c_{pc} values for $Ce_2@D_{5h}(6)$-C_{80} show that the Ce ions inside $D_{5h}(6)$-C_{80} do not circulate three dimensionally; instead, they move in a restricted way. Comparative analysis of the NMR spectra and computational calculations suggested that the encaged Ce ions circulate two dimensionally along a band of 10 contiguous hexagons inside a $D_{5h}(6)$-C_{80} cage, which was later confirmed by an X-ray crystallographic study.[20]

In contrast, all carbon signals of $Ce_2@D_2(10611)$-C_{72} and $Ce_2@D_{3h}(5)$-C_{78} exhibited much larger temperature-dependent shifts than those of $Ce_2@I_h(7)$-C_{80} and $Ce_2@D_{5h}(6)$-C_{80}. These observation is associated with the fact that the Ce ions are localized

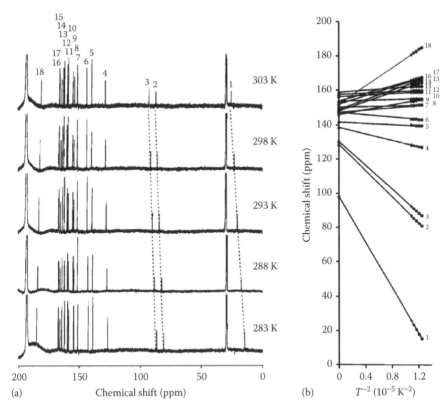

FIGURE 4.2 (a) ^{13}C NMR spectra of $Ce_2@C_{72}$ at 283–303K in CS_2 solution. (b) Line fitting plot for all carbon atoms of $Ce_2@C_{72}$. (Reprinted with permission from Yamada, M. et al., *J. Phys. Chem. A*, 112, 7627–7631, 2008. Copyright 2008 American Chemical Society.)

at the specific positions inside the $D_2(10611)$-C_{72} and $D_{3h}(5)$-C_{78} cages. Enhanced temperature-dependent shifts of the carbon atoms in adjacent pentagons in the non-IPR $Ce_2@D_2(10611)$-C_{72} or in metal-coordinated hexagonal rings in $Ce_2@D_{3h}(5)$-C_{78} were observed in the VT-^{13}C NMR spectra. The ^{13}C NMR spectrum of $Ce_2@D_2(10611)$-C_{72} is shown in Figure 4.2a, which indicates that the signals are very sensitive to temperature. Line-fitting plot of all the cage carbons (Figure 4.2b) revealed that the two paramagnetic Ce atoms interact strongly with the cage carbons and thus affect the chemical shifts dramatically. Therefore, the paramagnetic NMR shift analysis is a powerful tool to determine the localized position of Ce ions in carbon cages.

The paramagnetic effects caused by the buried 4f electrons are also observable in heteronuclear NMR measurements. Temperature dependence in ^{45}Sc NMR measurements was found in $CeSc_2N@I_h(7)$-C_{80}[15] and $Ho_xSc_{3-x}N@C_{80}$ isomers (I, II; x = 1, 2).[21] The ^{45}Sc NMR chemical shifts of $CeSc_2N@I_h(7)$-C_{80} at different temperatures were proportional to T^{-2}, suggesting that the pseudocontact term dominated the chemical shifts. In that case, dominant Fermi contact interaction is less likely, because it would only be possible via spin polarization transfer from Ce to the Sc nucleus though the Ce—N and N—Sc bonds. In $Ho_xSc_{3-x}N@C_{80}$, the Ho-induced paramagnetic shifts are also dominated by the pseudocontact term. However, the rather large difference between the simulated δ_{dia} values obtained by extrapolation to $T^{-2} = 0$ in the line-fitting plot of δ with respect to T^{-2} and the experimental values for $Sc_3N@C_{80}$ indicates that the Fermi contact term is probably not negligible in these systems.

It is noteworthy that the paramagnetic effect possesses a long-range nature and reaches the exohedral functional groups located outside of the carbon cages.[17,22,23] For instance, the ^{1}H NMR spectra of two isomers of $Ce@C_{2v}(9)\text{-}C_{82}(Ad)$ showed the corresponding proton signals in an extremely up-field region. In addition, all the proton signals shifted to the further up-field region with decreasing temperature. That observation was explained by Equation 4.2, because the θ values of all the protons are in the range of $\theta < 54.7°$. A similar trend was also found in the VT-^{1}H NMR spectra of $Ce_2@I_h(7)\text{-}C_{80}(Ad)$. Furthermore, the paramagnetic ^{1}H NMR shift analysis can distinguish two conformers of $Ce_2@I_h(7)\text{-}C_{80}[(Mes_2Si)_2CH_2]$ (Mes = mesityl) and metal atom positions in $Ce_2@I_h(7)\text{-}C_{80}[(CH_2)_2NTrt]$ (Trt = trityl) because of the anisotropic nature.[21] These results demonstrate clearly that the pseudocontact shift analysis is a useful means for NMR structure determination of EMFs and their derivatives in solution.

4.1.1.2   NMR Characterization of Carbide Cluster EMFs

Among the cluster EMFs that have been obtained so far, carbide cluster EMFs are particularly unique because of the encapsulation of two carbon atoms inside the cage along with one to three metal atoms.[24]

For the paramagnetic $Sc_3C_2@C_{80}$, the electrochemical reduction strategy was also applied to get its mono-anions suitable for NMR measurement.[25] The results lead to a surprising finding of the carbide structure because this compound had been long believed as $Sc_3@C_{82}$.

For other diamagnetic species, direct NMR characterization can give sufficient information about the cage symmetry. In addition, ^{13}C NMR spectroscopy can provide useful information about the metal–cage interactions and even dynamics of the metallic clusters inside the carbon cages. A highly temperature-dependent internal Sc_2C_2 cluster dynamics motion was observed by ^{13}C NMR spectroscopy for $Sc_2C_2@C_{2v}(5)\text{-}C_{80}$ (Figure 4.3).[26] The structure of this Sc_2C_{82} isomer was determined unambiguously by single crystal X-ray diffraction to be a carbide cluster EMF instead of $Sc_2@C_{82}$. However, it exhibits more ^{13}C NMR lines than expected for the $C_{2v}(5)\text{-}C_{80}$ cage when the temperature is below 335 K because the cluster is fixed in the cage and the metal atoms interact strongly with the specific cage carbons, engendering C_s symmetry for the whole molecule. At temperatures above 410 K, the cluster starts to rotate rapidly inside the cage, recovering the C_{2v} symmetry of the molecule because of the weak metal–cage interactions. Similar results were also observed in other carbide cluster EMFs, and thus, the temperature-sensitive motional behavior could be a signature of these molecules. Accordingly, these species are potential molecular thermometers useful in extreme conditions.

The signals of the encapsulated C_2-unit (acetylide unit) were also detectable using ^{13}C-enriched samples. The first report was on $Sc_2C_2@C_{3v}(8)\text{-}C_{82}$, $Sc_2C_2@D_{2d}(23)\text{-}C_{84}$, and $[Sc_3C_2@C_{80}]^-$, for which the C_2-signals were observed at 253.2, 249.2, and 328.3 ppm, respectively.[27] Temperature-dependent ^{13}C NMR spectroscopy was used

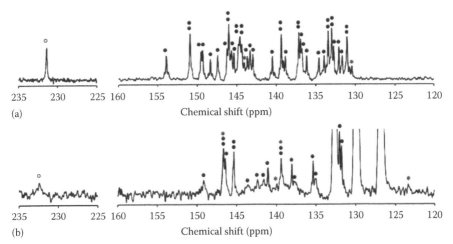

FIGURE 4.3 ^{13}C NMR spectra of $Sc_2C_2@C_{2v}(5)$-C_{80} measured (a) in CS_2 at 298 K and (b) in 1,2-dichlorobenzene at 413 K. The signal from the C_2 moiety of Sc_2C_2 is marked with an open circle; *indicates half intensity. (Reprinted with permission from Kurihara, H. et al., *J. Am. Chem. Soc.*, 133, 2382–2385, 2011. Copyright 2011 American Chemical Society.)

to study the dynamics of the C_2-unit in $Sc_2C_2@C_{3v}(8)$-C_{82}, $Sc_2C_2@D_{2d}(23)$-C_{84}, and $[Sc_3C_2@C_{80}]^-$ as well. The line-width of the signals is about five times wider than that of the cage carbon atoms, which increases with temperature rising due to spin-rotation relaxation. This confirmed that the C_2-unit rotates rapidly inside the cage and the motion is dependent on temperatures. A Sc_2C_{84} isomer is unambiguously identified as a new carbide cluster metallofullerene $Sc_2C_2@C_s(6)$-C_{82} using NMR spectroscopy.[28] The ^{13}C-nuclei signal of the internal C_2-unit was observed at 244.4 ppm with a 15% ^{13}C-enriched sample. Similar broadening and highly temperature-dependent shifts of the signals corresponding to dynamic motion of the internal Sc_2C_2 cluster were also revealed. For the remaining Sc_2C_{84} isomer, NMR and XRD results disclosed that $Sc_2C_2@C_{2v}(9)$-C_{82} is the correct structure. The signal of C_2-unit is observed at 242.7 ppm, and its motion was confirmed again to be temperature dependent.[29]

NMR spectroscopy has also been utilized to characterize $Y_2C_2@C_{2n}$ isomers. Earlier reports only focused on the characterization of cage symmetries.[30] Recently, the signals of the internal C_2-unit are also observed for some compounds which are not ^{13}C-enriched. Based on NMR results and calculations, Dorn and coworkers proposed that the Y_2C_2 cluster prefers a linear structure if the interior space is sufficiently large.[31] For example, in $Y_2C_2@C_{82}$ the cluster adopts a bent butterfly-like configuration, but it is nearly linear in $Y_2C_2@C_{100}$. The authors proposed a *nanoscale fullerene compression effect* on the structures of the internal carbide cluster.

Recently, a monometallic cyanide cluster EMF $YCN@C_s(6)$-C_{82} was reported by Yang and collaborators.[32] This discovery breaks the common belief that multiple (two to four) metal atoms are necessary to form the cluster which is suitable for being encapsulated inside EMFs. ^{13}C NMR spectra confirmed firmly the cage symmetry (Figure 4.4). Meanwhile, the signal from the internal CN unit was found at 292 ppm, which is downfield shifted as compared with the corresponding values of C_2-unit in

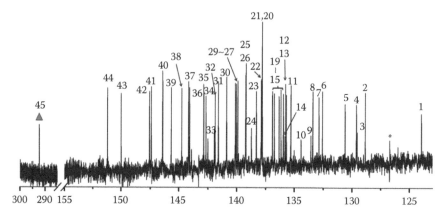

FIGURE 4.4 ^{13}C NMR spectrum of YCN@C_s(6)-C_{82}. The signal from the CN moiety is observed at 292 ppm (number 45).

carbide cluster EMFs. Finally, the molecular structure of this endohedral was unambiguously established by single-crystal X-ray crystallography.

4.1.1.3 NMR Characterization of Nitride Cluster EMFs

As one of the most promising EMF compounds, nitride cluster EMFs have attracted great attention.[33] In the very beginning, ^{13}C NMR spectroscopy has been used to characterize the cage symmetry of Sc$_3$N@C_{80}. In 2006, an extended ^{13}C NMR study was performed on Sc$_3$N@C_{80} at various temperatures, which focused on spin-lattice relaxation times to probe the orientational dynamics of the internal cluster and the Sc$_3$N@C_{80} molecule as well.[34] The measurements showed an activated behavior of the molecular reorientations over the full temperature range, with a similar behavior for the temperature dependence. Combined with spectral data from MAS NMR, the measurements can be interpreted to mean that the motion of the encapsulated Sc$_3$N cluster is independent of that of the C_{80} cage.

A more meaningful work was reported for the Y$_x$Lu$_{3-x}$N@I_h-(7)C_{80} and Lu$_x$Sc$_{3-x}$N@ series, which were systematically studied by ^{13}C NMR spectroscopy.[35] It was revealed that the two types of carbon atoms of the I_h(7)-C_{80} cage exhibited chemical downshifts when the size of the encaged nitride cluster shrinks. The chemical shift is in an ideal linear correlation with the π-orbital axis vector (POAV) pyramidalization angles of the carbon atoms, which are obtained by density functional theory (DFT) calculations. Furthermore, the shifts for pyrene-type carbon atoms are remarkable, which indicates strong interactions between the encaged metal atoms with these pyrene-type carbons. This is helpful in determining the coordination environment of the metal atoms with the cage framework. In this study, it was found that larger metal atoms tend to change their coordination sites from the pentagon/hexagon edge Sc$_3$N@C_{80} to the center of the hexagon. This change increases the pyramidalization of the pyrene carbon atoms.

The relaxation rates of carbon atoms in M$_3$N@C_{80} (M = Sc, Y, Lu) were investigated using ^{13}C NMR spectroscopy.[36] For endohedral nitride cluster fullerenes of I_h(7)-C_{80} cage symmetry, increased relaxation rates have been observed. The enlarged cage size increases the relaxation of the carbons, but the encapsulated metal atoms give an additional dipole–dipole interaction to the relaxation rate of the carbon atoms depending on their magnetic character. For the spherical shape of the fullerene cage, the study showed that the diffusion is fast as compared to the rotation of the molecule at high temperatures. Thus, only a minor deformation of the cage by the internal cluster is found and the cage shape is almost not influenced by the cluster.

4.1.2 Metal Nuclear NMR Spectroscopy

The non-zero nuclear spin of metal and nonmetal atoms in the cage of EMFs both could be engaged to study their molecular structure and dynamics by multinuclear NMR spectroscopy. There are several NMR spectroscopic studies that have been reported, including but not limited to ^{7}Li, ^{45}Sc, ^{89}Y, and ^{139}La.

4.1.2.1 ^{7}Li NMR Spectroscopy

In o-DCB/acetonitrile solution of $[Li^+@C_{60}]SbCl_6^-$ salt, a single ^{7}Li NMR signal at high field ($\delta = -10.5$ ppm with LiCl in D_2O as external reference) was detected. The π-electrons of the fullerene cage present shielding effect on the encapsulated Li^+ ion, which should be responsible for the up-field chemical shift.[37] It was found that the counter ion or exohedral functionalization changed the environment of the Li+ ion dramatically. For instance, the ^{7}Li signal of $[Li^+@C_{60}]PF_6^-$ in o-DCB solution was detected at $\delta = -11.2$ ppm, which is much more up-field shifted, whereas the signals in the [5,6] and [6,6] isomers of $[Li^+@PCBM]PF_6^-$ were observed at -10.7 and -12.3 ppm, respectively.[38] For the water-soluble Li@C_{60}-fullerenol, which is a mixture with a rough composition of Li@$C_{60}(OH)_{18}$, the ^{7}Li NMR signal is between -15 and -19 ppm.[39]

4.1.2.2 ^{45}Sc NMR Spectroscopy

^{45}Sc NMR spectroscopy is probably the most common metal NMR technique in EMF study, partially because of the high abundance of Sc-containing EMFs. Because the Sc atoms always interact strongly with the cage carbons and the interaction could be reflected by measuring the chemical shifts of the Sc atom, ^{45}Sc NMR spectroscopy becomes a powerful tool to probe the EMF internal states. In some cases, temperature-dependent ^{45}Sc NMR spectroscopy can be used to reveal the dynamics of encaged Sc atoms in spite of the fact that the large quadrupole moment of Sc nuclei ($I = 7/2$) would broaden ^{45}Sc NMR lines.

The first ^{45}Sc NMR spectroscopy report was on two isomers of $Sc_2C_2@C_{82}$ (at that time, these two isomers were thought to be $Sc_2@C_{84}$).[40] From 298 to 363 K, the ^{45}Sc NMR spectra of the isomer with the $C_s(6)$-C_{82} cage exhibited two signals of equal intensity with the distance of 50 ppm, indicating that the two Sc atoms are in different chemical environments. The two peaks merged gradually into a sharp one as the temperature increases. Actually, similar results were observed in the variable temperature ^{45}Sc NMR spectra of $Sc_2C_2@C_{2v}(5)$-C_{80} and $Sc_2C_2@C_s(6)$-C_{82} (renewed study) measured in ODCB-d_4.[26,28] These results confirmed that the Sc positions are highly sensitive to the temperatures. At low temperatures, the two Sc atoms are fixed and interact strongly with the cage carbons, but the coordination is activated by the energy introduced from the temperature increasing. The variable temperature

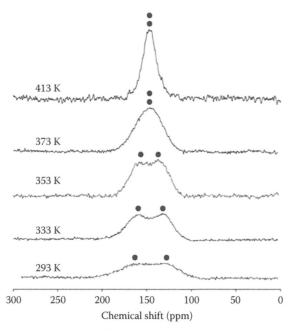

FIGURE 4.5 Variable temperature ^{45}Sc NMR spectra of $Sc_2C_2@C_{2v}(5)$-C_{80}. (Reprinted with permission from Kurihara, H. et al., *J. Am. Chem. Soc.* 133, 2382–2385, 2011. Copyright 2011 American Chemical Society.)

NMR study of $Sc_2C_2@C_{2v}(5)$-C_{80} showed that the two NMR signals coalesced into one when the temperature is over 353 K, and the estimated free energy of the inversion barrier was 63 kJ/mol (Figure 4.5).[26] In contrast, the $Sc_2C_2@C_{3v}(8)$-C_{82} isomer showed only one ^{45}Sc NMR line at 225 ppm in the measured temperature range from 238 to 433 K. Temperature dependence studies using line-width of the single lines in the solid state ^{45}Sc NMR spectra of $Sc_2C_2@C_{3v}(8)$-C_{82} and $Sc_2C_2@C_{82}$-$C_{3v}(8)$ resulted in activation energies of 6.4 and 6.6 kJ/mol, respectively.[29,41]

For the most abundant $Sc_3N@I_h(7)$-C_{80}, a single line of ^{45}Sc NMR spectrum was observed at approximately 200 ppm.[42] For the other nitride cluster EMFs with one- or hetero-metal atoms, the influence of the hetero-atom number is negligible in the nitride cluster series of $Lu_xSc_{3-x}N@C_{80}$ (x = 0–2), in which the ^{45}Sc chemical shifts lie in the range of 195–212 ppm.[43] Similarly, the ^{45}Sc NMR signal of $Sc_3N@C_{68}$ was first reported at approximately 90 ppm in 2000 and then updated to approximately 79 ppm recently.[33] This difference in value, compared to that of $Sc_3N@C_{80}$, could be an effect of cage size.

For sulfide cluster EMFs, $Sc_2S@C_{82}$ exhibits a single line at 290 ppm, indicative of two equivalent Sc atoms resulting from the fast rotation of the Sc_2S cluster at room temperature.[44] In contrast, the ^{45}Sc signal of $Sc_2S@C_{72}$ was shifted to 183 ppm, consistent with a cage size effect.[45] For the first cyanide $Sc_3NC@C_{80}$, two signals in a 2:1 intensity ratio were detected, which indicate two types of nonequivalent Sc atoms of the Sc_3NC cluster.[46] For $Sc_4O_2@C_{80}$, two types of Sc atoms were distinguished to present signals at 285 and 135 ppm, respectively. This indicates that the two types of Sc atoms have different valence states, which is consistent with DFT calculations.

4.1.2.3 ^{89}Y NMR Spectroscopy

Due to the long spin-lattice relaxation times of ^{89}Y isotope, ^{89}Y NMR spectroscopy has not been widely used to study the structures of EMFs. A representative work was reported by Dorn and coworkers in 2009.[47] They successfully obtained the ^{89}Y NMR spectra of $Y_3N@I_h(7)$-C_{80}, $Y_3N@C_s(51365)$-C_{84}, and $Y_3N@D_3(19)$-C_{86} and found that the cluster motion is highly dependent on the cage structures. As shown in Figure 4.6, although only one ^{89}Y signal was observed for $Y_3N@C_{80}$ or $Y_3N@C_{86}$, which indicates a rotating Y_3N cluster, three signals were found in $Y_3N@C_{84}$, which means that the cluster is fixed. Furthermore, it was found that the ^{89}Y chemical shift values were distributed over a large range: 191.63 ppm in $Y_3N@C_{80}$, 104.32/65.33/−19.53 ppm in $Y_3N@C_{84}$, and 62.65 ppm in $Y_3N@C_{86}$. The most up-field value of −19.53 ppm in $Y_3N@C_{84}$ should result from the Y atom coordinating with the fused-pentagon carbons.

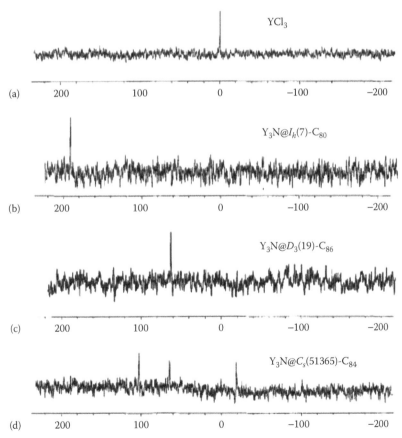

FIGURE 4.6 ^{89}Y NMR spectra of (a) YCl_3 (b) $Y_3N@C_{80}$ (c) $Y_3N@C_{86}$ (d) $Y_3N@C_{84}$. (Reprinted with permission from Fu, W. J. et al., *J. Am. Chem. Soc.*, 131, 11762–11769, 2009. Copyright 2009 American Chemical Society.)

4.1.2.4 ^{139}La NMR Spectroscopy

La-containing EMFs are among the earliest investigated species and have attracted great attention during past years.[48] ^{139}La NMR shows its great power in probing the structures and properties of this class of EMFs. In 1997, Akasaka and coworkers reported the ^{139}La NMR spectrum of La$_2$@I_h(7)-C$_{80}$ and observed only one peak at −403 ppm. Variable temperature NMR studies in the temperature range of 258–363 K revealed the free circulation of the two La atoms.[49] Later on, similar results were obtained in several derivatives of La$_2$@C$_{80}$. In detail, the ^{139}La chemical shift values of La$_2$@C$_{80}$(Dep$_2$Si)$_2$CH$_2$,[50] two isomers of La$_2$@C$_{80}$(Dep$_2$Si(CH$_2$)-CHtBp) (tBp = 4-*tert*-butylphenyl),[51] and cycloadduct of La$_2$@C$_{80}$(TCNEO)[52] were observed at −365, −392 and −397, and −323 ppm, respectively. In these compounds, the linewidth of the ^{139}La signal varies strongly with temperatures, showing that the metal atoms are moving inside the cage. In contrast, no signal boarding was observed in the La$_2$@C$_{80}$(CH$_2$)$_2$NTrt derivative (δ = −464 and −412 ppm for [6,6] and [5,6] isomers),[17] showing that the La atoms are steady inside the cage. In La$_2$@C$_{80}$(Ad), the two metal atoms are even collinear with the spiro carbon of the addend.[23]

Further investigations of a variety of La-EMFs and their derivatives showed that the properties of the internal La atom(s) depend strongly on the cage structure, size, and even the position of the addend. For instance, in the La@C$_{2n}$C$_6$H$_3$Cl$_2$ series, which was obtained by refluxing the soot in 1,2,4-trochlorobenzene, the ^{139}La NMR signal values go to downfield with the increase in cage size.[53–57] Table 4.1 shows that the values of the three isomers La@C$_2$(10612)-C$_{72}$C$_6$H$_3$Cl$_2$ lie between −603 and −619 ppm, and they decrease to −513 and −511 ppm for the two isomers of La@C$_{74}$C$_6$H$_3$Cl$_2$; to −493, −500, and −488 ppm for the three isomers of La@C$_{80}$C$_6$H$_3$Cl$_2$; and to −456 and −468 ppm for the two isomers of La@C$_{82}$C$_6$H$_3$Cl$_2$, respectively. The ^{139}La NMR chemical shift of [La@C$_{82}$]$^-$ appeared at −470 ppm[3].

TABLE 4.1
Summary of the ^{139}La NMR Chemical Shift Values of Typical La-EMFs

Compound	^{139}La Value/ppm	Specification	References
La@C$_{72}$C$_6$H$_3$Cl$_2$	−603 and −605	3 isomers	53
La@C$_{74}$C$_6$H$_3$Cl$_2$	−513 and −511	2 isomers	54,57
La@C$_{80}$C$_6$H$_3$Cl$_2$	−493, −500, and −488	3 isomers	55
La@C$_{82}$C$_6$H$_3$Cl$_2$	−456 and −468	2 isomers	56
[La@C$_{2v}$(9)-C$_{82}$]	−470	anionic form	3
La$_2$@C$_{72}$	−532	Non-IPR structure	58
La$_2$@C$_{76}$	−617.8	Non-IPR structure	59
La$_2$@C$_{80}$	−403		49
La$_2$@C$_{80}$(Dep$_2$Si)$_2$CH$_2$	−365		50
La$_2$@C$_{80}$(Dep$_2$Si(CH$_2$)-CHtBp)	−392 and −397	2 isomers	51
La$_2$@C$_{80}$(TCNEO)	−323		52
La$_2$@C$_{80}$(CH$_2$)$_2$NTrt	−464 and −412	2 isomers	23

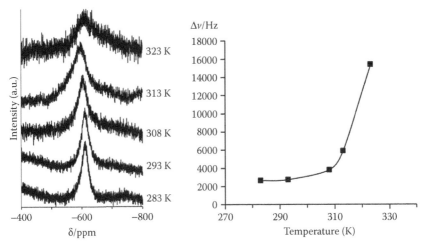

FIGURE 4.7   (Left) Variable-temperature ^{139}La NMR spectra of La$_2$@C_s(17490)-C$_{76}$. (Right) Plotting the line-width against temperature.

Di-EMFs containing two La atoms, for example, La$_2$@C_{72}, which contains two pairs of fused pentagons, show a ^{139}La NMR signal at −532 ppm, whose line-width is not obviously dependent on the temperature.[58] This confirms that the two La atoms coordinate strongly with the fused-pentagon carbons. Surprisingly, the ^{139}La NMR signal of La$_2$@C_{76}, which also contains two pairs of fused pentagons, is highly sensitive to temperature changes (Figure 4.7), demonstrating that the two La atoms adopt a motional behavior inside the cage.[59] The ^{139}La NMR chemical shift values of typical La-EMFs are summarized in Table 4.1.

4.2   X-RAY DIFFRACTION

Marilyn M. Olmstead, Kamran B. Ghiassi, and Alan L. Balch

Determination of single crystal structures of EMFs is the key to understanding their identity, and ultimately the chemical reactivity and properties of these recently discovered chemical species. Some misconceptions about these fascinating molecules stemmed from early disbelief that there could be interior contents at all, rather than external, adducts. Mass spectral data could not distinguish between interior and exterior species and those that comprised part of the fullerene cage. Theoretical calculations, especially in the early days, incorrectly predicted isomers of endohedral fullerenes, in part because of the failure to recognize encaged carbides and also the reluctance to expect adjacent pentagon pairs (non-IPR cages).[60,61] Powder diffraction could not provide sufficient data to unambiguously unravel the complicated spherical arrangement of carbon atoms. Single crystal X-ray crystallography has thus provided the necessary detailed information regarding how hexagons and pentagons are arranged, what encapsulated species exist, and the geometric details of their chemical bonding. Some of these facts still apply to present methods. For example, mass spectrometry cannot differentiate between cage carbons and carbides. Crystallography is able to corroborate the results of detailed ^{13}C analysis, a successful methodology for fullerenes that is outlined in Section 4.1.

As is true for empty cage fullerenes, endohedral fullerenes cannot be expected to crystallize by themselves in an ordered lattice. There are clearly fundamental problems. One is the rotational motion of spherical molecules which only have weak van der Waals forces between them. There is a lack of any favorable intermolecular potential energy that can possibly give rise to a repeat arrangement in three dimensions. Consequently, the crystals of pristine fullerenes are characterized by diffraction of spheres and not by diffraction of the individual carbons and other atoms that comprise their unit cells.[62] Typically, their crystals belong to the cubic crystal system with one of the closest packing arrangements.[3] In order to avoid these problems, the incorporation of solvent molecules and a variety of co-crystallizing agents have been employed in order to obtain the desired structural order. External derivatization of the fullerene cage has also been successful. This topic is discussed in Chapter 6. These various approaches have been only partially successful in simplifying crystal structure determination due to the residual disorder that occurs in cage rotation, circulation of interior groups, disorder due to crystallographic symmetry, and solvent molecule disorder in the structures. Thus, the challenge is still being met by practicing crystal engineers and crystallographers. Some examples of structures that are not externally modified and that contain solvates and co-crystallizing agents will be given in this section. Interesting geometric trends that have been pointed out by crystallographic results will also be presented.

4.2.1 Crystal Structures of Solvates of Endohedral Fullerenes

Very few structures with simple solvate molecules have been reported with endohedral fullerenes. By contrast, the empty cage fullerenes (particularly C_{60}) display crystal structures that contain benzene,[11] carbon disulfide,[10] bromobenzene,[63] bromoform,[64] pentane,[65] phenol,[66] thiophene, and so on.[67] Simple solvates may not have been reported for many endohedral fullerenes, because of a size mismatch. The larger co-crystallizing agents assist crystal growth and also increase the amount of crystalline material present in a typical crystallization. The marked success of crystallization with nickel octaethylporphyrin, Ni(OEP), has dictated its frequent choice and has produced by far the greatest number of endohedral fullerene structures.

Because $Sc_3N@I_h$-C_{80} is the third most abundant fullerene after C_{60} and C_{70}, and more material is available, it has been structurally characterized through the use of many crystallization methods, including with the solvent o-xylene. Crystals of $Sc_3N@I_h$-C_{80}•5(o-xylene) and $Lu_3N@I_h$-C_{80}•5(o-xylene) form upon cooling of the respective o-xylene solutions.[68] The Sc and Lu structures are isomorphous, although the Lu analog shows greater interior disorder. The Sc solvate gives a well-ordered structure at 90 K and is shown in Figure 4.8. Favorable $\pi \cdots \pi$ stacking interactions clearly aid in the production of the ordered crystal.

Short face-to-face contacts between carbons of the fullerene cage and o-xylene plane are in the range of 3.11–3.34 Å. Only one of the Sc sites shows minor disorder (20%). Based on the three major sites, the Sc—N distances are 1.9931(14), 2.0323(16), and 2.0526(14) Å; the N—Sc—N angles are 120.70(7), 119.21(6), and 118.47(7)°. There are only two different kinds of C—C bonds in the I_h cage. Those at 6:6 ring junctions average 1.417(6) Å in length and those at 6:5 ring junctions average 1.449(6) Å. The shortest Sc—C distances range from 2.145(4) to 2.205(4) Å. Further analysis has shown

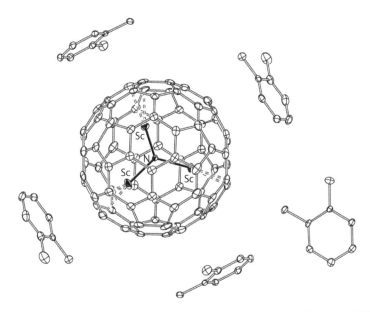

FIGURE 4.8   A view of the structure of $Sc_3N@I_h\text{-}C_{80}$•5(o-xylene) with thermal displacement parameters shown at the 50% probability level.

that the carbon atoms nearest to the Sc sites are *punched out*. Furthermore, pyramidalization angles are greatest for those carbon atoms nearest to the metals (both Sc and Lu). Interestingly, at 200 K, the structure suffers from disorder, displaying multiple cage orientations and circulation of the Sc atoms within the cage. Nevertheless, this behavior is reversible, similar to the structures of $2C_{60}$•$3CS_2$[69] and C_{60}•4benzene.[70]

4.2.2   CO-CRYSTALLIZATION METHODS FOR CRYSTAL GROWTH

The advantage of co-crystallization is that it is a simple and reliable method for crystal growth of fullerenes. It is only necessary to find a suitable host, usually with a delocalized π system, either curved or shape-compatible, that has the right solubility. The co-crystallization agent is able to add more mass to the sample and thereby create more crystals. This is a feature that is important in the field of endohedral fullerenes because the amount of fullerene material isolated from chromatographic separation is typically in the milligram or sub-milligram range. If the co-crystallization agent carries a heavier atom, such as a metalloporphyrin, phasing of the structure is easier with purely Patterson and/or dual space methods. The hope is that rotational disorder will be minimized if a suitable match between guest (fullerene) and host (co-crystallizing agent) is found. Even external modification of the fullerene cage cannot guarantee that this will always occur. Therefore, although the empty cage fullerene C_{60} has been co-crystallized with a variety of co-crystallization agents, the precious nature of endohedral fullerenes has dictated that porphyrins are almost always the agent of choice. As one example, the structure of $Tm_3N@I_h\text{-}C_{80}$ was obtained by co-crystallization with Ni(OEP) in benzene and led to a reliable

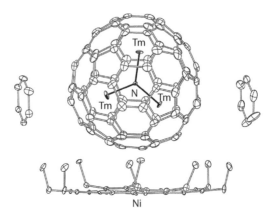

FIGURE 4.9   The structure of $Tm_3N@I_h\text{-}C_{80}\bullet Ni(OEP)\bullet 2benzene$ with thermal displacement parameters shown at the 50% probability level.

description of the structure interior contents.[71] This structure will be used as an example of the fullerene/metalloporphyrin co-crystal. The asymmetric unit of this structure consists of $Tm_3N@I_h\text{-}C_{80}\bullet Ni(OEP)\bullet 2benzene$. There are two orientations of the Tm_3N cluster with relative occupancies of 0.67/0.33. The structure is depicted with only the 0.67 occupancy cluster in Figure 4.9, in which it can be pointed out that the thermal ellipsoids for the fullerene are larger than the other atoms in the structure, particularly those of the porphyrin and cluster. These data were collected at 90 K. In our work, we have always found that the fullerene exhibits relatively large rotational thermal motion, with oblate spheroidal shapes (*pancakes*). This shape persists even at very low temperatures of 12 K using a He cryostat. Figure 4.9 draws attention to the guest–host arrangement of the fullerene and porphyrin. The association between these two species has been attributed to $\pi\cdots\pi$ interactions between the delocalized surfaces of the fullerene and of the porphyrin plane; van der Waals interactions between the hydrogen atoms of the enclosing ethyl groups and the fullerene; and, in some cases, donor–acceptor interactions between the two interactions. These species are certainly important, but do not give the whole story as to the stability of this particular structure. Other intermolecular interactions make a contribution, as shown in Figures 4.10 through 4.12. In Figure 4.10 the crystal structure in the *ac* plane is emphasized. In Figure 4.11, the back-to-back arrangement of porphyrins is shown, together with the shortest $Ni\cdots Ni$ contacts and $Ni\cdots C$ contacts. In the structures of these co-crystals, these are typical geometric arrangements and intermolecular values. Figure 4.12 portrays the intermolecular interactions that occur along the crystallographic *b* direction and thus expands the packing arrangement that is shown in Figure 4.10. In the next layer along *b*, intervening benzene molecules are π-stacked between fullerene cages. The second type of benzene molecule is located in an interstitial position and is weakly π-stacked with its inversion-related neighbor.

A general observation in the separation of endohedral fullerenes is the similarity in elution times for isomers of similar shapes and charges. For example, in the case of the trimetallic nitride C_{80} cages, two isomers of the C_{80} cage with point symmetries of I_h and D_{5h} elute together. In the arc-discharge method of endohedral fullerene

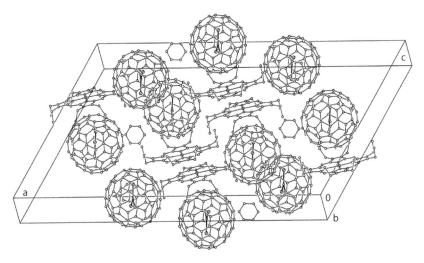

FIGURE 4.10   A view of the packing arrangement in the crystal structure of $Tm_3N@$ I_h-C_{80}•Ni(OEP)•2benzene.

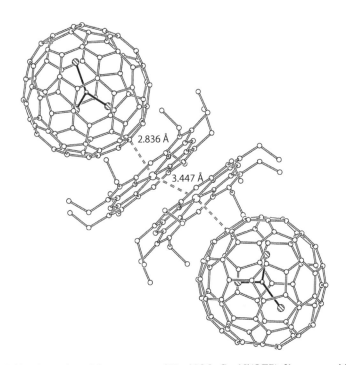

2.836 Å

3.447 Å

FIGURE 4.11   A portion of the structure of $Tm_3N@I_h$-C_{80}•Ni(OEP)•2benzene with selected intermolecular distances. There is a crystallographic center of symmetry midway between the two Ni(OEP) units.

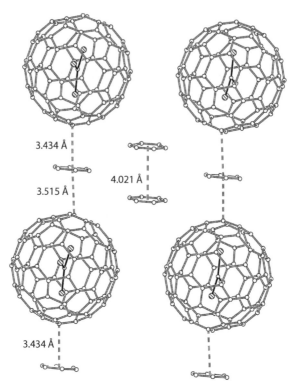

FIGURE 4.12 A view of the disposition of the two benzene molecules in the structure of $Tm_3N@I_h\text{-}C_{80}\bullet Ni(OEP)\bullet 2benzene$. The crystallographic b-axis is vertical in this drawing.

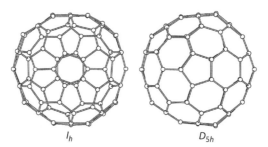

I_h D_{5h}

FIGURE 4.13 The two different C_{80} cages observed in endohedral fullerene chemistry.

synthesis, the I_h isomer is more abundant. For trimetallic nitrides, the cage has a 6− charge. Among many other examples is the cage of $Sm_3@I_h\text{-}C_{80}$, in which each Sm carries a 2+ charge and the cage similarly carries a 6− charge.[72] Figure 4.13 shows how the two cage isomers are related. The I_h cage can be obtained from the D_{5h} cage by rotation of the top half by 36° around its midpoint. If the two isomers are not well separated during the chromatographic stage, the isomers may crystallize in the same site in a crystal and lead to intractable disorder. Nevertheless, pure isomers of these two cages have been successfully obtained in some cases. In the I_h cage,

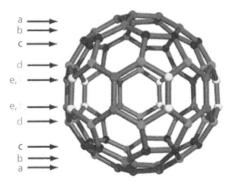

FIGURE 4.14   The six different types of carbon atoms in the D_{5h}-C_{80} cage.

there are two kinds of carbon atoms: those where three hexagons meet and those where two hexagons and a pentagon meet. In the D_{5h} cage, there are six different types of carbon atoms, illustrated in Figure 4.14. The difference in the two can be confirmed by ^{13}C NMR in enriched diamagnetic samples.

The crystal structure determination of Tm$_3$N@D_{5h}-C_{80}•Ni(OEP)•2benzene has also been carried out and it is similar to the I_h isomer, but it is not isostructural.[12] Although the I_h isomer crystallizes in the monoclinic space group $C2/c$ with Z = 8, the D_{5h} isomer crystallizes in the monoclinic space group $C2/m$ with Z = 4. The crystal packing in the two systems is almost identical. However, the D_{5h} structure suffers from the imposition of a crystallographic mirror plane on the fullerene. This mirror coincides with a porphyrin mirror plane but not with the fullerene mirror plane and results in disorder of the fullerene. In fact, this is a common occurrence in the porphyrin co-crystals, and is a major disadvantage. Nevertheless, with some effort, the disorder can usually be resolved and the structure determined. A third isomer of C$_{80}$ with a 6− cage has been reported. It is formed along with the more dominant isomers in the synthesis of Dy$_3$N@C_{80} and proposed to be the D_{5d}(1) isomer.[73] The isomer awaits confirmation by X-ray crystallography. Other cages cannot be precluded, particularly if the encaged contents have different geometric requirements. For example, metals or clusters not capable of donating six electrons to the cage, such as monometallic Sm,[74] Yb,[75] or La,[55] adopt the C_{2v}(3)-C_{80} cage. The carbide Sc$_2$C$_2$, utilizes C_{2v}(5)-C_{80}.[26] It is possible that non-IPR isomers may exist; there are seven IPR isomers of C$_{80}$ and 31917 non-IPR isomers.

A second co-crystallization agent that has experienced limited success to date is that of BEDT-TTF, bis(ethylenedithio)tetrathiafulvalene (ET) (Figure 4.15). First used in the co-crystallization of C$_{60}$,[76] and recently in that of C$_{70}$,[77] the molecule has greater

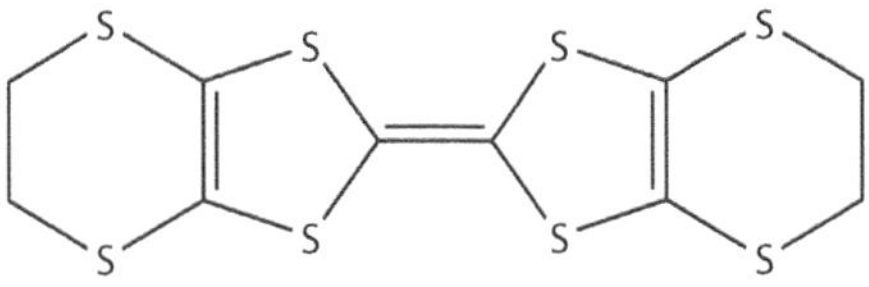

FIGURE 4.15   The structure of bis(ethylenedithio)tetrathiafulvalene (BEDT-TTF or ET, for short).

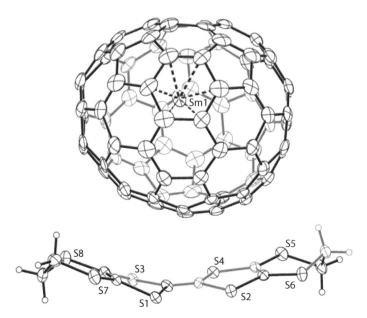

FIGURE 4.16 The co-crystal of Sm@C_{2v}(3)-C_{80} and ET. A half molecule of toluene in the structure is not shown.

flexibility in its curvature and may allow for greater degree of charge transfer than the porphyrin. In addition to Ni(OEP), it has been shown that ET co-crystallizes with Sm@ C_{2v}(3)-C_{80}.[74] In both the ET and Ni(OEP) co-crystals, the Sm atom occupies a range of sites in a disordered fashion (Figure 4.16). One particular site is dominant, slightly off the center of a hexagonal ring near the twofold axis. The distribution of Sm sites differs in the two co-crystals, and further structural studies are needed in order to determine if the distribution is controlled in some way by the co-crystallizing host.

4.2.3 STONE-WALES DISORDER

Another type of disorder can occur which relates to the existence of the so-called Stone-Wales transformation in the fullerene cage. In the original study,[78] it was proposed that a structural arrangement that involves rotation by 90° of a bond between two hexagons of C_{60} has a low-enough potential energy to yield multiple isomers of C_{60}. One such transformation is shown below. Successive transformations can yield additional isomers or trigger a return to the original isomer (Figure 4.17).

Many studies have suggested that Stone-Wales transformations readily occur in higher fullerenes both in bottom-up and top-down high temperature fullerene production, although the topic is controversial due to the large barrier for the process.[79] If the transformation has little impact on the overall cage curvature, the resultant isomer would be expected to elute with a similar retention time, be difficult to separate from its parent, and occupy the same site in the crystal. Thus, the concurrent presence of such isomers during crystallization adds confusion in the form of disorder to the crystal structure solution. Two isomers of the empty cage fullerene, C_{86},

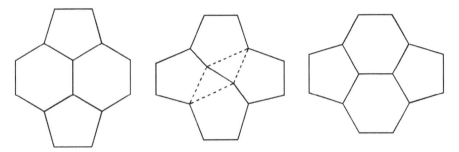

FIGURE 4.17   The Stone-Wales transformation.

were isolated and crystallized together. Analysis by X-ray crystallography showed that the $C_s(16)$-C_{86} isomer can be converted into both enantiomers of the $C_2(17)$-C_{86} isomer by two different single Stone-Wales transformations.[80] All three cages occur in the same crystallographic site. This unusual result is substantiated by several levels of theory that predict that these two isomers are the most thermodynamically and kinetically stable of the empty cages of 19 possible IPR isomers. Similarly, the endohedral fullerenes Sm@$C_2(5)$-C_{82}, Sm@$C_s(6)$-C_{82}, and Sm@$C_{3v}(7)$-C_{82} were formed in the carbon soot in the electric-arc vaporization of hollow graphite rods filled with Sm_2O_3 and graphite powder.[81] Although they were separated by lengthy chromatography, characterized by UV-Vis-NIR and laser-desorption ionization time-of-flight mass spectroscopy, crystallization of an inadequately purified sample yielded crystals of isomerically pure Sm@$C_2(5)$-C_{82} and Sm@$C_{3v}(7)$-C_{82} but a mixed crystal of 0.6667 Sm@$C_{3v}(7)$-C_{82}/0.3333 Sm@$C_s(6)$-C_{82}. Five different recrystallizations yielded the same result. It is clear that the latter two fullerene cages are closely related, only differing in a single Stone-Wales transformation. Synthesis of multiple isomers of Sm@C_{82} has been reported elsewhere, and different patterns of isomer formation occurred. In one case, the samarium source was a mixture of silicon carbide and Sm_2Co_{17} alloy.[82] In the other, $SmNi_2$ alloy was used.[83] In the latter two preparations, Sm@$C_{2v}(9)$-C_{82} was proposed to be formed, albeit not crystallographically characterized. A fifth isomer is unidentified. The crystal structure of the isomer Sm@$C_2(5)$-C_{82} as the Ni(OEP) co-crystal with chloroform and benzene solvates was reported from the Liu synthesis method.[84] It was shown that the chloroform/benzene solvate and the toluene solvate are essentially isostructural.[24] This outcome re-emphasizes the need to collect and structurally characterize as many different isomers as possible in order to be able to systematically synthesize endohedral fullerenes. We are far from being able to do this at this time.

4.2.4   The Residence Site for the Encaged Metal

A fundamental question that has many answers is, Where does the encapsulated metal atom reside inside the cage? The preference for a particular cage position of the metal atom is predicted by many theoretical studies. Monometallic endohedral fullerenes are attractive targets for these studies because the position might not be affected by electrostatic repulsion of other metals or by steric effects. Nevertheless,

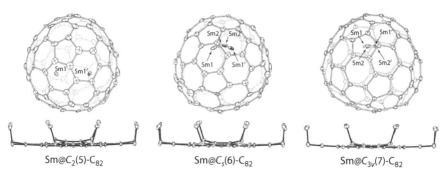

$Sm@C_2(5)-C_{82}$ $Sm@C_s(6)-C_{82}$ $Sm@C_{3v}(7)-C_{82}$

FIGURE 4.18   Views of the crystal structures of three isomers of $Sm@C_{82} \cdot Ni(OEP)$ showing mirror-related positions of the predominant Sm positions.

the tendency for there to be multiple metal sites in the crystal structures shows that there are shallow potential energy surfaces inside that can be populated to varying degrees. In the example given above for $Sm@C_{82}$ (Figure 4.18), there are three different outcomes. In the $C_2(5)$ isomer, the most populated site for Sm1 brings it near the twofold axis and under a hexagonal ring. In the $C_s(6)$ isomer, there are two nearly equal sites for Sm, Sm1, and Sm2, and both of these indicate a preference for the Sm to be close to a 5:6 ring junction. In the C_{3v} isomer, again, Sm1 and Sm2 are nearly equal in occupancy and reside close to the threefold axis and both are closest to a 5:6 ring junction. All three of these isomers crystallize with the relatively flat region of the cage in proximity to the flat porphyrin surface, and this may be a structure-directing interaction. Another ambiguity is due to the crystallographic mirror plane that gives rise to the primed atoms Sm1′ and Sm2′ shown in the Figure 4.18. Consequently, it is not particularly easy to define a rule for the metal location in monometallic endohedral fullerenes of this size. A similar disorder complicated the study of the metal atom position in a different isomer of C_{82} that occurs in Gd@ $C_{2v}(9)-C_{82}$, one of the monometallic endohedrals with a 3+ metal.[85] Although, by crystallographic mirror symmetry, the Gd could be located in either an η^6 fashion along the twofold axis or near a 5:6 bond junction away from the twofold axis, theoretical calculations showed the former to be the reasonable energy minimum.

In another study, a series of four Stone-Wales interconnected isomers of $Sm@C_{90}$ was identified by X-ray crystallography.[86] Three of these cages were rather spherical, being the isomers $C_2(40)$, $C_2(42)$, and $C_2(45)$. A fourth was more egg-shaped with symmetry $C_{2v}(46)$. All four cages possess a belt of 11 contiguous hexagons. In this belt, the metal atoms have a tendency to circulate, an effect that can be studied by temperature-dependent ^{13}C NMR. The large number of sites found for Sm in the crystal structures, even though carried out at 90 K, makes it impossible to reliably identify the favored positions in these cases.

A substantial amount of void space exists in the larger cages. The largest monometallic cages studied to date are those of C_{94}, where approximately 70% of the interior space is *empty*. Of the 134 possible IPR cages of C_{94}, only one has C_{3v} symmetry, and this is the cage adopted by the most abundant isomers for the divalent cations, Ca, Tm,[87] and Sm.[88] Both Ca and Sm crystallize as $M@C_{3v}(134)-C_{94} \cdot Ni(OEP) \cdot 2toluene$

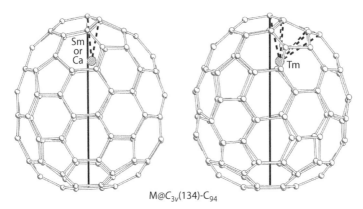

M@C_{3v}(134)-C_{94}

FIGURE 4.19   Three different metal atoms, Ca, Sm, and Tm, in structures of M@C_{3v}(134)-C_{94} favor sites near the same end of the threefold axis of the cage. The vertical line shows the position of the threefold axis.

in the monoclinic space group $C2/m$ and are isostructural. They also display the same cage interaction, shown in Figure 4.19.

In these structures, the Ca or Sm is near to one end of the threefold axis of the cage, in closest proximity to a 6:6 ring junction. The Sm⋯C [2.536(4) and 2.567(3) Å] distances are longer than the Ca⋯C [2.358(8) and 2.445(7) Å] distances. In the Tm structure, Tm@C_{3v}(134)-C_{94}•2Ni(OEP)•3.65benzene, the space group is tetragonal, $I4_1/a$, with a different packing arrangement. All seven Tm sites are also clustered near the same end of the threefold axis. The site with 40% occupancy is η^6 to a hexagon of the 6:6:6 junction that coincides with the threefold axis with a range of Tm⋯C distances of 2.43(4) to 2.59(2) Å.

4.2.5   NON-IPR CRYSTAL STRUCTURES

It has been reasonably proposed that the metal will be found in rather cupped, or highly curved, locations inside the cage such that it can achieve maximum interaction with valence orbitals or with negatively charged regions—in other words, preferring the most coordinatively saturated position, as in a metal complex. This is a useful paradigm. In addition, the residence of the metal inside the cage actually *causes* the nearest carbons to be more highly pyramidalized than other carbons, creating the *punched out* effect that has been noted. The examples of non-IPR endohedral fullerenes provide further evidence for the curvature distortion inherent in the metal–cage interaction. Non-IPR cages contain one or more sites where pentagons join pentagons rather than being isolated from one another by hexagons. If there are two abutting pentagons, this is called a pentalene unit in organometallic chemistry, and it is an 8π-electron anti-aromatic region of the cage. Theory predicts that negative charge is found at the 5:5 junction of two pentagons, and consequently, the metal cation experiences a strong electrostatic attraction at this point. Structurally, the two pentagons are remarkably folded at this junction and form a cavity that strongly encapsulates a metal. Wherever a pentalene exists, a metal is present at very near full occupancy. The first non-IPR endohedral fullerene to be characterized by single-crystal X-ray diffraction was that of $Sc_3N@D_3(6140)$-C_{68}.[89] There are no IPR cages having 68 carbons, but there are 6332 possible non-IPR cages comprised of pentagons and hexagons. An isomer with D_3 symmetry offers a perfect match for the Sc_3N cluster because there are three pentalene regions radiating at 120° around the three-fold axis. Only two such cages occur among the 6332 possibilities, one rather flat (6275) and the other more rounded one (6140) that was actually chosen. In the crystal structure, there is disorder caused by chirality because the compound is formed as a statistical (or racemic) mixture of two enantiomers. However, it is clear that each cage shares the central nitrogen and has an independent set of three Sc's. One such cage is depicted in Figure 4.20.

The sum of the angles around the central N is 359.7° and the Sc–N distances average 1.986 Å. The shortest Sc···C distances are those between Sc and the carbons at the 5:5 junction, with an average of 2.255 Å. Thus, in some respects, this geometry mimics that seen in $Sc_3N@I_h$-C_{80}, described above. However, the cage cannot be any smaller and still accommodate a planar Sc_3N unit. As shown in Figure 4.20b, the fold in the pentalenes allows for three distinctive cups that accommodate the three Sc's. Remarkably, all eight carbons of each pentalene unit are in close contact to the Sc's. The average Sc···C distance for 24 short contacts is 2.336 Å.

A second example of a non-IPR cage that encapsulates and matches a Sc cluster was revealed in the structure of $Sc_2S@C_s(10528)$-C_{72}.[45] The C_{72} cage has only one IPR isomer with D_{6d} symmetry. There are 11,189 possible non-IPR isomers. A single-crystal study established that the cage adopts an isomer with mirror symmetry, 10,528, that has an almost uncanny match to the expected bent geometry of Sc_2S (with a 4− charge). There are two essentially equal orientations of the fullerene in the crystal structure. In each case, the Sc's are found with short contacts to pentalene motifs. The structure is depicted for one orientation in Figure 4.21. The

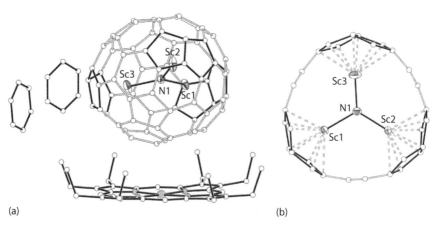

(a)   (b)

FIGURE 4.20   Results from the crystal structure determination of $Sc_3N@D_3(6140)$-C_{68}•Ni (OEP)•2benzene. The three pentalene regions are highlighted with solid lines. In (a) the porphyrin, fullerene, and two benzene arrangement is shown. View (b) is down the normal to the Sc_3N plane (some carbon atoms are omitted for clarity).

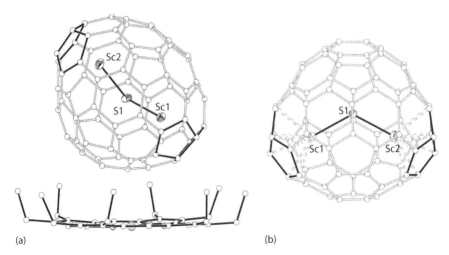

FIGURE 4.21   Views of the structure of $Sc_2S@C_s(10528)$-C_{72}. View (a) shows the disposition of the fullerene with respect to the co-crystallized Ni(OEP) (1.5 molecules of toluene in the structure are omitted for clarity). View (b) is down the normal to the Sc_2S plane. The pentalenes are drawn with filled bonds.

misshapen C_{72} cage results from Sc···C interactions: bumped outward where the metal is present, curved at the empty end. The S—Sc distances average 2.344 Å and the average Sc—S—Sc angle is 124.6°. The Sc's interact with all eight pentalene carbons with an average Sc···C distance of 2.339 Å. Those distances to the 5:5 junction are somewhat shorter with an average value of 2.259 Å. The results of this structure add evidence to the supposition that the shape of the cluster influences the selection of the most stable isomer. In comparison, it is striking that the structure of the non-IPR cage of $La_2@D_2(10611)$-C_{72} places two fused pentagons at opposite ends of the cage,[58] whereas the non-IPR $Sc_2S@C_{72}$ cage places them at an angle that conforms to the bent Sc—S—Sc geometry. At this time, it is not known if the cluster is acting as a template during the formation of the fullerene or if some other first-formed fullerene undergoes successive transformations to reach this more stable state.

An additional example of a non-IPR endohedral fullerene containing two patches of fused pentagons is found the structure $Gd_3N@C_2(22010)$-C_{78}.[90] Two of the Gd atoms are in close proximity to pentalene units while the third is η^6 to an opposing hexagon. The large size of Gd is likely to be the driving force in the occurrence of this unusual cage isomer; the majority of similar endohedral fullerenes adopt the IPR $D_{3h}(5)$-C_{78} cage.

A number of endohedral fullerenes containing only a single fused pentagon patch have been crystallographically characterized. Because of the single, folded pentagon–pentagon motif, they have a nose-like appearance and have been variously described as egg-shaped and also as looking like the Star Wars' *Millennium Falcon*. The first of these to be described was the structure of $Tb_3N@C_s(51365)$-C_{84},[91] and the same isomer has reappeared in the structures of the Gd and Tm analogs.[92] A possibly related structure is that of $Gd_3N@C_2(39663)$-C_{82}.[93]

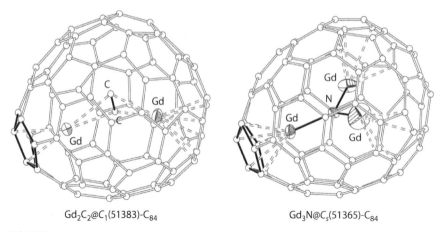

$Gd_2C_2@C_1(51383)-C_{84}$ $Gd_3N@C_s(51365)-C_{84}$

FIGURE 4.22 Views of the non-IPR, single pentalene, C_{84} cages of a carbide and a nitride that utilize the metal Gd^{3+}.

Recently, the first non-IPR endohedral carbide structure was reported, $Gd_2C_2@$ $C_1(51383)-C_{84}$.[94] It is interesting to compare the geometry of the carbide $Gd_2C_2@$ $C_1(51383)-C_{84}$, a 4− cage, and the trimetallic nitride $Gd_3N@C_s(51365)-C_{84}$, a 6− cage. As shown in Figure 4.22, different isomers are used for the different encaged clusters. It can be shown that the two cages are actually related by a single Stone-Wales transformation. The C_s isomer shows a more prominent *nose* than the C_1 cage. The Gd—N bond distances are fairly uniform, 2.177(8) Å for the Gd near the pentalene and 2.085(9) Å and 2.129(8) Å for the other two; the sum of the Gd—N—Gd angles is 359.7°. Interior Gd···C distances are similar and are 2.470(15) and 2.479(16) Å for those between Gd and the 5:5 junction. In comparison, the carbide structure has much shorter distances between Gd and the 5:5 junction of 2.300(19) and 2.383(14) Å but long distances from this Gd to the carbide carbons of 2.487(10) and 2.541(10) Å. The asymmetry suggests that there is competition between the pentalene and the carbide for Gd's charge, but, in the nitride, the nitrogen exerts the dominant bonding interaction.

4.2.6 ENDOHEDRAL CARBIDE STRUCTURES

Endohedral metal carbides are intriguing because of their possible connection to the formation of fullerenes by inclusion or extrusion of C_2 fragments in the fullerene plasma. But perhaps they include single carbon C^{4-} species, or larger C_n fragments. It is expected that more carbides will be discovered, but to date the best character-ized are acetylenic carbides of the type Sc_2C_2, Ti_2C_2, Y_2C_2, Gd_2C_2, and Lu_2C_2. Based on a simple electrostatic model in which the C_2 unit bears a 2− charge and the metal is 3+, a symmetrical planar structure might be expected, as seen in some organo-metallic examples.[95] However, encapsulation by a fullerene cage may not allow this arrangement due to the constraints of cage size, adduct formation, and other local-ized charge distributions. Thus, computations[96–100] have predicted a range of geom-etries, as depicted in Figure 4.23.

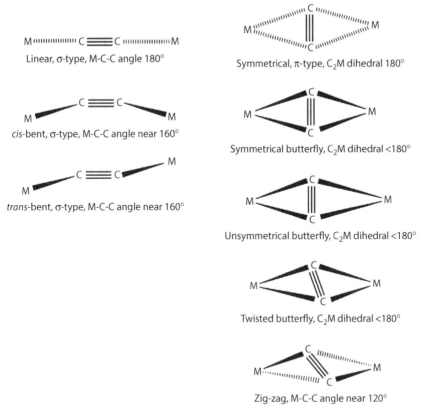

FIGURE 4.23   In the absence of encapsulation, the top left arrangement of the M_2C_2 unit is favorable, but, inside a fullerene, arrangements as shown on the right have been observed.

The largely σ-type linear arrangements shown on the left-hand column of Figure 4.23 have not been crystallographically confirmed inside fullerenes. Commonly, both orientational disorder in the cage and internal positional disorder occur, and therefore, rather high uncertainty is associated with the geometry of the M_2C_2 unit. Our understanding of the geometry is derived from the atom positions with the highest occupancies, but there are clear trends. As shown in Table 4.2, the larger metal···metal distances can be correlated with wider dihedral angles between the MC_2 planes. A cage that is large enough to accommodate a 180° dihedral angle has not yet been reported. The largest known carbide cluster endohedral, $Gd_2C_2@D_3(85)$-C_{92},[101] has a wide dihedral angle but is still a butterfly. The longest dimension of the D_3-C_{92} cage is 9.43 Å, and the dihedral angle range of 154–161° is in accord with typical distances between the metal and the carbons of the cage. As the cage size contracts to the longest dimension of approximately 8.3 Å in $Gd_2C_2@C_1(51383)$-C_{84}, repulsions between the metal and carbons require compression of the M_2C_2 unit, and more acute angles in the range of 123–124° are observed. By ^{13}C NMR and DFT calculations, the degree of compression can be estimated.[43] Additional crystal structures of carbides will be useful for the fine tuning of the agreement between experiment and

TABLE 4.2

Variation of M···M Distances versus M$_2$C Dihedral Angle

Endohedral	M$_2$C$_2$ Style	M···M Distance (Å)	Dihedral Angle (°)	References
Sc$_2$C$_2$@C_{2v}(5)-C$_{80}$	Zigzag/butterfly mixed	4.312	130.8, 127.0	102
Sc$_2$C$_2$@C_{3v}(8)-C$_{82}$	Symmetrical butterfly	3.981	132.5	103
Sc$_2$C$_2$@D_{2d}(23)-C$_{84}$	Twisted butterfly	4.468	150.2	48
Gd$_2$C$_2$@C_1(51383)-C$_{84}$	Unsymmetrical butterfly	4.106	123–124	39
Gd$_2$C$_2$@D_3(85)-C$_{92}$	Symmetrical butterfly	4.743	154–161	46

theory. One of the enduring fascinations of the endohedral carbide structures is the observation of the close interplay between steric effects and bending strain of the cluster. Another question remains to be addressed: will the encapsulation of carbide clusters end as the fullerene cage reaches some maximum size? Certainly there is a minimum size for encapsulation of the M$_2$C$_2$ moiety.

4.2.7 DIMETALLO ENDOHEDRAL FULLERENES

The dimetallo endohedral fullerenes clearly demonstrate that electrostatic repulsion rather than covalent bonding is dominant when the metals are not mediated by main group elements. As the cage size increases, the metal–metal distance increases, and, significantly, the longest cage axis often is used for the vector between the metals. A few examples are given in Table 4.3.

4.2.8 CONCLUSION

In this brief survey, some of the achievements as well as some of the pitfalls of crystal structure determination of the endohedral fullerenes have been highlighted. Polarizable solvates and co-crystallization agents utilize the delocalized and electrostatic surfaces of the endohedral fullerenes in order to insert themselves into the crystalline lattice. Disorder is prevalent. Many types of disorders exist, including orientational disorder, positional disorder from circulation of the encaged moieties, and disorder with respect to crystallographic symmetry. The search for a recognizable

TABLE 4.3

Variation of M···M Distances versus Cage Dimensions

Endohedral	Cage Style	M···M Distance (Å)	Cage Dimensions (Å)	References
Sm$_2$@D_2(35)-C$_{88}$	Tubular	4.216, 4.207	8.915 × 8.518 × 7.241	49
Sm$_2$@C_1(21)-C$_{90}$	Quasi-spherical	4.147	9.058 × 7.511	104
Sm$_2$@D_3(85)-C$_{92}$	Tubular	4.619, 4.604	9.574 × 8.263	49
La$_2$@D_5(450)-C$_{100}$	Tubular	5.744	10.083 × 8.024	105
Sm$_2$@D_{3d}(822)-C$_{104}$	Tubular	5.832	10.841 × 8.266	106

model can doomed by isomer impurity. In practical terms, our experience has shown that high-resolution data are essential for discrimination of individual carbon atoms when more than one cage orientation exists at a given crystallographic site. The position of a carbon atom from each cage can actually overlap or be much less separated than a normal C—C bond. Therefore, a data resolution of 0.6 Å is more in agreement with the C···C separations than the usual IUCr-suggested value of 0.83 Å for small molecule crystallography. In order to obtain good counting statistics at this resolution, the use of synchrotron radiation is preferred. Of course, high-quality crystals are always acceptable in laboratory sources. It is well known that temperature has a major effect on bond distances as a consequence of thermal motion. With endohedral fullerenes, shallow potential wells can be occupied at elevated temperatures. Thus, caution must be exercised in comparing endohedral fullerene structures when they are carried out at different temperatures. Also, in general, only the major sites should be identified as having the characteristic geometry. As an example, the structure of $Sc_3N@$ I_h-C_{80}•5(o-xylene) at 90 K displays Sc—N bond distances of 2.023 Å (average of three values) whereas the distance shrinks to 1.994 Å (average of nine values) at 200 K.[68]

ACKNOWLEDGMENTS

The authors are grateful for the financial support of the National Science Foundation (grants CHE-1305125 and CHE-1011760 to Alan L. Balch and Marilyn M. Olmstead) and for beam time at the Advanced Light Source, supported by the Director, Office of Science, Office of Basic Energy Sciences of the U.S. Department of Energy under contract number DE-AC02-05CH11231.

REFERENCES

1. Diederich, F.; Whetten, R. L.; Thilgen, C.; Ettl, R.; Chao, I.; Alvarez, M. M. Fullerene isomerism—isolation of C_{2v}-C_{78} and D_3-C_{78}. *Science* **1991**, *254*, 1768–1770.
2. Fowler, P. W.; Manolopoulos, D. E. *An Atlas of Fullerenes*. Oxford Press: Clarendon, Australia, 1995.
3. Akasaka, T.; Wakahara, T.; Nagase, S.; Kobayashi, K.; Waelchli, M.; Yamamoto, K.; Kondo, M. et al. La@C_{82} anion. An unusually stable metallofullerene. *J. Am. Chem. Soc.* **2000**, *122*, 9316–9317.
4. Feng, L.; Wakahara, T.; Tsuchiya, T.; Maeda, Y.; Lian, Y. F.; Akasaka, T.; Mizorogi, N.; Kobayashi, K.; Nagase, S.; Kadish, K. M. Structural characterization of Y@C_{82}. *Chem. Phys. Lett.* **2005**, *405*, 274–277.
5. Akasaka, T.; Wakahara, T.; Nagase, S.; Kobayashi, K.; Waelchli, M.; Yamamoto, K.; Kondo, M. et al. Structural determination of the La@C_{82} isomer. *J. Phys. Chem. B* **2001**, *105*, 2971–2974.
6. Wakahara, T.; Kobayashi, J.; Yamada, M.; Maeda, Y.; Tsuchiya, T.; Okamura, M.; Akasaka, T. et al. Characterization of Ce@C_{82} and its anion. *J. Am. Chem. Soc.* **2004**, *126*, 4883–4887.
7. Wakahara, T.; Okubo, S.; Kondo, M.; Maeda, Y.; Akasaka, T.; Waelchli, M.; Kako, M. et al. Ionization and structural determination of the major isomer of Pr@C_{82}. *Chem. Phys. Lett.* **2002**, *360*, 235–239.
8. Kodama, T.; Ozawa, N.; Miyake, Y.; Sakaguchi, K.; Nishikawa, H.; Ikemoto, I.; Kikuchi, K.; Achiba, Y. Structural study of three isomers of Tm@C_{82} by ^{13}C NMR spectroscopy. *J. Am. Chem. Soc.* **2002**, *124*, 1452–1455.

9. Kirbach, U.; Dunsch, L. The existence of stable Tm@C_{82} isomers. *Angew. Chem. Int. Ed.* **1996**, *35*, 2380–2383.

10. Xu, W.; Niu, B.; Shi, Z. J.; Lian, Y. F.; Feng, L. Sm@C_{2v}(3)-C_{80}: Site-hopping motion of endohedral Sm atom and metal-induced effect on redox profile. *Nanoscale* **2012**, *4*, 6876–6879.

11. Plant, S. R.; Ng, T. C.; Warner, J. H.; Dantelle, G.; Ardavan, A.; Briggs, G. A. D.; Porfyrakis, K. A bimetallic endohedral fullerene: PrSc@C_{80}. *Chem. Commun.* **2009**, 4082–4084.

12. Yamada, M.; Wakahara, T.; Lian, Y. F.; Tsuchiya, T.; Akasaka, T.; Waelchli, M.; Mizorogi, N.; Nagase, S.; Kadish, K. M. Analysis of lanthanide-induced NMR shifts of the Ce@C_{82} anion. *J. Am. Chem. Soc.* **2006**, *128*, 1400–1401.

13. Kobayashi, K.; Nagase, S. Structures and electronic states of M@C_{82} (M = Sc, Y, La and lanthanides). *Chem. Phys. Lett.* **1998**, *282*, 325–329.

14. Yamada, M.; Wakahara, T.; Tsuchiya, T.; Maeda, Y.; Akasaka, T.; Mizorogi, N.; Nagase, S. Spectroscopic and theoretical study of endohedral dimetallofullerene having a Non-IPR fullerene cage: Ce_2@C_{72}. *J. Phys. Chem. A* **2008**, *112*, 7627–7631.

15. Yamada, M.; Wakahara, T.; Tsuchiya, T.; Maeda, Y.; Kako, M.; Akasaka, T.; Yoza, K.; Horn, E.; Mizorogi, N.; Nagase, S. Location of the metal atoms in Ce_2@C_{78} and its bis-silylated derivative. *Chem. Commun.* **2008**, 558–560.

16. Yamada, M.; Nakahodo, T.; Wakahara, T.; Tsuchiya, T.; Maeda, Y.; Akasaka, T.; Kako, M. et al. Positional control of encapsulated atoms inside a fullerene cage by exohedral addition. *J. Am. Chem. Soc.* **2005**, *127*, 14570–14571.

17. Yamada, M.; Okamura, M.; Sato, S.; Someya, C. I.; Mizorogi, N.; Tsuchiya, T.; Akasaka, T.; Kato, T.; Nagase, S. Two regioisomers of endohedral pyrrolidi-nodimetallofullerenes M_2@I_h-C_{80}(CH_2)(2)NTrt (M = La, Ce; Trt = trityl): Control of metal atom positions by addition positions. *Chem. Eur. J.* **2009**, *15*, 10533–10542.

18. Wang, X. L.; Zuo, T. M.; Olmstead, M. M.; Duchamp, J. C.; Glass, T. E.; Cromer, F.; Balch, A. L.; Dorn, H. C. Preparation and structure of CeSc$_2$N@C_{80}: An icosahedral carbon cage enclosing an acentric CeSc$_2$N unit with buried *f* electron spin. *J. Am. Chem. Soc.* **2006**, *128*, 8884–8889.

19. Zhang, L.; Popov, A. A.; Yang, S. F.; Klod, S.; Rapta, P.; Dunsch, L. An endohedral redox system in a fullerene cage: The Ce based mixed-metal cluster fullerene Lu_2CeN@C_{80}. *Phys. Chem. Chem. Phys.* **2010**, *12*, 7840–7847.

20. Feng, L.; Suzuki, M.; Mizorogi, N.; Lu, X.; Yamada, M.; Akasaka, T.; Nagase, S. Mapping the metal positions inside spherical C_{80} cages: Crystallographic and theoretical studies of Ce_2@D_{5h}-C_{80} and Ce_2@I_h-C_{80}. *Chem. Eur. J.* **2013**, *19*, 988–993.

21. Zhang, Y.; Popov, A. A.; Schiemenz, S.; Dunsch, L. Synthesis, isolation, and spectroscopic characterization of Holmium-based mixed-metal Nitride clusterfullerenes: Ho_xSc$_{3-x}$N@C_{80} (x = 1, 2). *Chem. Eur. J.* **2012**, *18*, 9691–9698.

22. Takano, Y.; Aoyagi, M.; Yamada, M.; Nikawa, H.; Slanina, Z.; Mizorogi, N. et al. Anisotropic magnetic behavior of anionic Ce@C_{82} carbene adducts. *J. Am. Chem. Soc.* **2009**, *131*, 9340–9346.

23. Yamada, M.; Someya, C.; Wakahara, T.; Tsuchiya, T.; Maeda, Y.; Akasaka, T.; Yoza, K. et al. Metal atoms collinear with the spiro carbon of 6,6-open adducts, M_2@C_{80}(Ad) (M = La and Ce, Ad = adamantylidene). *J. Am. Chem. Soc.* **2008**, *130*, 1171–1176.

24. Lu, X.; Akasaka, T.; Nagase, S. Carbide cluster metallofullerenes: Structure, properties, and possible origin. *Acc. Chem. Res.* **2013**, *47*, 1627–1635.

25. Iiduka, Y.; Wakahara, T.; Nakahodo, T.; Tsuchiya, T.; Sakuraba, A.; Maeda, Y.; Akasaka, T. et al. Structural determination of metallofullerene Sc$_3$C$_{82}$ revisited: A surprising finding. *J. Am. Chem. Soc.* **2005**, *127*, 12500–12501.

26. Kurihara, H.; Lu, X.; Iiduka, Y.; Mizorogi, N.; Slanina, Z.; Tsuchiya, T.; Akasaka, T.; Nagase, S. $Sc_2C_2@C_{80}$ rather than $Sc_2@C_{82}$: Templated formation of unexpected $C_{2v}(5)$-C_{80} and temperature-dependent dynamic motion of internal Sc_2C_2 cluster. *J. Am. Chem. Soc.* **2011**, *133*, 2382–2385.

27. Yamazaki, Y.; Nakajima, K.; Wakahara, T.; Tsuchiya, T.; Ishitsuka, M. O.; Maeda, Y.; Akasaka, T.; Waelchli, M.; Mizorogi, N.; Nagase, S. Observation of ^{13}C NMR chemical shifts of metal carbides encapsulated in fullerenes: $Sc_2C_2@C_{82}$, $Sc_2C_2@C_{84}$, and $Sc_3C_2@C_{80}$. *Angew. Chem. Int. Ed.* **2008**, *47*, 7905–7908.

28. Lu, X.; Nakajima, K.; Iiduka, Y.; Nikawa, H.; Mizorogi, N.; Slanina, Z.; Tsuchiya, T.; Nagase, S.; Akasaka, T. Structural elucidation and regioselective functionalization of an unexplored carbide cluster metallofullerene $Sc_2C_2@C_s(6)$-C_{82}. *J. Am. Chem. Soc.* **2011**, *133*, 19553–19558.

29. Lu, X.; Nakajima, K.; Iiduka, Y.; Nikawa, H.; Tsuchiya, T.; Mizorogi, N.; Slanina, Z.; Nagase, S.; Akasaka, T. The long-believed $Sc_2@C_{2v}(17)$-C_{84} is actually $Sc_2C_2@C_{2v}(9)$-C_{82}: Unambiguous structure assignment and chemical functionalization; *Angew. Chem. Int. Ed.* **2012**, *51*, 5889–5892.

30. Inoue, T.; Tomiyama, T.; Sugai, T.; Okazaki, T.; Suematsu, T.; Fujii, N.; Utsumi, H.; Nojima, K.; Shinohara, H. Trapping a C_2 radical in endohedral metallofullerenes: Synthesis and structures of $Y_2C_2@C_{82}$ (Isomers I, II, and III). *J. Phys. Chem. B* **2004**, *108*, 7573–7579.

31. Zhang, J. Y.; Fuhrer, T.; Fu, W. J.; Ge, J. C.; Bearden, D. W.; Dallas, J.; Duchamp, J. et al. Nanoscale fullerene compression of an Yttrium Carbide cluster. *J. Am. Chem. Soc.* **2012**, *134*, 8487–8493.

32. Yang, S.; Chen, C.; Liu, F.; Xie, Y.; Li, F.; Jiao, M.; Suzuki, M. et al. An improbable monometallic cluster entrapped in a popular fullerene cage: $YCN@C_s(6)$-C_{82}. *Sci. Rep.* **2013**, *3*, 1487.

33. Stevenson, S.; Fowler, P. W.; Heine, T.; Duchamp, J. C.; Rice, G.; Glass, T.; Harich, K.; Hajdu, E.; Bible, R.; Dorn, H. C. Materials science—A stable non-classical metallofullerene family. *Nature* **2000**, *408*, 427–428.

34. Gorny, K. R.; Pennington, C. H.; Martindale, J. A.; Phillips, J. P.; Stevenson, S.; Heinmaa, I.; Stern, R. Molecular orientational dynamics of the endohedral fullerene $Sc_3N@C_{80}$ as probed by ^{13}C and [45]Sc NMR; eprint arXiv:cond-mat/0604365v1 **2006**, http://adsabs.harvard.edu/abs/2006cond.mat..4365G.

35. Yang, S. F.; Popov, A. A.; Dunsch, L. Carbon Pyramidalization in fullerene cages induced by the endohedral cluster: Non-scandium mixed metal nitride cluster fullerenes. *Angew. Chem. Int. Ed.* **2008**, *47*, 8196–8200.

36. Klod, S.; Zhang, L.; Dunsch, L. The role of the cluster on the relaxation of endohedral fullerene cage carbons: A NMR spin-lattice relaxation study of an internal relaxation reagent. *J. Phys. Chem. C* **2010**, *114*, 8264–8267.

37. Aoyagi, S.; Nishibori, E.; Sawa, H.; Sugimoto, K.; Takata, M.; Miyata, Y.; Kitaura, R. et al. A layered ionic crystal of polar $Li@C_{60}$ superatoms; *Nat. Chem.* **2010**, *2*, 678–683.

38. Matsuo, Y.; Okada, H.; Maruyama, M.; Sato, H.; Tobita, H.; Ono, Y.; Omote, K.; Kawachi, K.; Kasama, Y. Covalently chemical modification of lithium ion-encapsulated fullerene: Synthesis and characterization of $[Li^+@PCBM]PF_6^-$. *Org. Lett.* **2012**, *14*, 3784–3787.

39. Ueno, H.; Nakamura, Y.; Ikuma, N.; Kokubo, K.; Oshima, T. Synthesis of a lithium-encapsulated fullerenol and the effect of the internal lithium cation on its aggregation behavior. *Nano Res.* **2012**, *5*(8), 558–564.

40. Miyake, Y.; Suzuki, S.; Kojima, Y.; Kikuchi, K.; Kobayashi, K.; Nagase, S.; Kainosho, M.; Achiba, Y.; Maniwa, Y.; Fisher, K. Motion of scandium ions in Sc_2C_{84} observed by [45]Sc solution NMR. *J. Phys. Chem.* **1996**, *100*, 9579–9581.

41. Okimoto, H.; Hemme, W.; Ito, Y.; Sugai, T.; Kitaura, R.; Eckert, H.; Shinohara, H. Solid-state ^{13}C and ^{45}Sc NMR studies on endohedral scandium-carbide metallofullerenes: A motional dynamics of Sc atoms in fullerenes. *NANO* **2008**, *3*(1), 21–25.

42. Stevenson, S.; Rice, G.; Glass, T.; Harich, K.; Cromer, F.; Jordan, M. R.; Craft, J. et al. Small-bandgap endohedral metallofullerenes in high yield and purity. *Nature* **1999**, *401*, 55–57.

43. Yang, S.; Popov, A. A.; Chen, C.; Dunsch, L. Mixed metal Nitride clusterfullerenes in cage isomers: $Lu_xSc_{3-x}N@C_{80}$ (x = 1, 2) as compared with $M_xSc_{3-x}N@C_{80}$ (M = Er, Dy, Gd, Nd); *J. Phys. Chem. C* **2009**, *113*, 7616–7623.

44. Dunsch, L.; Yang, S.; Zhang, L.; Svitova, A.; Oswald, S.; Popov, A. A. Metal sulfide in a C_{82} fullerene cage: A new form of endohedral clusterfullerenes; *J. Am. Chem. Soc.* **2010**, *132*, 5413–5421.

45. Chen, N.; Beavers, C. M.; Mulet-Gas, M.; Rodriguez-Fortea, A.; Munoz, E. J.; Li, Y. Y.; Olmstead, M. M.; Balch, A. L.; Poblet, J. M.; Echegoyen, L. $Sc_2S@C_s(10528)$-C_{72}: A dimetallic sulfide endohedral fullerene with a non isolated Pentagon rule cage. *J. Am. Chem. Soc.* **2012**, *134*, 7851–7860.

46. Wang, T. S.; Feng, L.; Wu, J. Y.; Xu, W.; Xiang, J. F.; Tan, K.; Ma, Y. H. et al. Planar quinary cluster inside a fullerene cage: Synthesis and structural characterizations of $Sc_3NC@C_{80}$-I_h. *J. Am. Chem. Soc.* **2010**, *132*, 16362–16364.

47. Fu, W. J.; Xu, L. S.; Azurmendi, H.; Ge, J. C.; Fuhrer, T.; Zuo, T. M.; Reid, J.; Shu, C. Y.; Harich, K.; Dorn, H. C. ^{89}Y and ^{13}C NMR cluster and Carbon cage studies of an Yttrium metallofullerene family, $Y_3N@C_{2n}$ (n = 40–43). *J. Am. Chem. Soc.* **2009**, *131*, 11762–11769.

48. Chai, Y.; Guo, T.; Jin, C. M.; Haufler, R. E.; Chibante, L. P. F.; Fure, J.; Wang, L. H.; Alford, J. M.; Smalley, R. E. Fullerenes with metals inside. *J. Phys. Chem.* **1991**, *95*, 7564–7568.

49. Akasaka, T.; Nagase, S.; Kobayashi, K.; Walchli, M.; Yamamoto, K.; Funasaka, H.; Kako, M.; Hoshino, T.; Erata, T. ^{13}C and ^{139}La NMR studies of $La_2@C_{80}$: First evidence for circular motion of metal atoms in endohedral dimetallofullerenes. *Angew. Chem. Int. Ed.* **1997**, *36*, 1643–1645.

50. Wakahara, T.; Yamada, M.; Takahashi, S.; Nakahodo, T.; Tsuchiya, T.; Maeda, Y.; Akasaka, T. et al. Two-dimensional hopping motion of encapsulated La atoms in silylated $La_2@C_{80}$. *Chem. Commun.* **2007**, 2680–2682.

51. Yamada, M.; Minowa, M.; Sato, S.; Kako, M.; Slanina, Z.; Mizorogi, N.; Tsuchiya, T.; Maeda, Y.; Nagase, S.; Akasaka, T. Thermal carbosilylation of endohedral dimetallofullerene $La_2@I_h$-C_{80} with Silirane. *J. Am. Chem. Soc.* **2010**, *132*, 17953–17960.

52. Yamada, M.; Minowa, M.; Sato, S.; Slanina, Z.; Tsuchiya, T.; Maeda, Y.; Nagase, S.; Akasaka, T. Regioselective cycloaddition of $La_2@I_h$-C_{80} with tetracyanoethylene oxide: Formation of an endohedral dimetallofullerene adduct featuring enhanced electron-accepting character; *J. Am. Chem. Soc.* **2011**, *133*, 3796–3799.

53. Wakahara, T.; Nikawa, H.; Kikuchi, T.; Nakahodo, T.; Rahman, G. M. A.; Tsuchiya, T.; Maeda, Y. et al. $La@C_{72}$ having a non-IPR carbon cage. *J. Am. Chem. Soc.* **2006**, *128*, 14228–14229.

54. Nikawa, H.; Kikuchi, T.; Wakahara, T.; Nakahodo, T.; Tsuchiya, T.; Rahman, G. M. A.; Akasaka, T. et al. Missing metallofullerene $La@C_{74}$. *J. Am. Chem. Soc.* **2005**, *127*, 9684–9685.

55. Nikawa, H.; Yamada, T.; Cao, B. P.; Mizorogi, N.; Slanina, Z.; Tsuchiya, T.; Akasaka, T.; Yoza, K.; Nagase, S. Missing metallofullerene with C_{80} cage. *J. Am. Chem. Soc.* **2009**, *131*, 10950–10954.

56. Akasaka, T.; Lu, X.; Kuga, H.; Nikawa, H.; Mizorogi, N.; Slanina, Z.; Tsuchiya, T.; Yoza, K.; Nagase, S. Dichlorophenyl derivatives of $La@C_{3v}(7)$-C_{82}: Endohedral metal induced localization of pyramidalization and spin on a Triple-Hexagon junction. *Angew. Chem. Int. Ed.* **2010**, *49*, 9715–9719.

57. Lu, X.; Nikawa, H.; Kikuchi, K.; Mizorogi, N.; Slanina, Z.; Tsuchiya, T.; Akasaka, T.; Nagase, S. Radical derivatives of insoluble La@C_{74} : X-ray structures, metal positions, and isomerization. *Angew. Chem. Int. Ed.* **2011**, *50*, 6356–6359.

58. Lu, X.; Nikawa, H.; Nakahodo, T.; Tsuchiya, T.; Ishitsuka, M. O.; Maeda, Y.; Akasaka, T. et al. Chemical understanding of a non-IPR metallofullerene: Stabilization of encaged metals on fused-pentagon bonds in La$_2$@C_{72}. *J. Am. Chem. Soc.* **2008**, *130*, 9129–9136.

59. Suzuki, M.; Mizorogi, N.; Yang, T.; Uhlik, F.; Slanina, Z.; Zhao, X.; Yamada, M. et al. La$_2$@C_s(17490)-C_{76}: A new non-IPR dimetallic metallofullerene featuring unexpectedly weak Metal-Pentalene interactions. *Chem. Eur. J.* **2013**, *19*, 17125–17130.

60. Rodriguez-Fortea, A.; Balch, A. L.; Poblet, J. M. Endohedral metallofullerenes: A unique host-guest association. *Chem. Soc. Rev.* **2011**, *40*, 3551–3563.

61. Xie, Y.-P.; Lu, X.; Akasaka, T.; Nagase, S. New features in coordination chemistry: Valuable hints from X-ray analyses of endohedral metallofullerenes. *Polyhedron* **2013**, *52*, 3–9.

62. Fleming, R. M.; Hessen, B.; Siegrist, T.; Kortan, A. R.; Marsh, P.; Tycko, R.; Dabbagh, G.; Haddon, R. C. Crystalline fullerenes. Round pegs in square holes. In *Fullerenes*; Hammond, G., Kuck, V. J., Eds.; ACS Symposium Series, American Chemical Society: Washington, DC, **1992**; pp. 25–39.

63. Korobov, M. V.; Mirakian, A. L.; Avramenko, N. V.; Valeev, E. F.; Neretin, I. S.; Slovokhotov, Y. L.; Smith, A. L.; Oloffson, G.; Ruoff, R. S. C_{60}•bromobenzene solvate: Crystallographic and thermochemical studies and their relationship to C_{60} solubility in romobenzene. *J. Phys. Chem. B.* **1998**, *102*, 3712–3717.

64. Hardie, M. J.; Torrens, R.; Raston, C. L. Characterisation of a new 1:1 (C_{60})(CHBr$_3$) intercalation complex. *Chem. Commun.* **2003**, 1854–1855.

65. Oszlanyi, G.; Bortel, G.; Faigel, G.; Pekker, S.; Tegze, M. Dynamic origin of the ortho-rhombic symmetry of C_{60}-n-pentane. *Solid State Commun.* **1994**, *89*, 417–419.

66. Schulz-Dobrick, M.; Panthöfer, M.; Jansen, M. Supramolecular arrangement of C_{60} and phenol into a square packing arrangement of π···π interacting and Hydrogen-bonded rods in C_{60}•5C$_6$H$_5$OH. *Eur. J. Inorg. Chem.* **2005**, 4064–4069.

67. Bond, A. D. C_{60} thiophene disolvate. *Acta Crystallogr. Sect. E. Struct. Rep. Online* **2003**, *59*, o1992–o1993.

68. Stevenson, S.; Lee, H. M.; Olmstead, M. M.; Kozikowski, C.; Stevenson, P.; Balch, A. L. Preparation and crystallographic characterization of a new endohedral, Lu$_3$N@C_{80}•5(o-xylene), and comparison with Sc$_3$N@C_{80}•5(o-xylene). *Chem. Eur. J.* **2002**, *8*, 4528–4535.

69. Olmstead, M. M.; Jiang, F.; Balch, A. L. 2C$_{60}$•3CS$_2$: Orientational ordering accompanies the reversible phase transition at 168 K. *Chem. Commun.* **2000**, 483–484.

70. Olmstead, M. M.; Balch, A. L.; Lee, H. M. An order-disorder phase transition in the structure of C_{60}•4benzene. *Acta Crystallogr. Sect. B Struct. Sci.* **2012**, *68*, 66–70.

71. Zuo, T.; Olmstead, M. M.; Beavers, C. M.; Balch, A. L.; Wang, G.; Yee, G. T.; Shu, C. et al. Preparation and structural characterization of the I_h and the D_{5h} isomers of the endohedral fullerenes Tm$_3$N@C_{80}: Icosahedral C_{80} cage encapsulation of a trimetallic nitride magnetic cluster with three uncoupled Tm^{3+} ions. *Inorg. Chem.* **2008**, *47*, 5234–5244.

72. Xu, W.; Feng, L.; Calvaresi, M.; Liu, J.; Liu, Y.; Niu, B.; Shi, Z.; Lian, Y.; Zerbetto, F. An experimentally observed trimetallofullerene Sm$_3$@I_h-C_{80}: Encapsulation of three metal atoms in a cage without a nonmetallic mediator. *J. Am. Chem. Soc.* **2013**, *135*, 4187–4190.

73. Yang, S.; Dunsch, L. Expanding the number of stable isomeric structure of the C_{80} cage: A new fullerene Dy$_3$N@C_{80}. *Chem. Eur. J.* **2006**, *12*, 413–419.

74. Yang, H.; Wang, Z.; Jin, H.; Hong, B.; Liu, Z.; Beavers, C. M.; Olmstead, M. M.; Balch, A. L. Isolation and crystallographic characterization of Sm@C_{2v}(3)-C_{80} through cocrystal formation with Ni^{II}(octaethylporphyrin) or bis(ethylenedithio)tetrathiafulvalene. *Inorg. Chem.* **2012**, *52*, 1275–1284.

75. Lu, X.; Lian, Y.; Beavers, C. M.; Mizorogi, N.; Slanina, Z.; Nagase, S.; Akasaka, T. Crystallographic X-ray analyses of Yb@C_{2v}(3)-C_{80} reveal a feasible rule that governs the location of a rare earth metal inside a medium-sized fullerene. *J. Am. Chem. Soc.* **2011**, *133*, 10772–10775.

76. Izuoka, A.; Tachikawa, T.; Sugawara, T.; Suzuki, Y.; Konno, M.; Saito, Y.; Shinohara, H. An X-ray crystallographic analysis of a (BEDT-TTF)$_2$$C_{60}$ charge-transfer complex. *J. Chem. Soc. Chem. Commun.* **1992**, 1472–1473.

77. Ghiassi, K. B.; Olmstead, M. M.; Balch, A. L. Orientational variation, solvate composition and molecular flexibility in well-ordered cocrystals of the fullerene C_{70} with bis(ethylenedithio)tetrathiafulvalene. *Chem. Commun.* **2013**, *49*, 10721–10723.

78. Stone, A. J.; Wales, D. J. Theoretical studies of icosahedral C_{60} and some related species. *Chem. Phys. Lett.* **1986**, *128*, 501–503.

79. Dunk, P. W.; Kaiser, N. K.; Hendrickson, C. L.; Quinn, J. P.; Ewels, C. P.; Nakanishi, Y.; Sasaki, Y.; Shinohara, H.; Marshall, A. G.; Kroto, H. W. Closed network growth of fullerenes. *Nature Chem.* **2012**, *3*, 855.

80. Wang, Z.; Yang, H.; Ziang, A.; Liu, Z.; Olmstead, M. M.; Balch, A. L. Structural similarities in C_s(16)-C_{86} and C_2(17)-C_{86}. *Chem. Commun.* **2010**, *46*, 5262–5264.

81. Yang, H.; Jin, H.; Wang, X.; Liu, Z.; Yu, M.; Zhao, F.; Mercado, B. Q.; Olmstead, M. M.; Balch, A. L. X-ray crystallographic characterization of new soluble endohedral fullerenes utilizing the popular C_{82} Bucky cage. Isolation and structural characterization of Sm@C_{3v}(7)-C_{82}, Sm@C_s(6)-C_{82}, and Sm@C_2(5)-C_{82}. *J. Am. Chem. Soc.* **2012**, *134*, 14127–14136.

82. Okazaki, T.; Lian, Y. F.; Gu, Z. N.; Suenaga, K.; Shinohara, H. Isolation and spectroscopic characterization of Sm-containing metallofullerenes. *Chem. Phys. Lett.* **2000**, *320*, 435–440.

83. Liu, J.; Shi, Z. J.; Gu, Z. N. The cage and metal effect: Spectroscopy and electrochemical survey of a series of Sm-containing high metallofullerenes. *Chem. Asian J.* **2009**, *4*, 1703–1711.

84. Xu, W.; Niu, B.; Feng, L.; Shi, Z.; Lian, Y. Access to an unexplored chiral C_{82} cage by encaging a divalent metal: Structural elucidation and electrochemical studies of Sm@C_2(5)-C_{82}. *Chem. Eur. J.* **2012**, *18*, 14246–14249.

85. Suzuki, M.; Lu, X.; Sato, S.; Nikawa, H.; Mizorogi, N.; Slanina, Z.; Tsuchiya, T.; Nagase, S.; Akasaka, T. Where does the metal cation stay in Gd@C_{2v}(9)-C_{82}? A single-crystal X-ray diffraction study. *Inorg. Chem.* **2012**, *51*, 5270–5273.

86. Yang, H.; Jin, H.; Zhen, H.; Wang, Z.; Liu, Z.; Beavers, C. M.; Mercado, B. Q. M.; Olmstead, M. M.; Balch, A. L. Isolation and crystallographic identification of four isomers of Sm@C_{90}. *J. Am. Chem. Soc.* **2011**, *133*, 6299–6306.

87. Che, Y.; Yang, H.; Wang, A.; Jin, H.; Liu, Z.; Lu, C.; Zuo, T. et al. Isolation and structural characterization of two very large, and largely empty, endohedral fullerenes: Tm@C_{3v}-C_{94} and Ca@C_{3v}-C_{94}. *Inorg. Chem.* **2009**, *48*, 6004–6010.

88. Jin, H.; Yang, H.; Yu, M.; Liu, Z.; Beavers, C. M.; Olmstead, M. M.; Balch, A. L. Single samarium atoms in large fullerene cages. Characterization of two isomers of Sm@C_{92} and four isomers of Sm@C_{94} with the X-ray crystallographic identification of Sm@C_1(42)-C_{92}, Sm@C_2(24)-C_{92}, and Sm@C_{3v}(134)-C_{94}. *J. Am. Chem. Soc.* **2012**, *134*, 10933–10941.

89. Olmstead, M. M.; Lee, H. M.; Duchamp, J. C.; Stevenson, S.; Marciu, D.; Dorn, H. C.; Balch, A. L. $Sc_3N@C_{68}$: Folded Pentalene coordination in an endohedral fullerene that does not obey the isolated pentagon rule. *Angew. Chem. Int. Ed.* **2003**, *42*, 900–903.

90. Beavers, C. M.; Chaur, M. N.; Olmstead, M. M.; Echegoyen, L.; Balch, A. L. Large metal ions in a relatively small fullerene cage: The structure of $Gd_3N@C_2$(22010)-C_{78} departs from the isolated pentagon rule. *J. Am. Chem. Soc.* **2009**, *131*, 11519–11524.

91. Beavers, C. M.; Zuo, T.; Duchamp, J. C.; Harich, K.; Dorn, H. C.; Olmstead, M. M.; Balch, A. L. $Tb_3N@C_{84}$: An improbable, egg-shaped endohedral fullerene that violates the isolated pentagon rule. *J. Am. Chem. Soc.* **2006**, *128*, 11352–11353.

92. Zuo, T.; Walker, K.; Olmstead, M. M.; Melin, F.; Holloway, B. C.; Echegoyen, L.; Dorn, H. C. et al. New egg-shaped fullerenes: Non-isolated pentagon structures of $Tm_3N@C_s(51365)$-C_{84} and $Gd_3N@C_s(51365)$-C_{84}. *Chem. Commun.* **2008**, 1067–1069.

93. Mercado, B. Q.; Beavers, C. M.; Olmstead, M. M.; Chaur, M. N.; Walker, K. W.; Holloway, B. C.; Echegoyen, L.; Balch, A. L. Is the isolated Pentagon rule merely a suggestion for endohedral fullerenes? The structure of a second egg-shaped endohedral fullerene—$Gd_3N@C_s(39663)$-C_{82}. *J. Am. Chem. Soc.* **2008**, *130*, 7854–7855.

94. Zhang, J., Bowles, F. L; Bearden, D. W.; Ray, W. K.; Fuhrer, T. ; Ye, Y.; Dixon, C. et al. A missing link in the transformation from asymmetric to symmetric metallofullerene cages implies a top-down fullerene formation mechanism. *Nature Chem.* **2013**, *5*, 880–885.

95. a. Song, H.-B.; Wang, Q.-M.; Zhang, Z.-Z.; Mak, T. C. W. A novel luminescent copper(I) complex containing an acetylenediide-bridged, butterfly-shaped tetranuclear core. *Chem. Commun.* **2001**, 1658–1659. b. St. Clair, M.; Schaefer, W. P.; Bercaw, J. E. Reactivity of permethylscandocene derivatives with acetylene. Structure of acetylene diylbis(permethylscandocene), η^5-$(C_5Me_5)_2Sc$-$C\equiv C$-$Sc(\eta^5$-$C_5Me_5)_2$. *Organometallics* **1991**, *10*, 525–527. c. Ogawa, H.; Onitsuka, K.; Joh, T.; Takahashi, S. Synthesis, characterization, and molecular structure of μ-ethynediyl complexes $[X(Pr_3)MC\equiv CM(PR_3)_2X]$ (M = Pd, Pt; R = Me, Et, *n*-Bu; X = Cl, I). *Organometallics* **1998**, *7*, 2257–2260. d. Bruce, M. L.; Snow, M. R.; Tiekink, E. R. T.; Williams, L. The first example of a μ_4-η^1, η^2-acetylide dianion: Preparation and X-ray structure of $Ru_4(\mu_4$-η^1,η^2-$C_2)$ $(\mu$-$PPh_2)_2(CO)_{12}$. *J. Chem. Soc. Chem. Commun.* **1986**, 701–702. e. Blau, R. J.; Chisholm, M. H.; Folting, K.; Wang, R. J. Synthesis, X-ray crystal structure, and N.M.R. studies of *trans*-bis(dimethylphenylphosphine)bis[{dicarbido(4-)σ-Pt, η^2-W^2}-pentakis(t-butoxy)ditungsten(M=M)]platinum(II). *J. Chem. Soc. Chem. Commun.* **1985**, 1582–1584.

96. Popov, A. A.; Yang, S.; Dunsch, L. Endohedral fullerenes. *Chem. Rev.* **2013**, *113*, 5989–6113.

97. Lu, X.; Feng, L.; Akasaka, T.; Nagase, S. Current status and future developments of endohedral metallofullerenes. *Chem. Soc. Rev.* **2012**, *41*, 7723–7760.

98. Zhang, J.; Fuhrer, T.; Fu, W.; Ge, J.; Bearden, D. W.; Dallas, J.; Duchamp, J., Walker, K.; Champion, H.; Azurmendi, H.; Harich, K.; Dorn, H. C. Nanoscale fullerene compression of an yttrium carbide cluster. *J. Am. Chem. Soc.* **2012**, *134*, 8487–8493.

99. Maki, S.; Nishibori, E.; Terauchi, I.; Ishihara, M.; Aoyagi, S.; Sakata, M.; Takata, M.; Umemoto, H.; Inoue, T.; Shinohara, H. A structural diagnostics diagram for metallofullerenes encapsulating metal carbides and nitrides. *J. Am. Chem. Soc.* **2013**, *135*, 918–923.

100. Garcia-Borrìs, M.; Osuna, S.; Luis, J. M.; Swart, M.; Solà, M. The exohedral Diels-Alder reactivity of the titanium carbide endohedral metallofullerene $Ti_2C_2@D_{3h}$-C_{78}; Comparison with D_{3h}-C_{78} and $M_3N@D_{3h}$-C_{78} (M = Sc and Y) reactivity. *Chem. Eur. J.* **2012**, *18*, 7141–7154.

101. Yang, H.; Lu, C.; Liu, Z.; Jin, H.; Che, Y.; Olmstead, M. M.; Balch, A. L. Detection of a family of gadolinium-containing endohedral fullerenes and the isolation and crystallographic characterization of one member as a metal carbide encapsulated inside a large fullerene cage: $(Gd_2C_2)@D_3(85)$-C_{92}. *J. Am. Chem. Soc.* **2008**, *130*, 17296–17300.

102. Kurihara, H.; Lu, X.; Iiduka, Y.; Nikawa, H.; Mizorogi, N.; Slanina, Z. Chemical understanding of carbide cluster metallofullerenes: A case study on $Sc_2C_2@C_{2v}(5)$-C_{80} with complete X-ray crystallographic characterizations. *J. Am. Chem. Soc.* **2012**, *134*, 3139–3144.

103. Kurihara, H.; Lu, X.; Iiduka, Y.; Nikawa, H.; Hachiya, M.; Mizorogi, N. Slanina, Z.; Tsuchiya, T.; Nagase, S.; Akasaka, T. X-ray structures of $Sc_2C_2@C_{2n}$ (n = 40–42): In depth understanding of the core-shell interplay in carbide cluster metallofullerenes. *Inorg. Chem.* **2012**, *51*, 746–750.

104. Yang, H.; Jin, H.; Hong, B.; Liu, Z.; Beavers, C. M.; Zhen, H.; Wang, Z.; Mercado, B. Q.; Olmstead, M. M.; Balch, A. L. Large endohedral fullerenes containing two metal ions: $Sm_2@D_2(35)$-C_{88}, $Sm_2@C_1(21)$-C_{90}, and $Sm_2@D_3(85)$-C_{92} and their relationship to endohedral fullerenes containing two gadolinium ions. *J. Am. Chem. Soc.* **2011**, *133*, 16911–16919.

105. Beavers, C. M.; Jin, H.; Yang, H.; Wang, Z.; Wang, H.; Ge, H.; Liu, Z.-Y.; Mercado, B. Q.; Olmstead, M. M.; Balch, A. L. Very large, soluble endohedral fullerenes in the series La_2C_{92} to La_2C_{134}: Isolation and crystallographic characterization of a $La_2@D_5(450)$-C_{100}. *J. Am. Chem. Soc.* **2011**, *133*, 15338–15341.

106. Mercado, B. Q.; Jiang; A.; Yang, H.; Wang, Z.; Jin, H.; Liu, Z.; Olmstead, M. M.; Balch, A. L. Isolation and structural characterization of the molecular nanocapsule, $Sm_2@D_{3d}(822)$-C_{104}. *Angew. Chem. Int. Ed.* **2009**, *48*, 9114–9116.

5 Intrinsic Properties of Endohedral Metallofullerenes

Song Wang, Taishan Wang, Tao Wei,
Chunru Wang, and Shangfeng Yang

CONTENTS

5.1   INTRODUCTION

Endohedral metallofullerenes (EMFs) have been attracting wide interest all over the world over the past two decades [1]. To date, a wealth of EMFs have been synthesized and isolated, including conventional EMFs ($M_x@C_{2n}$, $x = 1$–3) and several different types of cluster fullerenes such as nitride cluster fullerenes (NCFs) [2], carbide cluster fullerenes (CCFs) [1], oxide cluster fullerenes (OCFs) [1], sulfide cluster fullerenes (SCFs) [1], hydrocarbide cluster fullerenes [1], carbonitride cluster fullerenes [1], and cyanide cluster fullerenes [1]. EMFs exhibit unique properties resulting from the electron transfer from the encapsulated species (metal ions or clusters) to the carbon cage. Therefore, EMFs not only inherit the properties of the carbon cage and the encapsulated species but also exhibit intriguing properties which are derived from the intramolecular interaction between the carbon cage and the encapsulated species [1,3].

An exhaustive review on all aspects of EMFs was recently made by Popov et al. [1]. Several other reviews containing the general properties of EMFs or specific ones such as chemical or electrochemical properties have also been published in the past five years, including those by Echegoyen et al. in 2009 [4]; Lu et al. in 2013, 2012, and 2011 [3,5,6]; and Popov and Dunsch in 2011 [7]. In this chapter, we focus on the intrinsic properties of EMFs, including their electronic, electrochemical, optical, and magnetic properties.

5.2   ELECTRONIC PROPERTIES

5.2.1   UV-vis-NIR Absorption Property

UV-vis-NIR absorption spectroscopy is an important technique to characterize EMFs. The electronic absorption of EMFs mainly results from $\pi \rightarrow \pi^*$ excitation of the carbon cage, which is primarily dependent on the structure and charge state of carbon cage. The optical band gap of EMFs is the energy of the lowest energy excitation, which corresponds to the HOMO-LUMO gap and with which the electronic stability of EMF can be evaluated [1]. Practically, it is difficult to assign the lowest energy excitation of EMF because the absorption bands are often broad and some bands overlap. Thus, the optical band gap ($E_{g, \text{optical}}$, in electron volts) is often estimated from the absorption onset (λ_{onset}, in nanometers) according to the equation $E_{g, \text{optical}} = 1240/\lambda_{\text{onset}}$. For $E_{g, \text{optical}}$, a borderline of 1.0 eV is often used to distinguish large and small band gap EMFs, which indicates the stability of EMFs in some cases. For instance, the majority of NCFs have large $E_{g, optical}$ exceeding 1 eV, whereas $E_{g, optical}$ of mono-metallofullerenes (mono-EMFs) is generally quite small [8,9].

The absorption spectroscopy is highly structure sensitive. Because the electronic absorption of EMFs mainly results from $\pi \rightarrow \pi^*$ excitation of the carbon cage, UV-vis-NIR absorption spectra of EMFs with the same carbon cage isomer in the same formal charge state are very similar. For instance, the absorption spectra of a series of mono-EMFs $M^{III}@C_{82}$-$C_{2v}(9)$ (M = Y, La, Ce, Pr, Nd, Gd, Tb, Dy, Ho, Er, Lu) are quite similar and almost independent on the encapsulated metal [9]. However, when the isomeric structure or charge state of carbon cage changes, the absorption spectrum varies markedly [1]. A good example is that the absorption spectra of the divalent metal-based mono-EMFs, $M^{II}@C_{82}$-$C_{2v}(9)$ (M = Ca, Sm, Tm, Eu, Yb), which are very similar, but quite different from those of $M^{III}@C_{82}$-$C_{2v}(9)$ because of the different charge state of the same C_{82}-$C_{2v}(9)$ cage [9]. For the di-metallofullerenes (di-EMFs), $M_2@C_{80}$-$I_h(7)$ (M = La, Ce, Pr), the absorption spectra are quite similar and featureless, and the absorption spectra are monotonically decreasing absorption coefficient with increasing wavelength [10–12].

Given that UV-vis-NIR absorption spectra of EMFs are very similar for the same carbon cage isomer in the same formal charge state, UV-vis-NIR absorption spectroscopy can be used as an indirect technique for structure elucidation of newly isolated EMFs when the cage structures of analogous EMFs with other metals are known. This solution has been frequently used in preliminary determination of the cage structures of carbide cluster fullerenes (CCFs) [1]. Figure 5.1 shows UV-vis-NIR spectra of three isomers of $Y_2C_2@C_{82}$ in carbon disulfide solution reported by Shinohara et al. Obviously, the absorption spectra of $Y_2C_2@C_{82}$ (I, II, III) are different from each other, indicating that $Y_2C_2@C_{82}$ (I, II, III) possess different cage isomers [13]. Later on, the cage isomers of $Y_2C_2@C_{82}$ (I, II, III) were confirmed by ^{13}C NMR spectra to be $C_s(6)$, $C_{2v}(9)$, and $C_{3v}(8)$, respectively [12]. Of note, the absorption spectra of $Y_2C_2@C_{82}$ (I, II, III) are similar to those of $(Sc_2C_2)@C_{82}$ (I, II, III), $Dy_2C_2@C_{82}$ (I, II, III), and $Er_2C_2@C_{82}$ (I, II, III) reported previously for the same label of isomer [14–16]. More recently, Yang et al. proposed that the newly isolated Tb_2C_{84} was actually a CCF with an isomeric structure of C_{82}-C_s (6) based on the high resemblance of its UV-vis-NIR spectrum with those of $Y_2C_2@C_s(6)$-C_{82} and $Sc_2C_2@C_s(6)$-C_{82}, and such an assumption was confirmed by an X-ray crystallographic study [17]. Interestingly, UV-vis-NIR absorption spectra of $(Er_2C_2)@C_{82}$ (I, II, III) were found to be similar to those of the di-EMFs $Er_2@C_{82}$ (I, II, III) with the same cages [18].

For NCFs, the encapsulated M_3N cluster transferred six electrons to the cage, and the UV-vis-NIR absorption spectra mainly depend on the isomer structures of the sixfold negatively changed cage [19]. Comparing the UV-vis-NIR spectra of $Sc_3N@C_{80}$-$I_h(7)$ and $Sc_3N@C_{80}$-$D_{5h}(6)$ reported by Dunsch et al. (Figure 5.2), calculated from the absorption onsets of 820 and 920 nm for these two isomers, respectively, the optical band gaps ($E_{g, optical}$) of $Sc_3N@C_{80}$-$I_h(7)$ and $Sc_3N@C_{80}$-$D_{5h}(6)$ were calculated to be 1.51 and 1.35 eV, respectively [20]. In particular, it was found that a series of C_{80}-$I_h(7)$-based NCFs encapsulating three homogenous metals $M_3N@C_{80}$-$I_h(7)$ (M = Sc, Y, Gd-Lu) exhibited quite similar spectra without NIR absorptions [20–22]. However, it should be noted that in some cases, the similarity of UV-vis-NIR absorption spectra

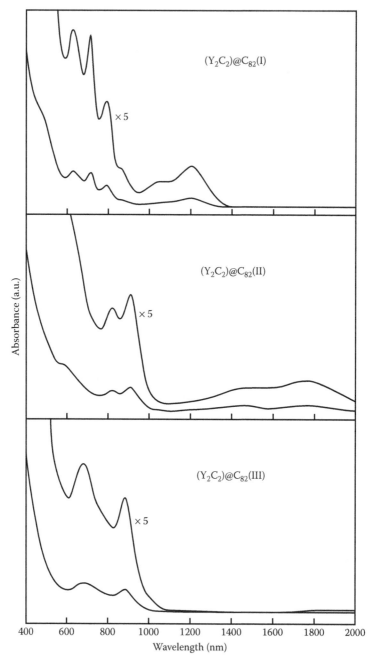

FIGURE 5.1 UV-vis-NIR spectra of $Y_2C_2@C_{82}$ (I, II, III) in CS_2 solvent. (Reprinted with permission from Inoue, T. et al., *J. Phys. Chem. B.* 108, 7573–7579, 2004. Copyright 2004 American Chemical Society.

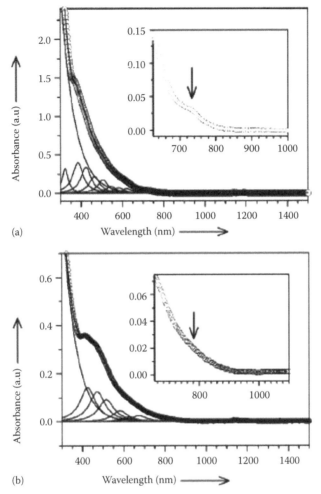

(a)

(b)

FIGURE 5.2 UV-vis-NIR spectra of $Sc_3N@C_{80}$ (I) (a) and $Sc_3N@C_{80}$ (II) (b) in toluene. Lorentzian fit data are given by solid lines. The onset range is represented with expanded intensity scale as inset. The lowest energetic transitions are indicated by arrows. (From Krause, M. and Dunsch, L.: *ChemPhysChem.* 2004. 5. 1445–1449. Copyright Wiley-VCH Verlag GmbH & Co. KGaA. Reproduced with permission.)

of a same type of EMFs with the same carbon cage becomes obscure because of the significant perturbation of the encapsulated metal atoms contributing to the frontier orbitals. A clear example is reported for mixed metal nitride clusterfullerenes NCFs (MMNCFs) [23,24]. Figure 5.3 compares the UV-vis-NIR spectra of $Lu_xSc_{3-x}N@C_{80}$ [I, $I_h(7)$] (x = 0–3) and $Lu_xSc_{3-x}N@C_{80}$ [II, $D_{5h}(6)$] (x = 0–2), isolated by Yang et al. [24]. For $Lu_xSc_{3-x}N@C_{80}$ (I, x = 0–3), while a doublet peak with absorption maxima at 686/659 nm is observed in the spectrum of $Lu_3N@C_{80}$ (I), this peak is slightly redshifted to 691/666 nm in $Lu_2ScN@C_{80}$ (I) and 706/685 nm in $LuSc_2N@C_{80}$ (I). However, in $Sc_3N@C_{80}$ (I), this feature is dramatically changed to a broad single band with an absorption maximum at 735 nm. Interestingly, $Lu_xSc_{3-x}N@C_{80}$ (I; x = 1, 2)

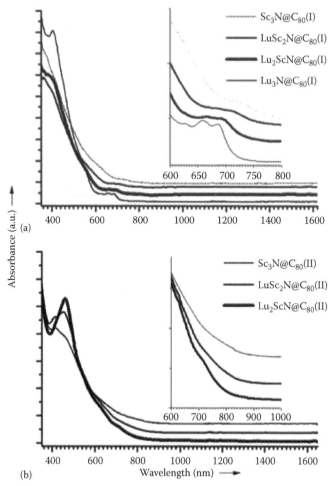

FIGURE 5.3 UV-vis-NIR spectra of $Lu_xSc_{3-x}N@C_{80}\text{-}I_h(7)$ ($x = 0–3$) (a) and $Lu_xSc_{3-x}N@$ $C_{80}\text{-}D_{5h}(6)$ ($x = 0–2$) (b) dissolved in toluene. The insets show the enlarged spectral region. (Reprinted with permission from Yang, S. F. et al., *J. Phys. Chem. C*, 113, 7616–7623, 2009. Copyright 2009 American Chemical Society.)

exhibits intermediate electronic absorption properties in comparison with $Sc_3N@C_{80}$ (I) and $Lu_3N@C_{80}$ (I), with $LuSc_2N@C_{80}$ (I) resembling $Sc_3N@C_{80}$ (I) and $Lu_2ScN@$ C_{80} (I) being closer to $Lu_3N@C_{80}$ (I) (see Figure 5.3a) [22]. Likewise, a significant change of the electronic spectrum occurs in $Lu_xSc_{3-x}N@C_{80}$ (II). Clearly, the absorption peak at 714 nm observed in the spectrum of $Lu_2ScN@C_{80}$ (II) is redshifted to approximately 727 nm of $LuSc_2N@C_{80}$ (II) and turns out to be nondetectable in $Sc_3N@C_{80}$ (II). Thus, the overall absorption spectrum of $LuSc_2N@C_{80}$ (II) exhibits an intermediate feature in comparison with $Lu_2ScN@C_{80}$ (II) and $Sc_3N@C_{80}$ (II) with a higher resemblance to $Sc_3N@C_{80}$ (II) (see Figure 5.3b). The obvious dependence of UV-vis-NIR absorption spectra of MMNCFs based on the same $C_{80}\text{-}I_h(7)$ on the encapsulated metal atoms was further revealed in the comparative study of several $M_xSc_{3-x}N@C_{80}$ (I) (x = 1,2; M = Sc, Lu, Er, Dy, Gd, Nd), indicating that, although their electronic spectra are dominated by the C_{80}-I_h(7) carbon cage, a significant perturbation of the C_{80} cage is dependent on the size of the encaged metal [24]. Another representative example demonstrating the influence of the encapsulated metals on the electronic absorption property of EMFs is the comparison of the UV-vis-NIR absorption spectra of three different types of EMFs, $Sc_3N@C_{78}$-D_{3h}(5) [25], $La_2@C_{78}$-D_{3h}(5) [26], and $Ti_2C_2@$ C_{78}-D_{3h}(5) [27]. Although these three EMFs have the same C_{78}-D_{3h}(5) cage and same number of electron transfer (six electrons) from the encapsulated cluster to the C_{78} cage, their UV-vis-NIR absorption spectra are dramatically different. This phenomenon was interpreted by their substantial difference on the metal–cage interactions, which dramatically change the energy distribution of frontier molecular orbitals (MO) and hence the excitation spectra [1].

For SCFs, the UV-vis-NIR absorption spectrum of $Sc_2S@C_{82}$ shows a strikingly high resemblance to that of $Sc_2C_2@C_{82}$-C_{3v}(8), enabling the assignment of C_{82}-C_{3v}(8) to the cage isomer of $Sc_2S@C_{82}$ [28], which was later confirmed by X-ray crystallographic study [29]. Both C_{82} cages are in the formal -4 charge state and the encapsulated cluster both contains Sc2 moiety: no wonder these two EMFs have almost identical electronic absorption properties.

5.2.2 High-Energy Spectroscopy

Charge transfer from the encaged species to the carbon cage is determinative for the stability of EMFs with specific isomers. The electronic structure of EMFs and the formal oxidation state (valency) of the encapsulated metal atoms resulted from charge transfer from the encaged species to the carbon cage were obtained by high-energy spectroscopies such as X-ray absorption and photoemission spectroscopies. Herein, we focus on EMFs in versatile forms including mono-EMF and cluster fullerenes based on Sc, Y, La, Ce, Gd, Dy, Tm, Ca, and Ti metals commonly studied in literatures, and more details on high-energy spectroscopic study of EMFs based on other lanthanide metals can be referred in the most recent review by Popov et al. [1].

5.2.2.1 Sc-EMFs

An early ultraviolet photoelectron spectroscopic (UPS) study of $Sc@C_{82}$ reported in 1999 by Hino et al. pointed out that two electrons were transferred from Sc to C_{82} cage, that is, Sc atom took the +2 valence state [30]. However, according to the more recent experimental and computational results, the +3 valence state of Sc atom within $Sc@C_{82}$ seems more reasonable [31]. X-ray spectroscopy (XPS) and UPS were performed on $Sc_2C_2@C_{82}$ CCF [32]. The Sc 2p XPS core level of $Sc_2C_2@C_{82}$ was located between those of the Sc metal and the oxide (Sc_2O_3), suggesting that Sc atoms take an intermediate valency [32]. A more recent and detailed study of the valency of the two Sc ions of $Sc_2C_2@C_{82}$ by high-energy spectroscopy revealed unambiguously that the Sc ions were in the valence state of +3. Combined with the calculation results, the effective valency of the Sc ions was determined to be 2.6 [33]. The electronic structure of $Sc_3N@C_{80}$ NCF was also studied by photoemission and X-ray absorption spectroscopy (XAS) [34]. Photoemission from the Sc 2p levels gave a first rough estimation of the Sc valency from the binding energy, which was close to a formally trivalent Sc^{3+} state. XAS at the Sc $L_{2,3}$ edge showed a more intense Sc L_3 peak and new shoulders appeared on the high energy sides of both main Sc peaks compared to $Sc_2C_2@C_{82}$ [34]. Combining the XAS of $Sc_3N@C_{80}$ at the Sc $L_{2,3}$ edge with atomic multiplet calculations, an effective Sc valence of 2.4 was determined [34]. Furthermore, Sc 3d absorption edge in XAS of $K_xSc_3N@C_{80}$, which was prepared by doping $Sc_3N@C_{80}$ with K intercalation, showed no changes in the line shape and in the peak intensities. This makes it clear that the effective Sc valence was not altered upon K intercalation [34]. More recently, XPS studies of $Sc_3N@C_{78}$ and $Sc_3N@C_{80}$ were also reported. For $Sc_3N@C_{78}$, the binding energy of Sc 2p supported a much less oxidation state than +3, possibly +1 state [35]. More recently, UPS and XPS study of a non-IPR (IPR = isolated pentagon rule) NCF $Sc_3N@C_{68}$ was reported, suggesting an oxidation state of Sc atoms in the cage of +2 or +3 [36]. Of note, the XPS peak positions of Sc $2p_{3/2}$ of $Sc_3N@C_{80}$, $Sc_3N@C_{78}$, and $Sc_3N@C_{68}$ shifted toward lower binding energy in sequence (Figure 5.4), suggesting the electron population on the Sc atoms increases with the cage size contraction [36].

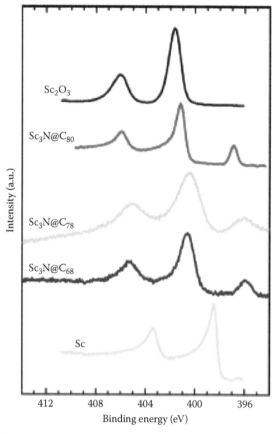

FIGURE 5.4 The Sc 2p XPS of $Sc_3N@C_{68}$, $Sc_3N@C_{78}$, and $Sc_3N@C_{80}$ together with Sc_2O_3 and Sc metal. (Reprinted from *Chem. Phys.*, 421, Hino, S. et al., 39–43. Copyright 2013, with permission from Elsevier B.V.)

5.2.2.2 Y-EMFs

The UPS spectral feature of Y@C_{82} valance band was particularly similar to those EMFs with a trivalent encapsulated atom [37], and later on, the +3 valence state of the encapsulated Y atom was confirmed by Y 3d XPS [38]. XPS spectrum of potassium-intercalated $K_{2.3}$Y@C_{82} was also studied, indicating that the Y atom remained close to +3 valence state, which did not change by the intercalation [38]. UPS study of a CCF Y_2C_2@C_{82}-C_{3v} (8) in comparison with that of di-EMF Y_2@C_{82}-C_{3v} (8) has also been reported [39]. The UPS of these two EMFs basically resembled each other, although minute differences between them have been detected. Simulating the spectrum with a $C_{3v}(8)$-C_{82}^{4-} structure, the UPS spectrum of Y_2C_2@C_{82}-$C_{3v}(8)$ has been well reproduced [39], indicating that the electronic structure of Y_2C_2@C_{82} is $(Y_2C_2)^{2+}$@C_{82}^{4-}, which was substantially different to that of $(Y^{3+})_2$@C_{82}^{6-} proposed for Y_2@C_{82}-$C_{3v}(8)$ [39].

5.2.2.3 La-EMFs

La@C_{82} was the first extracted EMF and extensively studied because of its highest yield in the standard EMFs synthesis [40]. The UPS spectra of La@C_{82} and empty C_{82} were compared, revealing a critical difference in the region just below the Fermi level [37]. In the UPS of La@C_{82}, two new peak components at 0.9 and 1.6 eV with the intensity ratio of 1:2 originated from the transfer of three electrons from La to C_{82}, affording an electronic structure of La^{3+}@C_{82}^{3-} [37]. The electronic structure of potassium-doped La@C_{82} was also reported, and the UPS spectrum showed that the La atom within $K_{5.7}$La@C_{82} remained close to the +3 valence state and was well screened from the outside chemical environment of its hosting cage [37]. For di-EMF La_2@C_{80}, the valence of La was studied by K-edge X-ray absorption fine-structure (XANES) spectrum, indicating that the absorption spectrum was very broad and the threshold energy was consistent with La in La_2O_3. Thus, the valence of La atoms in La_2@C_{80} was +3 as well [41].

5.2.2.4 Ce-EMFs

The valence of Ce atom in Ce@C_{82} was first studied by XPS [42]. In the region of the Ce $3d_{3/2}$ and $3d_{5/2}$ core levels, there was no characteristic peak at a binding energy of ~914 eV correlated with the $3d_{3/2}$ peak of Ce^{4+}, so the existence of Ce^{4+} in Ce@C_{82} was ruled out [42]. The peak position and shape of the Ce 3d XPS peak of Ce@C_{82} were rather similar to those of cerium trihalides, suggesting that the oxidation state of Ce was +3 [42]. In 2003, Shibata et al. confirmed the Ce^{3+} valence state in Ce@C_{82} by Ce L_{III}-edge XANES. Besides, XPS study of di-EMF Ce_2@C_{80} indicated that the valency of Ce atom within Ce_2@C_{80} was +3 too [43]. Another Ce-based di-EMF, Ce_2@C_{72}, was studied by the XAS resonant photoelectron spectroscopy (ResPES), which was measured at the Ce $M_{4,5}$ ($3d \rightarrow 4f$) and $N_{4,5}$ (mainly $4d \rightarrow 4f$) edges, and the main groups of features and multiplet shapes were found to be characteristic for a trivalent state of Ce [1]. More recently, the valency of Ce atom within $CeSc_2N$@C_{80} NCF was studied by XPS [44]. The Ce region of the spectrum was quite similar to those of Ce@C_{82} and Ce_2@C_{80} with the two prominent features for the $3d_{3/2}$ and $3d_{5/2}$ levels. Based on the comparison with cerium trihalides, it was confirmed that the Ce atom encaged in C_{80} was in the form of a Ce^{3+} ion [44].

5.2.2.5   Gd-EMFs

UPS study of $Gd@C_{82}$ supported that three electrons were transferred from the metal atom to the C_{82} cage [45]. It was similar to that of $La@C_{82}$, which exhibited a SOMO-derived low-energy feature with the spectral onset of 0.3 eV below Fermi level [45]. The varying-temperature EXAFS of $Gd@C_{82}$ from 10 to 300 K was later studied [46]. From the energy position and comparison with GdF_3, it was confirmed that Gd ion was present in a trivalent state [46]. The fine-structure analysis of Gd $M_{4,5}$ edge taken from the $Gd@C_{82}$ microcrystal was also studied [47]. Compared to the reference spectrum of Gd_2O_3, the results indicated that Gd^{3+} ion was encapsulated in Gd@ C_{82} [47]. The ResPES study of the Gd $N_{4,5}$ edge was carried out too, and the binding energy of the $4f^{-1}$ band and the lack of resonance at low binding energies were interpreted as an evidence of charge transfer from Gd to the C_{82} cage, which left Gd in the Gd^{3+} valence state [48]. The valence photoelectron spectra and pre-edge and resonance regions in Gd (4d $\rightarrow$ 4f) XAS spectra of $Gd@C_{82}$ and various Gd@ $C_{82}(OH)_x$ derivatives (x = 12, 20, 22, 26) were also studied, revealing that the fine structures of electronic configurations of metallic atoms can be modulated through chemical modifications [49]. The electronic structure of $Gd@C_{60}$ was measured by UPS, ResPES along with DFT+U calculations, indicating good agreement between the experimental photoemission spectra and theoretically calculated ones [50]. The UPS showed that Gd 4f states exhibited enhanced correlation energies that have values larger than those normally observed in metallic gadolinium and gadolinium compounds GdX (X = N, P, As, and Bi). The ResPES taken from 6 eV below the Fermi level showed prominent resonant intensity features, which indicated strong hybridization between the Gd valence states and the fullerene cage [50].

5.2.2.6   Dy-EMFs

XPS spectrum of a Langmuir–Blodgett (LB) film of $Dy@C_{82}$ has been reported [51]. The comparison of the XPS pattern of $Dy@C_{82}$ LB film and $DyCl_3$ in the core-level region of Dy 4d indicates a striking similarity with similar peak separations and intensity distributions, and the valence of Gd in $Gd@C_{82}$ was determined to be +3 [51]. In another study, the Dy L_{III}-edge XANES spectrum of $Dy@C_{82}$ was compared to Gd_2O_3, and the edge energies were 7792 eV for both samples, implying that the valence of the Dy atom in $Dy@C_{82}$ was +3 [52]. The XANES spectrum of potassium-doped $Dy@C_{82}$ ($K_xDy@C_{82}$) showed a peak at an energy level at 6 eV lower than that for Dy^{3+} [53]. This peak was assigned to Dy^{2+}, and thus, it indicated that doping can change the valence state of encapsulated metal atom [53]. So far, $K_xDy@C_{82}$ is the only report that doping potassium can change the valence of the encapsulated metal atom [1], providing a method to control the electronic structure of EMF by metal doping into its solid [53]. Besides, the XANES spectrum of $Dy@C_{60}$ showed the same threshold energies E_0, implying that the valence of the Dy atom is +3 in $Dy@C_{60}$ as in Dy@ C82 [54]. On the other hand, a detailed high-energy spectroscopic study of $Dy_3N@C_{80}$ was reported by Shiozawa et al. [55]. Using X-ray valence-band, Dy 4d photoemission, and Dy 4d-edge absorption spectroscopies, the effective valency of the encapsulated

Dy ions was accurately determined to be 2.8, which was in between those of the metal ions in $Tm_3N@C_{80}$ and $Sc_3N@C_{80}$ (see Figure 5.5) [55]. Likewise, based on a detailed analysis of the resonant photoemission across the Dy 4d-4f edge, the effective Dy valency within $Dy_3N@C_{80}@SWCNTs$ was estimated to be about 3.0 [56]. Of note, the electronic structure of potassium intercalated $Dy_3N@C_{80}$ was also investigated using X-ray and UV photoemission spectroscopy, revealing that the valence state of the trivalent Dy ions was not affected by the additional charge in the fullerene cage [57].

5.2.2.7 Tm-EMFs

The valence of the Tm ion in mono-EMF $Tm@C_{82}$-C_{3v} (8) was studied by using XPS/valence-band photoelectron spectroscopy (VB-PES) and electron energy loss spectroscopy (EELS) in 1997 [58]. The Tm 4d core level photoemission spectrum was similar to the calculated one for Yb^{3+} in Yb_2O_3 and different from the spectrum of Tm metal, providing the evidence for a $4f^{13}$ ground state correlated with Tm^{2+} in $Tm@C_{82}$ [58]. The characteristic multiplet structures in 4f photoemission were measured to further probe quantitatively the divalent state of Tm [58]. Further evidence for the divalent state of Tm was obtained from Tm 4d excitation spectrum measured by EELS in transmission, indicating clearly a $4f^{13}$ ground state with no indication of trivalent Tm [58]. Therefore, $Tm@C_{82}$ represented the first divalent rare-earth metal-based EMF [58]. The valence of Tm in another isomer of $Tm@C_{82}$ with $C_s(6)$-C_{82} cage was also studied by the same methods, revealing +2 valence state for Tm ion as well. This result demonstrated that the isomeric structure of the host carbon cage did not significantly affect the valence of the encapsulated Tm ion [59]. The electronic structure of the potassium intercalated endohedral mono-metallofullerene $Tm@C_{82}$ was also studied by valence band photoemission, pointing out that the valency of the Tm ions did not change upon the intercalation process [60]. On the contrary, Tm 4d photoemission spectrum of $Tm_3N@C_{80}$ revealed that the multiplet structures of $Tm_3N@C_{80}$ were similar to that of Tm^{3+} but the measured Tm 4d binding energy in $Tm_3N@C_{80}$ (177.7 eV) was significantly higher than that in $Tm@C_{82}$ (172.8 eV). Consequently, the effective valency of Tm ion in $Tm_3N@C_{80}$ was determined as 2.9 [61]. Therefore, photoemission spectroscopy provided the direct experimental proof for a +3 formal valence state of Tm atom within $Tm_3N@C_{80}$, which is different to that within $Tm@C_{82}$ (+2 formal valence state) [58]. Besides, as expected, potassium intercalation of $Tm_3N@C_{80}$ did not change the valence state of Tm atoms [62].

5.2.2.8 Ca-EMFs

UPS study of the isomers III and IV of $Ca@C_{82}$ indicated that the photoelectron onset energies of these two isomers were 0.7 and 0.8 eV below the Fermi level [63]. This was a rather high shift compared to those EMFs based on trivalent metals. Combining the UPS results with the *ab initio* calculation, the charge state of Ca was confirmed as +2 [63].

5.2.2.9 Ti-EMFs

EELS spectrum of $Ti_2C_2@C_{78}$ (first identified erroneously as $Ti_2@C_{80}$) was measured by Shinohara et al. [64]. First, a comparison of the $L_{2,3}$-edge EELS of $Ti_2C_2@C_{78}$ with that of TiO_2 indicated that the Ti atoms encapsulated in $Ti_2C_2@C_{78}$ were not in +4 oxidation state [64]. Moreover, comparing the L_3-edge EELS of $Ti_2C_2@C_{78}$

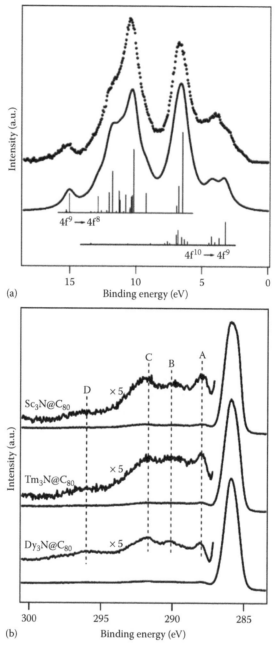

(a)

(b)

FIGURE 5.5 (a) Comparison between X-ray valence-band photoemission spectrum of $Dy_3N@C_{80}$ subtracted by the secondary photoelectron background (solid circles) and calculated spectrum (solid line); (b) C is core-level photoemission spectrum of $Dy_3N@C_{80}$ together with those of $Sc_3N@C_{80}$ and $Tm_3N@C_{80}$, which were shifted by 0.7 and 0.5 eV to higher binding energy, respectively. (Reprinted with permission from Shiozawa, H. et al., *Phys. Rev. B*, 72, 195409, 2005. Copyright 2005 by the American Physical Society.)

with those of Ti_2O_3, TiO, and TiC, a valence state lower than +2 was proposed for the Ti atoms within $Ti_2C_2@C_{78}$ [64]. It should be noted further that the UPS spectrum of $Ti_2C_2@C_{78}$ differs considerably from the spectra of $La_2@C_{78}$ and $Sc_3N@C_{78}$, although they have the same C_{78}-D_{3h} (5) cage. According to DFT calculations, the difference in the electronic structure between $Ti_2C_2@C_{82}$ and $La_2@C_{80}$ was due to the hybridization of the wave functions between Ti atoms and C atoms on the carbon cage [35].

5.3 ELECTROCHEMICAL PROPERTIES

5.3.1 CONVENTIONAL EMFS

5.3.1.1 M@C$_{2n}$

$M@C_{82}$ is the most abundant special member among the conventional mono-EMFs $M@C_{2n}$ family. For the trivalent metal-based $M@C_{82}$ EMFs, because three electrons transferred from capsulated metal ions to the carbon cage, an open-shell electron structure was formed. This led to relatively small electrochemical gaps [the difference between the first oxidation ($E_{ox, 1}$) and first reduction ($E_{red, 1}$) potentials]. Compared to the empty fullerenes, $E_{ox, 1}$ of mono-EMFs is much more negative. The first systematic electrochemical study of $M^{III}@C_{82}$-C_{2v}(9) (M = Y, La, Ce, Gd) was reported by Suzuki et al. in 1996, indicating clearly that $M^{III}@C_{82}$ EMFs with the same carbon cage in the same formal charge state exhibited similar redox properties (Figure 5.6A) [65]. For instance, cyclic voltammogram of $Y@C_{82}$ showed one reversible oxidation and four reversible reductions, which were close to those of La@

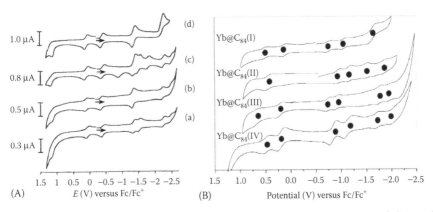

FIGURE 5.6 (A) Cyclic voltammograms of (a) La@C$_{82}$, (b) Y@C$_{82}$, (c) Ce@C$_{82}$, and (d) Gd@C$_{82}$ measured in *o*-DCB at a scan rate of 20 mV/s. (Reprinted from *Tetrahedron*, 52, Suzuki, T. et al., 4973, Copyright 1996, with permission from Elsevier Science Ltd.) (B) CV curves of Yb@C$_{84}$ (I), Yb@C$_{84}$ (II), Yb@C$_{84}$ (III), and Yb@C$_{84}$ (IV) (from top to bottom). Each redox step is marked with a solid dot to aid comparison. The third reduction of Yb@C$_{84}$ (I) is a two-electron process but is marked with only one solid dot. (Reprinted with permission from Lu, X. et al., *J. Am. Chem. Soc. 132*, 5896–5905, 2010. Copyright 2010 American Chemical Society.)

C_{82} and $E_{ox, 1}$ and $E_{red, 1}$ of Y@C_{82}; they exhibited only a few tens of mV shifts compared to that of La@C_{82} [65]. Such a small shift was probably due to the open-shell electronic structure that correlates with the ion radii of the metals and their third ionization potentials [65].

The electrochemical properties of M^{II}@C_{2n} (M = Ca, Sm, Eu, Tm, Yb) were also studied by cyclic voltammetry [1]. These mono-EMFs have a carbon cage in the closed-shell electronic state due to the two-electron transfer; as a result, they have higher oxidation potentials compared to those of M^{III}@C_{2n}. For the same mono-EMFs with different cage isomers, their electrochemical properties generally have a difference to a small extent [1,5]. For example, cyclic voltammogram of four isomers of Yb@C_{84}, namely, Yb@C_{84} (I, II, III, IV), looks similar in general (Figure 5.6B) [66]. Specifically, Yb@C_{84} (I, III, IV) exhibited two reversible oxidation steps and four reversible reduction processes, and the gap between the third and the second reduction steps was generally larger compared to that between the first and second ones, whereas the latter was mutually very close as well as the third and fourth reduction [66]. Particularly, the third and fourth reduction steps of Yb@C_{84} (I) are combined into a two-electron process [66]. For these three isomers, the difference on the first oxidation potential and first reduction potential is only tens of mV. However, Yb@C_{84} (II) has only one oxidation step along with four reversible reduction processes, and the oxidation and first reduction potential were observably shifted compared to the other three isomers. As a result, Yb@C_{84} (II) has the highest redox potentials (both $E_{red, 1}$ and $E_{ox, 1}$) and the largest electrochemical band gap [66].

5.3.1.2 M_2@C_{2n}

The first report on the electrochemical study of di-EMF M_2@C_{2n} was La$_2$@C_{80}-I_h(7). It exhibited two reversible reduction and two reversible oxidation steps in o-dichlorobenzene (o-DCB) at room temperature [67]. Later on, similar results were reported for Ce$_2$@C_{80}-I_h(7). Their oxidation potentials and reduction potentials are almost equal, only tens of mV shifted [68]. When the isomeric structure of C_{80} changes to D_{5h}(6), the first oxidation potentials of M_2@C_{80}-D_{5h}(6) (M = La, Ce) have approximately 330 mV negatively shifted, whereas the reduction potentials were close to those of the M_2@C_{80}-I_h(7) isomer [69]. When the cage size decreases to C_{78}, the first oxidation potential of La$_2$@C_{78}-D_{3h}(5) was +0.26 V [70], much smaller than that of La$_2$@C_{80}-I_h(7) (0.56 V) but almost equal to that of La$_2$@C_{72}-D_2(10611) (0.24 V) [71]. The first reduction potential of La$_2$@C_{78}-D_{3h}(5) was −0.40 V, which was more negative than that of La$_2$@C_{80}-I_h(7) (−0.31 V), suggesting that it is harder to be reduced [70]. Contrarily, the first reduction potential of La$_2$@C_{78}-D_{3h}(5) was more positive than that of La$_2$@C_{72}-D_2(10611) (−0.68 V), and thus, La$_2$@C_{78}-D_{3h}(5) is easier to be reduced than La$_2$@C_{72}-D_2(10611) [71].

The electrochemical properties of an aza-di-metallofullerene Gd$_2$@C_{79}N-I_h(7) have also been reported recently by Dorn et al. [72]. Gd$_2$@C_{79}N-I_h(7) exhibited one reversible oxidation accompanied by two reduction steps: the first reduction peak was reversible with a half-wave potential ($E_{1/2, red, 1}$) at 0.96 V, whereas the second reduction peak was irreversible with peak potential being at 1.98 V. The first reversible reduction potential of Gd$_2$@C_{79}N was significantly negatively shifted compared to that of $La_2@C_{80}$-$I_h(7)$ (−0.31 V). For this reason, it can be assigned to the low-lying SOMO [72].

5.3.1.3 $Sm_3@C_{80}$

Conventional EMFs encapsulating three metals (tri-EMFs) have been rarely reported, and so far only one structure, $Sm_3@C_{80}$-$I_h(7)$, has been experimentally isolated and characterized including the electrochemical study [73]. The cyclic voltammetry (CV) and differential pulse voltammetry (DPV) profiles of $Sm_3@C_{80}$-$I_h(7)$ indicated two reversible oxidation steps accompanied by two reduction steps, with the first reduction being reversible and the second one being irreversible. The first oxidation potential of $Sm_3@C_{80}$-$I_h(7)$ is 0.30 V and much lower than those of $La_2@C_{80}$-$I_h(7)$ (0.56 V), indicating a stronger electron-donating ability. The first reduction potential of $Sm_3@C_{80}$-$I_h(7)$ (−0.83 V) is much more negative than that of $La_2@C_{80}$-$I_h(7)$ (−0.31 V). The electrochemical potential gap of $Sm_3@C_{80}$-$I_h(7)$ is 1.13 V, which is a little larger than that of $La_2@C_{80}$-$I_h(7)$ (0.87 V). This comparison study therefore shows that the endohedral cluster composition indeed has a significant influence on the electrochemical behavior of EMFs [73].

5.3.1.4   Derivatives of Conventional EMFs

Chemical derivatization of EMFs may significantly change their electrochemical properties. For the derivatives of $M^{III}@C_{2n}$ mono-EMFs with the open-shell structures, two different cases should be considered: one is the derivitization of $M^{III}@C_{2n}$ by the groups forming a single bond to the carbon cage and the other is the groups bonding to the fullerene cage via two single bonds [1,5]. The former leads to the diamagnetic state and consequently dramatic change on electrochemical properties due to the transformation of the open-shell structure (pristine EMFs) to the closed-shell one (derivatives), whereas the latter retains paramagnetic. For example, the redox potentials of Bingel monoadducts of $La@C_{82}$ have been reported [74]. The first oxidation potential of three isomers of the monoadducts shifted anodically by 0.16 ~ 0.31 V relative to $E_{ox, 1}$ for $La@C_{82}$, whereas their reduction waves shifted cathodically by 0.24 ~ 0.41 V relative to $E_{red, 1}$ for $La@C_{82}$. These three isomers of the Bingel monoadducts of $La@C_{82}$ have almost identical electrochemical gaps (1.04, 1.06, and 1.08 V, respectively), which are larger than the $La@C_{82}$ (0.49 V). Contrarily, another isomer of the Bingel monoadducts of $La@C_{82}$ exhibited two reduction steps and one oxidation step, among which the first oxidation potential (0.08 V) was almost identical to that of $La@C_{82}$ (0.07 V), whereas its first reduction potential (−0.28 V) shifted positively by 0.14 V relative to that of $La@C_{82}$. These results suggested that this monoadduct was a cycloadduct retaining the paramagnetic property of $La@C_{82}$ [74].

The redox properties of derivatives of di-EMFs were studied widely [1,5]. It has been revealed that chemical modifications on the carbon cage can greatly change the redox behaviors of di-EMF. The redox potentials of the derivatives of di-EMFs are all shifted negatively with respect to those of the pristine di-EMFs [1]. For instance, the reduction potentials of $La_2@C_{80}(Mes_2Si)_2CH_2$ (Mes = mesityl) with two isomers shifted cathodically by 450 and 390 mV, and the oxidation potentials also shifted cathodically by 620 and 590 mV, as compared to those of the pristine $La_2@C_{80}$ [75]. Replacing La with Ce atom, both oxidation and reduction potentials of the silyl adducts of $M_2@C_{80}$ were cathodically shifted by 640 and 340 mV compared to those of $Ce_2@$ C_{80} [76]. Besides, three isomers of the bis-carbene adduct of $La_2@C_{80}$-$I_h(7)$ with Ad (Ad = adamentylidene) and CClPh groups exhibited negative shifts of reduction and oxidation potentials by 0.10 ~ 0.17 and 0.08 ~ 0.11 V relative to those of $Ce_2@C_{80}$, which were much smaller than the case for the silyl adducts, suggesting that the extent of the shifts upon derivatization strongly depends on the nature of the addends [1,77].

5.3.2   NCFs

5.3.2.1   Pristine NCFs

Electrochemical properties of NCFs have been extensively studied since 2004, when the first electrochemical study of pure $Sc_3N@C_{80}$-$I_h(7)$ was reported by Dunsch et al. [20] The cyclic voltammogram of $Sc_3N@C_{80}$-$I_h(7)$ showed two reduction steps with formal redox potentials of –1.24 and –1.62 V, and one oxidation step at $E_{1/2}$(Ox 1) = 0.62 V. Accordingly, the electrochemical gap of $Sc_3N@C_{80}$-$I_h(7)$ is 1.86 V, which is significantly larger than that of $La_2@C_{80}$ with (0.87 V) [20]. Another key feature of the redox property of $Sc_3N@C_{80}$-$I_h(7)$ is that the reductions are electrochemically irreversible at moderate voltammetric scan rates, and this is dramatically different to mono- and di-EMFs [78]. Because the reductive waves of the voltammogram were chemically irreversible at the scan rates of 100 mV/s, cyclic voltammetry studies of $Sc_3N@C_{80}$-$I_h(7)$ at various scan rates were further performed to probe the kinetics of the first anion formation. No significant change was observed when the scan rates were smaller than 100 mV, whereas increasing the scan rate resulted in more chemically reversible behavior. Consequently, the first reduction step became electrochemically reversible when the scan rate reached 6 V/s, whereas the minimum scan rates of 10 and 20 V/s were required for achieving reversible second and third reduction steps, respectively. Such a scan rate-dependent behavior of $Sc_3N@C_{80}$-$I_h(7)$ indicates clearly that its electrochemical reduction is electrochemically irreversible, but chemically reversible [78].

Electrochemical properties of other non-Sc $M_3N@C_{80}$-$I_h(7)$ (M = Y, Nd, Pr, Gd, Tb, Dy, Ho, Er, Tm, and Lu) NCFs were also studied [1]. Compared to $Sc_3N@C_{80}$-$I_h(7)$, the first reduction potentials of non-Sc $M_3N@C_{80}$-$I_h(7)$ NCFs are cathodically shifted (Figure 5.7A), whereas the first oxidation potentials of non-Sc $M_3N@C_{80}$-$I_h(7)$ NCFs (+0.65 ~ 0.70 V) are somewhat more positive than that of $Sc_3N@C_{80}$-$I_h(7)$ (approximately +0.6 V) and usually reversible [79]. Similar to $Sc_3N@C_{80}$-$I_h(7)$, it has been revealed that the reductions of non-Sc $M_3N@C_{80}$-$I_h(7)$ NCFs are electrochemically irreversible but chemically reversible as well [4,6]. To explain this phenomenon, a double-square reaction scheme was proposed, involving the charge-induced reversible rearrangement of the monoanion resulted from the change of the endohedral cluster in terms of shape or position inside the cage [80].

Upon the formation of the mixed-metal NCFs $A_xB_{3-x}N@C_{80}$-$I_h(7)$, the redox property observably changes compared to the homogeneous NCFs. A striking example is the electrochemical study of Ti-based mixed-metal NCFs $TiM_2N@C_{80}$-$I_h(7)$ (M = Sc, Y) by Popov, Yang, and Dunsch et al. [81,82]. Cyclic voltammetry of $TiSc_2N@$ C_{80}-$I_h(7)$ exhibits one reversible oxidation at +0.16 V and three reversible reduction steps at –0.94, –1.58, –2.21 V at a small scan rate of 20 mV/s (Figure 5.7B) [81]. Compared to $Sc_3N@C_{80}$-$I_h(7)$, the first oxidation potential of $TiSc_2N@C_{80}$-$I_h(7)$

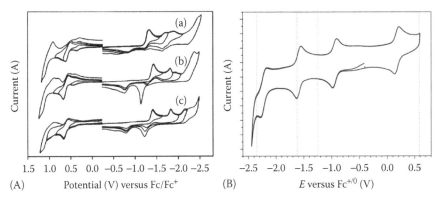

FIGURE 5.7   (A) Cyclic voltammogram of (a) $Sc_3N@C_{80}$, (b) $Y_3N@C_{80}$, and (c) $Er_3N@$ C_{80} in o-dichlorobenzene, 0.1 M $TBAPF_6^-$, 100 mV/s scan rate. (Reprinted with permission from Cardona, C. M. et al., *J. Am. Chem. Soc.*, 128, 6480, 2006. Copyright 2006 American Chemical Society.) (B) Cyclic voltammetry of $TiSc_2N@C_{80}$ measured in o-DCB solution (room temperature, $TBABF_4$ as supporting electrolyte) at a scan rate of 20 mV s^{-1}. Dotted vertical bars denote reversible redox potentials of $Sc_3N@C_{80}$. (Reprinted with permission from Popov, A. A. et al., *ACS Nano*, 4, 4857, 2010. Copyright 2010 American Chemical Society.)

shifted cathodically by 0.46 V and the first reduction potential shifted anodically by 0.3 V to the anode. As a result, the electrochemical gap of $TiSc_2N@C_{80}$-$I_h(7)$ is 1.10 V, which is much smaller than that of $Sc_3N@C_{80}$-$I_h(7)$ (1.86 V). Particularly, the most remarkable change for the redox property of $TiSc_2N@C_{80}$-$I_h(7)$ compared to $Sc_3N@C_{80}$-$I_h(7)$ is that the three reduction steps of $TiSc_2N@C_{80}$-$I_h(7)$ all become electrochemically reversible under the same conditions. To interpret this dramatic difference, it was proposed that for $TiSc_2N@C_{80}$-$I_h(7)$, both oxidation and reduction occur at the endohedral cluster, changing the valence state of Ti from Ti^{III} to Ti^{II} in $TiSc_2N@C_{80}^-$ (reduction) and Ti^{IV} in $TiSc_2N@C_{80}^+$ (oxidation), whereas these processes take place on the carbon cage instead for $Sc_3N@C_{80}$-$I_h(7)$ [81]. A similar so-called endohedral electrochemistry was later observed for $TiY_2N@$ C_{80}-$I_h(7)$ as well, which, however, exhibited negative shifts of around 0.16 V for both first oxidation and reduction potentials relative to those of $TiSc_2N@C_{80}$-$I_h(7)$ [82]. These comparative results reveal the strong influence of the encaged group-III metal (Sc or Y) on the electronic property of $TiM_2N@C_{80}$ (M = Sc,Y) [79]. Later on, a similar endohedral oxidation behavior was unveiled for $CeM_2N@C_{80}$-$I_h(7)$ (M = Lu, Y) as well [83,84].

The isomerization of the carbon cage remarkably affects the redox properties of EMFs. In 2005, Echegoyen et al. studied the electrochemical property of the mixture of I_h and D_{5h} isomers of $Sc_3N@C_{80}$ and found that the first oxidation potential of $Sc_3N@C_{80}$-$D_{5h}(6)$ is 270 mV less positive than that of $Sc_3N@C_{80}$-$I_h(7)$ isomer [78]. Besides, other $M_3N@C_{80}$ (M = Dy, Tm, and Lu) NCFs show some similarity on the cathodic shift of the oxidation potential of $M_3N@C_{80}$-$D_{5h}(6)$ isomer relative to $M_3N@C_{80}$-$I_h(7)$ isomer [1].

The electrochemical properties of EMFs depend not only on the isomeric structure of the cage but also on the cage size [85,86]. For instance, the electrochemical study of $Gd_3N@C_{2n}$ ($2n = 80-88$) NCF family (Figure 5.8) revealed that the first reduction step was roughly the same for the series of $Gd_3N@C_{2n}$ except $Gd_3N@C_{82}$, for which the first reduction step is cathodically shifted [85]. On contrary, the carbon cage size has a prominent effect on the oxidation process of $Gd_3N@C_{2n}$: When the carbon cage size increases, the oxidation potential decreased [85]. Of note, $Gd_3N@C_{88}$ exhibits reversible oxidation and reduction waves with the lowest first oxidation potential and HOMO–LUMO gap within the family [85]. A similar situation was observed in other $M_3N@C_{2n}$ ($M = Nd$, Pr, and Ce) NCF families as well [85]. Based on these results, it was concluded that the LUMO of NCFs is highly localized on the encapsulated metal nitride cluster, whereas the HOMO is localized on the carbon cage [85].

5.3.2.2 Derivatives of NCFs

Compared to the pristine NCFs, derivatization can change the reversibility of the redox processes or the redox potentials. Up to now, derivatives of $M_3N@C_{80}\text{-}I_h(7)$ NCFs have been studied mostly [1,4]. Because of the high symmetry of the $C_{80}\text{-}I_h(7)$ cage, there are only two isomers for the mono-cycloadducts of $M_3N@$ $C_{80}\text{-}I_h(7)$ with the addition taking place on either the [6,6] (hexagon/hexagon edge) or [5,6] (pentagon/hexagon edge) site which is not open and is affected by the metal size [1,4]. Electrochemical studies of these NCF derivatives provide helpful evidence on determining the [5,6] or [6,6] isomer. For example, according to the

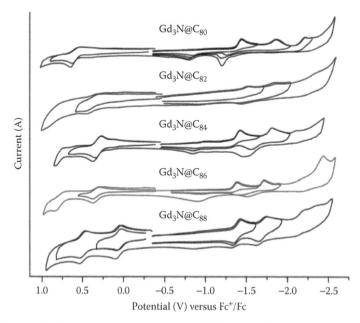

FIGURE 5.8 Cyclic voltammograms of the $Gd_3N@C_{2n}$ family measured in o-DCB at 100 mV/s. (Reprinted from *Tetrahedron*, 64, Manuel, N. C. et al., 11387–11393, Copyright 2008, with permission from Elsevier Science Ltd.)

comparative electrochemical study of Prato cycloadducts of $M_3N@C_{80}$ (M = Sc, Y, Er), the reduction steps of the [6,6]-pyrrolidinofullerene monoadducts of $Y_3N@C_{80}$ and $Er_3N@C_{80}$ were irreversible, similar to the pristine NCFs, whereas the electrochemical behavior of the [5,6]-pyrrolidinofullerene derivatives was quite different with reversible reductions [79]. Likewise, the cyclic voltammogram of the [5,6] Diels-Alder monoadduct of $Sc_3N@C_{80}$ showed three one-electron reversible reductions, whereas electrochemically irreversible reduction steps were observed in the cyclic voltammograms of the [6,6]-methanofullerene derivatives of $Y_3N@C_{80}$ and $Er_3N@C_{80}$ [79]. These results reveal that [5,6] adducts of NCFs exhibit reversible cathodic electrochemical properties, although [6,6] adducts have irreversible cathodic behavior [4].

The influence of the addends on the redox property of NCFs is very complex. For example, the reduction potential of [5,6]-pyrrolidine derivative of $Sc_3N@C_{80}$ is usually 0.10–0.15 V more positive than that in $Sc_3N@C_{80}$-$I_h(7)$, whereas a more positive shift, +0.23 V, is for the [6,6] pyrrolidine (N-trityl) derivative of $Sc_3N@C_{80}$ [87]. The reduction and oxidation potentials of methano-$Sc_3N@C_{80}$ and methano-$Lu_3N@C_{80}$ adducts are slightly negatively shifted, whereas for $Y_3N@C_{80}$ and $Gd_3N@C_{80}$, the shifts are positive but also small [1]. Similar potential shifts were observed by other derivatives of NCFs, such as bis-silylated derivative $Sc_3N@C_{80}(Mes_2Si)_2CH_2$ and trifluoromethyl derivative $Sc_3N@C_{80}(CF_3)_x$ [1].

5.3.3 CARBIDE CLUSTER FULLERENES

Few electrochemical studies of CCFs have been reported [1,5]. Among the CCF family, $Sc_3C_2@C_{80}$ is quite special because of its paramagnetic properties [88]. The cyclic voltammogram of $Sc_3C_2@C_{80}$-$I_h(7)$ shows one reversible oxidation step at –0.03 V and three reversible reduction steps at –0.50, –1.64, and –1.84 V, affording an electrochemical gap of 0.47 V [88]. Another paramagnetic CCF, $Lu_3C_2@C_{88}$, exhibits the redox properties resembling EMFs with a closed-shell structure, and its electrochemical gap (1.65 V) is much larger than that of $Sc_3C_2@C_{80}$-$I_h(7)$ [89]. Compared to $Sc_3C_2@C_{80}$, $M_2C_2@C_{2n}$ CCFs have closed-shell structures, and thus, their electrochemical properties are remarkably different. For instance, for $Sc_2C_2@C_{82}$-$C_{3v}(8)$, the first oxidation step at 0.47 V is reversible, whereas the first reduction process at –0.94 V is obviously irreversible followed by two reversible reduction steps [90]. The redox potentials of $Sc_2C_2@C_{82}$-$C_{3v}(8)$ are somewhat close to those of diamagnetic metallofullerenes such as $Sc_3N@C_{80}$ and $La_2@C_{80}$ [90]. The electrochemical gap of $Sc_2C_2@C_{82}$-$C_{3v}(8)$ is 1.41 V, and in fact, the electrochemical gaps of all $M_2C_2@C_{2n}$ isolated till now are larger than 1 V [5]. As for the only quaternary CCF reported so far, $Sc_4C_2@C_{80}$-$I_h(7)$ has an electrochemical gap of 1.56 eV, which is the largest among all Sc-based CCFs [5,91].

Derivatization of CCFs induces an influence on their electrochemical properties to a certain degree. For example, pyrrolidine addition to $Sc_2C_2@C_{82}$-$C_s(6)$ led to negative shifts for both oxidation and reduction steps, and so did the adamantylidene derivative of $Sc_2C_2@C_{80}$-$C_{2v}(5)$ [92,93]. Different addition sites affect the redox potentials of pyrrolidine monoadducts of $Sc_2C_2@C_{82}$-$C_{2v}(9)$ as well [94].

5.3.4 Sulfide Cluster Fullerenes

$Sc_2S@C_{82}$-$C_{3v}(8)$ represents the first SCF isolated by Dunsch et al. in 2010, and its electrochemical property was studied at that time [28]. The cyclic voltammogram of $Sc_2S@C_{82}$-$C_{3v}(8)$ shows one reversible oxidation step at 0.48 V and one irreversible reduction step at –1.03 V, followed by two reversible steps at –1.16 and –1.61 V [28]. Interestingly, a high resemblance of the redox properties between $Sc_2S@C_{82}$-$C_{3v}(8)$ and $Sc_2C_2@C_{82}$-$C_{3v}(8)$ was observed, and this can be understood by their similarity on the electronic configuration [28]. Soon after the isolation of $Sc_2S@C_{82}$-$C_{3v}(8)$, another isomer, $Sc_2S@C_{82}$-$C_s(6)$, was isolated by Echegoyen et al., for which the first oxidation potential shifted cathodically to +0.39 V and the reduction steps became fairly reversible (Figure 5.9) [95]. Interestingly, the first reduction (–0.98 V) and oxidation (0.39 V) potentials of $Sc_2S@C_{82}$-$C_s(6)$ are very similar to those of $Sc_2O@C_{82}$-$C_s(6)$ (–0.96 and 0.35 V, respectively) and $Sc_2C_2@C_{82}$-$C_s(6)$ (–0.93 and 0.42 V, respectively) [29]. Nevertheless, obvious difference on the second reduction step among these three different types of

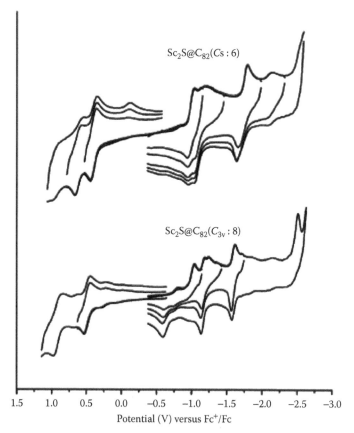

FIGURE 5.9 Cyclic voltammograms of $Sc_2(\mu_2$-$S)@C_{3v}(8)$-C_{82} and $Sc_2(\mu_2$-$S)@C_s(6)$-C_{82}, in $(n$-$Bu_4N)(PF_6)/o$-DCB with ferrocene as the internal standard. Scan rate, 100 mV s^{-1}. (From Chen, N. et al., *Chem. Commun.*, 2010, 46: pp. 4818–4820. Copyright 2010 The Royal Society of Chemistry.)

cluster fullerenes with the same cage can be seen: the second reduction step of $Sc_2S@$ $C_{82}-C_s(6)$ was very close to the first one (-0.98 and -1.12 V, respectively), whereas the gap between the first and the second reduction steps exceeded 0.3 V for $Sc_2O@$ $C_{82}-C_s(6)$ and $Sc_2C_2@C_{82}-C_s(6)$ [29,92]. This phenomenon was interpreted by their geometrical differences that the Sc_2S cluster is much more constrained by the cage than the Sc_2O cluster, which affects the interactions between the cage and the cluster and then the electrochemical property [29].

Recently, several new SCFs have been isolated with intriguing electrochemical properties [96–98]. The cyclic voltammogram of $Sc_2S@C_{72}-C_s(10528)$ shows a reversible first oxidation at $+0.64$ V, and the three reduction steps are reversible with the first reduction observed at -1.14 V [96]. Accordingly, the electrochemical gap of $Sc_2S@C_{72}-C_s(10528)$ is 1.78 V, which is larger than those of $Sc_2S@C_{82}-$ $C_{3v}(8)$ and $Sc_2S@C_{82}-C_s(6)$ [96]. The cyclic voltammogram of $Sc_2S@C_{70}-C_2(7892)$ shows a reversible first oxidation as well but all four reduction steps are electrochemically irreversible, and its electrochemical gap (1.57 V) is smaller than that of $Sc_2S@C_{72}-C_s(10528)$ [97]. More recently, the cyclic voltammogram of a Ti-only SCF, $Ti_2S@D_{3h}-C_{78}$, has been reported [96]. Its redox behavior is quite complex, exhibiting three reversible reductions at -0.92, -1.53, and -1.80 V, two additional irreversible reductions at -2.29 and -2.41 V, and two reversible oxidation steps at $+0.23$ and $+0.65$ V along with one quasi-reversible oxidation at $+1.03$ V [88]. The electrochemical gap of $Ti_2S@D_{3h}-C_{78}$ is calculated to be 1.15 V. These features are significantly different from all $Sc_2S@C_{2n}$ SCFs discussed above [98].

5.3.5 Oxide, Carbonitride, and Cyanide Cluster Fullerenes

With the discovery of several new cluster fullerenes, electrochemical study has been implemented as a routine characterization technique. For oxide clusterfullerene (OCF) $Sc_4O_2@C_{80}-I_h(7)$, two reversible oxidation and two reversible reduction steps were observed at low scan rate (20 mV/s), whereas a third reduction step could also be observed at higher scan rates of 1.0 V/s or using square-wave voltammetry [99]. The first oxidation potential of $Sc_4O_2@C_{80}-I_h(7)$ is 0.00 V, which is substantially more negative than the first oxidation potentials of the majority of EMFs with a diamagnetic state of the carbon cage, but is instead comparable to that those of $Sc_2@$ $C_{82}-C_{3v}(8)$ (0.05 V), $CeLu_2N@C_{80}-I_h(7)$ (0.01 V), $TiSc_2N@C_{80}-I_h(7)$ (0.16 V) and $TiY_2N@C_{80}-I_h(7)$ (0.00 V) [99]. On the other hand, the first reduction potential of $Sc_4O_2@C_{80}-I_h(7)$ was observed at -1.10 V, which is close to those of several EMFs with the endohedral reduction step such as $TiM_2N@C_{80}$ (M = Sc, Y). The electrochemical gap is 1.10 V of $Sc_4O_2@C_{80}-I_h(7)$, almost two times smaller than those of $M_3N@C_{80}$ (M = Sc, Y, Gd-Lu) NCFs [99]. It was concluded that both the oxidation and reduction of $Sc_4O_2@C_{80}-I_h(7)$ are reversible endohedral redox processes, and this fact is dramatically different to the cases of $M_3N@C_{80}$ (M = Sc, Y, Gd-Lu) NCFs for which the oxidation and reduction are localized on the carbon cage [99].

As the only carbonitride fullerene reported so far, $Sc_3NC@C_{80}-I_h$ exhibits a reversible first oxidation step at 0.6 V and a reversible first reduction step at -1.05 V followed by another reduction irreversible step at -1.68 V [100]. Of note, the first oxidation potential of $Sc_3NC@C_{80}-I_h$ is close to that of $Sc_3N@C_{80}-I_h$ (0.62 V),

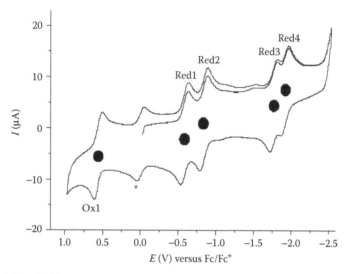

FIGURE 5.10 Cyclic voltammogram of YCN@C_{82} in *o*-dichlorobenzene (*o*-DCB) solution with ferrocene (Fc) as the internal standard and tetrabutylamonium hexafluorophosphate (TBAPF$_6$) as supporting electrolyte. Each redox step is marked with a number and a solid dot to aid comparison. (Reprinted by permission from Macmillan Publishers Ltd. *Sci. Rep.*, Yang, S. F. et al., 3, 1487, 2013, copyright 2013.)

whereas its first reduction potential (−1.05 V) is shifted cathodically by approximately 0.20 V relative to that of Sc$_3$N@C_{80}-I_h (−1.24 V); consequently, the electrochemical gap of Sc$_3$NC@C_{80}-I_h (1.65 V) is 0.21 V smaller than that of Sc$_3$N@C_{80}-I_h (1.86 V) [100].

Very recently, a monometallic cyanide cluster fullerene, YCN@C_{82}-C_s(6), was discovered by Yang et al, and its redox property was studied by cyclic voltammetry (Figure 5.10) [101]. YCN@C_{82}-C_s(6) exhibits one reversible oxidation step at +0.56 V and four reversible reduction steps at −0.59, −0.84, −1.76, and −1.92 V, respectively [99]. Interestingly, the first two reduction steps as well as the last two reduction steps are mutually close, whereas there is an abrupt large separation (0.92 V) between the second and third reduction steps [101]. Such a characteristic reduction behavior of YCN@C_{82}-C_s(6) highly resembles that of Yb@C_{82}-C_s(6), implying their similarity on the electronic configuration featuring a formal two-electron transfer from the encapsulated moiety to C_{82} cage [101].

5.4 OPTICAL PROPERTIES

5.4.1 Luminescence Properties of Er-EMFs

Although the luminescence properties of lanthanide ions are one of the widely used properties, for lanthanide-based EMFs the luminescence of endohedral lanthanide ions can hardly be detected because the majority of lanthanide ions emit in the visible range, where fullerenes intensively absorb light [1]. So far, only a near-IR (NIR) $^4I_{13/2} \rightarrow {}^4I_{15/2}$ luminescence of Er^{3+}-based EMFs was studied.

5.4.1.1 $Er_2@C_{82}$ and $Er_2C_2@C_{82}$

In the first luminescence study of Er-EMFs reported by Hoffman et al. in 1996, a fullerene extract mixture containing Er-EMFs was excited by an argon ion laser (514 nm), NIR emission from Er^{3+}-doped fullerenes in the region of 1500–1650 nm was observed [102]. These transitions correspond to the expected energies of the Er transitions between the $^4I_{13/2} \rightarrow {}^4I_{15/2}$ levels [102]. Later on, the same group found that different sample forms affected the transitions: samples obtained by extracting metallofullerenes with toluene exhibited relatively broad emission lines with the major feature appearing at 1520 nm, whereas the emission lines are shifted to higher energy for samples from films sublimated directly from soot mixture [103]. Based on the energy dynamics analysis between $Er_2@C_{82}$ and $Er@C_{82}$, it has been concluded that the emission signal resulted from Er^{3+} ions in $Er_2@C_{82}$ EMF, whereas in $Er@C_{82}$, emission should be quenched by the lower-energy C_{82}^{3-} state [103]. Several years later, this group made further study on the sample sublimed at low temperature and proposed that the emission of the Er–EMF mixture may be attributed to $Er@C_{60}$ [104].

With the availability of a purified $Er_2@C_{82}$ sample whose isomeric structure was unknown, Ding et al. studied the luminescence property of $Er_2@C_{82}$ in the frozen CS_2 solution at 30K as well as the hydrogenated derivative $Er_2@C_{82}H_n$ [105]. They concluded that the fluorescence arose from the $^4I_{13/2} \rightarrow {}^4I_{15/2}$ transition of Er^{3+} within $Er_2@C_{82}$, which was weak and the lifetime was less than 10 μs because of quenching by the overlapping electric state of the fullerene cage [105]. Upon hydrogenation, the fluorescence intensity of $Er_2@C_{82}H_n$ increased markedly while the lifetime was too short to measure. The increased fluorescence was interpreted as follows: the hydrogenation raised the energy of the lowest electronic state of the fullerene cage so there was no overlap with the Er^{3+} $^4I_{13/2}$ state [105]. In another report by Macfarlane et al., emission spectra of $Er_2@C_{82}$ in a glass (decalin) and crystalline (frozen CS_2) media were measured at 1.6K [106]. The Er emission spectrum of $Er_2@C_{82}$ consists of a well-resolved eight-line pattern near 1.55 μm at 1.6K. The similarity of spectra measured in glassy and polycrystalline media along with the very narrow inhomogeneous line width (<1 cm^{-1}) proved that the two Er ions in the fullerene cage were well shielded from the solvent environment [106]. However, at a resolution of 0.5 nm in frozen CS_2, the spectrum shows additional splittings, which is because of the exchange interaction between the two Er ions [106].

A more systematic photoluminescence studies of $Er_2@C_{82}$ (I, II, III) and $(Er_2C_2)@C_{82}$ (I, II, III) (isomers I, II, III correspond to $C_s(6)$, $C_{2v}(9)$, and $C_{3v}(8)$, respectively) in CS_2 solution at room temperature were reported by Shinohara et al. in 2007 [18]. These photoluminescence (PL) spectra of all EMFs correspond well with $^4I_{13/2} \rightarrow {}^4I_{15/2}$ emission whereas their intensities differ from each other [17]. In specific, enhanced NIR PL at 1520 nm from Er^{3+} ions in $Er_2@C_{82}$ (I, III) and $(Er_2C_2)@C_{82}$ (I, III) was observed. $(Er_2C_2)@C_{82}$ (III) exhibits the strongest PL intensity which is about 150 times stronger than those of $(Er_2C_2)@C_{82}$ (I, II) [18]. Such a difference was interpreted by their difference on the absorption property that the absorption onset of $(Er_2C_2)@C_{82}$ (III) was found at 1250 nm. There was a broad absorption band at 1500 nm in $(Er_2C_2)@C_{82}$ (I, II) leading to the quenching of Er^{3+} emission by the carbon cage of $(Er_2C_2)@C_{82}$ (I, II), similar results were obtained for $Er_2@C_{82}$ (I, II, III) as well [18]. The low-temperature (3.5K) PL spectra of $Er_2@C_{82}$ (III) and $(Er_2C_2)@C_{82}$ (III) in bisphenol A polycarbonate consist of about eight principal lines which correspond to $^4I_{13/2} \rightarrow {}^4I_{15/2}$ emission of Er^{3+} and show no significant difference between $Er_2@C_{82}$ (III) and $(Er_2C_2)@C_{82}$ (III) (Figure 5.11) [18]. In another study on the photoluminescence spectra of $Er_2@C_{82}$ (I) and $Er_2C_2@C_{82}$ (I) measured at 5 K by Plant et al. in 2009, the characteristic emission in the 1.5–1.6 µm region corresponds to the $^4I_{13/2}(1) \rightarrow {}^4I_{15/2}(n)$ transitions of the Er^{3+} ion for both EMFs [107]. Besides, a remarkable difference between these two EMFs was identified: the spectrum of $Er_2C_2@C_{82}$ (I) was acuminated with narrow and well-resolved spectral lines, whereas that of $Er_2@C_{82}$ (I) was broad without clearly defined emission lines and the principal emission was blueshifted with respect to that of $Er_2C_2@C_{82}$ (I) [107].

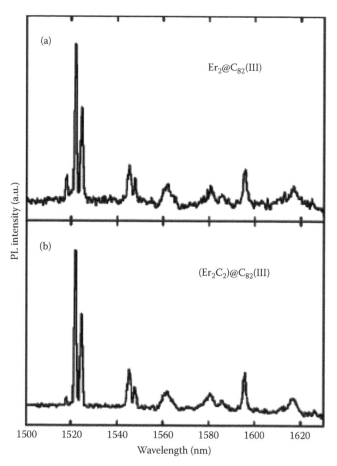

FIGURE 5.11   PL spectra of (a) $Er_2@C_{82}$ (III) and (b) $(Er_2C_2)@C_{82}$ (III) in bisphenol A polycarbonate thin film at 3.5K. (Reprinted with permission from Ito, Y. et al., *ACS Nano*, 1, 456, 2007. Copyright 2007 American Chemical Society.)

5.4.1.2 $Er_xSc_{3-x}N@C_{80}$

In 2001, Macfarlane et al. reported the first fluorescence study of $Er_3N@C_{80}$, $Er_2ScN@C_{80}$, and $ErSc_2N@C_{80}$ between 1.6 K and room temperature (Figure 5.12) [108]. At 1.6 K, the fluorescence spectra exhibited a characteristic eight-line pattern corresponding to the $^4I_{13/2}(1) \rightarrow {}^4I_{15/2}(n)$ transitions of Er^{3+} [108]. A similar doublet structure at around 1500 nm was observed in $ErSc_2N@C_{80}$ and $Er_2ScN@C_{80}$, whereas $Er_3N@C80$ showed a single line [108]. With the temperature increasing to 77 K, *hot* bands due to transitions from the first thermally excited component of the upper state [$^4I_{13/2}(2)$] were observed, and the intensity of the doublet components in $ErSc_2N@C_{80}$ and $Er_2ScN@C_{80}$ redistributed [108]. The measured emission decay lifetimes of $Er_xSc_{3-x}N@C_{80}$ at 1.6K were all in the range of 1.1–1.6 μs, correlating with a quantum efficiency for emission of ~10^{-4}, which is much lower than that of Er^{3+} in glassy or crystalline hosts [108]. The authors attributed such a low quantum efficiency to the quenching via interactions with the high-frequency vibrations of the carbon cage [108].

Later on, Jones et al. reported the detection of $Er_3N@C_{80}$ luminescence by direct excitation of the Er^{3+} f-f transitions using a near-IR tunable diode laser in 2006 [109]; this technique is different to the previous study which was carried out by photoexcitation of the fullerene cage states leading to relaxation via the ionic states [1]. On the basis of the direct non-cage-mediated optical interaction with the Er ion, the excitation at wavelength of the strongest emission line in the luminescence spectrum [$^4I_{13/2}(1) \leftrightarrow {}^4I_{15/2}(1)$, 1519 nm] led to the observation of the second-strongest luminescence line of the [$^4I_{13/2}(1) \rightarrow {}^4I_{15/2}(4)$, 1546 nm] [109]. Furthermore, by using a similar method, each principal emission line was observed, and an excitation spectrum of the $^4I_{13/2}$ manifold at 5 K in frozen CS_2 was obtained [109]. It was found that the splitting of the $^4I_{13/2}(n)$ manifold (around 250 cm^{-1}) was comparable to that observed for the lower energy $^4I_{15/2}$ manifold (350cm^{-1}) [109]. In a follow-up study, the same group measured the photoluminescence of $Er_3N@C_{80}$ in frozen CS_2 at 4.2 K at a range of magnetic fields from 0 T to 19.5 T. As the magnetic field increased, a Zeeman-like splitting of the observed spectrum linear with the field was observed and the principal peak exhibited a splitting in transition energy of 1.8 cm^{-1}/T [110].

The photoluminescence properties of Er-mixed-metal NCFs have been comprehensively studied recently. Low-temperature photoluminescence and photoluminescence excitation spectroscopy of $ErSc_2N@C_{80}$ and $Er_2ScN@C_{80}$ fullerenes measured by Tiwari et al in 2007 revealed at least two metastable configurations of the Er^{3+} ion within the cage [111]. The configuration-selective crystal-field splitting of the ground state $^4I_{15/2}$ and first excited state $^4I_{13/2}$ manifolds of Er^{3+} ions were determined by photoluminescence excitation spectroscopy [111]. Later on, the same group reported the variable temperature photoluminescence study of $ErSc_2N@C_{80}$ and $Er_2ScN@C_{80}$ in the range of 5–80K [112]. New emission peaks arise from thermally populated crystal-field levels of the excited state when the temperature was above 20K [112]. In $ErSc_2N@C_{80}$, the photoluminescence peak area shows that the $ErSc_2N$ cluster inside the C_{80} cage rearranges with the temperature increase [112]. In 2008, Morton et al. reported the photoluminescence study of $ErSc_2N@C_{80}$, revealing a change in the relative populations of the two $ErSc_2N$ configurations upon the application of 532-nm

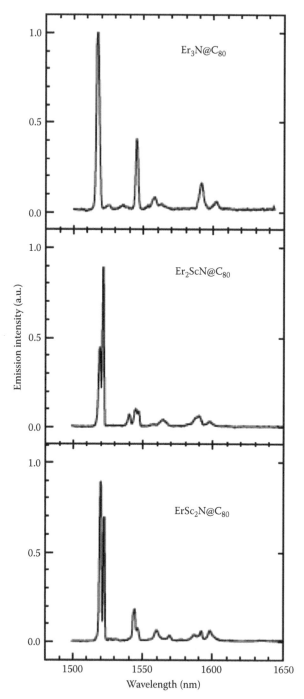

FIGURE 5.12 Fluorescence spectra of $Er_xSc_{3-x}N@C_{80}$ at 1.6K. (Reprinted from *Chem. Phys. Lett.*, 343, Macfarlane, R. M. et al., 229, Copyright 2001, with permission from Elsevier Science B.V.)

illumination [113]. In order to compare the optical properties of Er^{3+} encapsulated in a fullerene cage and Er^{3+} dispersed in single crystal, Dantelle et al. compared the photoluminescence properties of several Er-EMFs including $Er_2C_2@C_{82}$ (I, III) and $Er_{3-x}Sc_xN@C_{80}$ ($x = 0$, 1 and 2), and Er^{3+}-doped β-PbF_2 single crystals [114]. Low-temperature photoluminescence measurements showed that the line width of Er^{3+} in a carbon cage was similar to the one of Er^{3+} in β-PbF_2 single crystals, and the quantum efficiency of Er^{3+} emission at 1.5 μm in fullerenes is four orders of magnitude lower than that for Er^{3+} in β-PbF_2 crystals, and this was interpreted that molecular vibrations of the fullerene cage may induce rapid non-radiative decay processes [114].

The photoluminescence property of a dithiolane derivative of $Er_3N@C_{80}$ was also studied at 5 K in frozen CS_2 by Gimenez-Lopez et al. in 2010 [115]. The photoluminescence spectra comprise a series of sharp lines corresponding to the $^4I_{13/2}(1) \rightarrow {}^4I_{15/2}(n)$ transitions of Er^{3+} [115]. Although the exohedral functionality does not quench the luminescence of $Er_3N@C_{80}$, variations in the relative intensities of the photoluminescence peaks were observed; besides, the emission lines of $Er_3N@C_{80}$ are mono-component, whereas those of derivatized $Er_3N@C_{80}$ are further split [115]. The reasons for these observations are that the functional group reduces the C_{80} cage symmetry and therefore perturbs the local environment of each Er^{3+} inside the cage but all Er^{3+} of $Er_3N@C_{80}$ are equivalent [115].

5.4.2 NONLINEAR OPTIC PROPERTIES

The third-order nonlinear optical property of $Dy@C_{82}$ in CS_2 at low concentration (<0.1 g/l) was studied by Du et al. in 1998 [116]. The second-order hyperpolarizability at 532 nm was determined, which is consistent with the theoretical result. The measured polarizability of $Dy@C_{82}$ was much higher than that of empty fullerenes [116]. The resonant enhancement mechanism and the electron transfer from the endohedral metal atom to the carbon cage contributed to the large nonlinear optical response of $Dy@C_{82}$ [116]. A similar finding on the nonlinear property of $Er_2@C_{82}$ was reported by Dorn et al. in the same year, revealing that the nonlinear response of $Er_2@C_{82}$ (III) in CS_2 is two to three orders of magnitude larger than that of empty-cage fullerenes. The molecular susceptibility of $Er_2@C_{82}$ (III) is higher than that of empty fullerene as well [117]. The mechanism for the increase of the third-order nonlinear optical susceptibility is also concluded by the charge transfer from the encapsulated metal to the carbon cage [117]. In 2001, Yaglioglu et al. studied the nonlinear optical properties of $Gd_2@C_{80}$ thin films and solutions and found that the third-order nonlinearity was large, negative, and strongly dependent on pulse duration and wavelength [118].

In 2004, Shinohara et al. compared the nonlinear optical response of $Dy@C_{82}$ (I), $Dy_2@C_{82}$(I), and $Er_2@C_{92}$ (IV) using the optical Kerr effect technique [119]. $Dy@C_{82}$ (I) exhibited the largest second polarizability followed by C_{82}-C_{2v}, whereas the second polarizability response of $Dy_2@C_{82}$ (I) was the weakest of all three [119]. A similar phenomenon was observed for C_{92}-C_2 and $Er_2@C_{92}$(IV) as well [119]. Thus, encapsulating one atom into the fullerene cage enhances the second hyperpolarizability, which decreases by encapsulating a second atom [119]. In a theoretical study by Hu et al. in 2008, it was shown that the third-order nonlinear optical polarizabilities and the two-photon absorption cross sections of $M@C_{82}$ (M = Sc, Y, La) EMFs increased with the increase in the atomic radius of encapsulated metal M in the order of Sc $\rightarrow$ Y $\rightarrow$ La [120].

5.4.3 Transient Absorption Dynamics, Singlet Oxygen Generation, and Nanoplasmon Excitations

Akasaka et al. studied the transient spectroscopic properties of $La@C_{82}$ and $La_2@C_{80}$ EMFs on the nanosecond timescale in 2000 [121], revealing a two-step decay with characteristic lifetimes of 83 ns/2.9 µsec for $La@C_{82}$ and 150 ns/~40 µsec for $La_2@$ C_{80}, respectively. The slow decay of $La@C_{82}$ was assigned to the lowest excited state with quartet spin multiplicity [121]. Contrarily, according to the transient absorption studies of $La@C_{82}$ and $La_2@C_{80}$ derivatives by Guldi et al., the lifetimes of the first excited states (doublet and singlet, respectively) did not exceed 100 ps, whereas the lifetimes of the quartet and triplet states evolved after inter-system crossing were at the order of nanoseconds [122–124].

Transient absorption dynamics of $Gd_2@C_{80}$ was measured by Yaglioglu et al. in 2001 [125]. Two-photon resonance and subsequent resonance absorption and bleaching of photo-induced optical transitions resulted in the size and pulse-duration dependence of the nonlinear response at 800 nm, which was explained by a theoretical model combining two-photon and excited-state absorptions [125].

Photodynamics of $Li^+@C_{60}$ and photo-induced electron-transfer reactions of $Li^+@C_{60}$ were studied by Fukuzumi et al. In 2012 [126], no emission of $Li^+@C_{60}$ was observed by photoirradiation of a deaerated benzonitrile (PhCN) solution at room temperature. However, the emission spectrum of $Li^+@C_{60}$ could be detected in a deaerated 2-methyltetrahydrofuran glass at 77 K. The ultrafast photodynamics for the intersystem crossing from the singlet to the triplet excited state of $Li^+@C_{60}$ was observed by transient absorption spectra. The rate constant of singlet excited state of $Li^+@C_{60}$ undergoing intersystem crossing to the triplet excited state in PhCN is $8.9 \times 10^8\,s^{-1}$, which is larger than that of pristine C_{60}. The triplet lifetime of $Li^+@C_{60}$ (48 µs) is comparable to that of C_{60} (49 µs) [126].

The first single molecule photoemission study of $Y_3N@C_{80}$ was reported by Bharadwaj and Novotny in 2010 [127]. The photoluminescence spectrum of $Y_3N@$ C_{80} in xylene excited at 633 nm is represented by a peak centered around 710 nm corresponding to the $(HOMO-LUMO)_0$ emission. The transition lifetime is 240 ns, which is surprisingly shorter than those of typical rare-earth metal ions (Er^{3+}, Yb^{3+}, Y^{3+}, and so forth) in solid-state hosts but longer than the lifetime of typical organic dyes [125]. Confocal photoluminescence microscopy was used to study the emission of single $Y_3N@C_{80}$ molecules spin-coated on different transparent surfaces [127]. The photophysical properties of the single $Y_3N@C_{80}$ molecules were found to be very sensitive to the local dielectric environment, and the best results were achieved with poly(methyl methacrylate) [127]. The intrinsic quantum yield of $Y_3N@C_{80}$ at 635 nm was estimated less than 0.05 [127]. However, the photoluminescence efficiency of $Y_3N@C_{80}$ can be plasmon enhanced by two orders of magnitude by coupling it to a gold nanoparticle [127]. Very recently, a strongly emitting liquid-crystalline derivative of $Y_3N@$ C_{80} was synthesized, exhibiting outstanding luminescence properties in the near-IR region [128]. The emitting excited state of liquid-crystalline derivative of $Y_3N@C_{80}$

is extremely long-lived (16 ms at 298 K, which is 20 times higher than $Y_3N@C_{80}$, and 13 ms at 77 K) [128].

Shinohara et al. reported in 2001 that endohedral metallofullerenes $M@C_{82}$ (M = Dy, Gd, La) and $Dy_2@C_{2n}$ (2n = 84,94) could efficiently generate singlet oxygen under photolytic conditions [129]. The yield of singlet oxygen production was dependent on the type and the number of the encapsulated metal atoms [129]. A similar singlet oxygen generation ability under irradiation in the visible range was reported for $Sc_3N@C_{80}$ by Phillips et al. in 2009 [130]. Although EMFs can quench the singlet oxygen, the total quenching rate constants of singlet oxygen by $La@C_{82}$, $Ce@C_{82}$, and $Ce_2@C_{80}$ were measured by Yanagi et al. The $^1O_2$ quenching capabilities of C_{82} and EMFs were found to be comparable to that of β-carotene, a well-known strong $^1O_2$ quencher, and a combination of energy and charge transfer processes was proposed for the $^1O_2$ quenching [131].

A study on the optical properties of $M@C_{82}$ EMFs (M = Gd, Ce, La, and Y) in dimethylformamide (DMF) solution by Alidzhanov et al. in 2010 has revealed that all these mono-EMFs luminesced in the visible region, and enhanced Raman scattering has also been observed [132]. The luminescence spectrum was described by a superposition of three to four Gaussian spectral profiles, whose energy positions and intensities nonmonotonically depended on the energy of excitation quanta. Interestingly, visible luminescence and Raman scattering enhancement were observed only when nanosized anionic complexes of EMFs were formed in solution, resulting in nanoplasmon excitations in the cluster of EMFs [132].

5.5 MAGNETIC PROPERTIES

The special electronic structures and internal metal atoms in EMFs result in various magnetic properties, such as paramagnetism, ferromagnetism, and antiferromagnetism. These magnetic properties have promising applications in quantum information processing, memory devices, medical imaging, single-molecule magnet (SMM), and spintronics. $La@C_{82}$ is the first paramagnetic EMF studied by electron spin resonance (ESR) spectroscopy [133]. Subsequently, the paramagnetic properties of $Sc@C_{82}$ [134], $Y@C_{82}$ [135], $Sc_3C_2@C_{80}$ [136], etc. were investigated. For these paramagnetic species, each of them has an unpaired electron localized on certain site. For example, the $La@C_{82}$ has its spin delocalizing on the fullerene cage, but the spin in $Sc_3C_2@C_{80}$ distributes on the C_2 core unit [137]. For their ESR studies, the hyperfine coupling constants (hfcc) and g-factor were analyzed in detail to disclose the spin-center and kinds of metal nuclei. Furthermore, the electron spin manipulation based on these paramagnetic species has been realized by changing temperatures and exohedral modification. These paramagnetic species have wide applications in chemistry, physics, material, and biology.

For many lanthanide ions in EMFs, such as Gd, Eu, Ho, Tb, and Dy, they have unpaired f-electrons, which also make the molecule paramagnetic [9]; especially for $Gd@C_{82}$ and $Gd_3N@C_{80}$, the seven unpaired f-electrons in each Gd^{3+} result in super-paramagnetic properties, which are studied as excellent contrast agents for magnetic resonance imaging (MRI) [138,139]. Moreover, the solid states of these f-electron-based paramagnetic species were also studied by superconducting quantum interference device (SQUID) and X-ray magnetic circular dichroism (XMCD) measurements [140]. By means of these techniques, the transformation of magnetic properties is investigated, in which the magnetic susceptibility and effective magnetic moment are important parameters for the magnetism of EMFs.

Herein, we introduce the magnetic properties of EMFs. The spin-centers, hyperfine couplings, and spin modulations are discussed based on $Sc@C_{82}$, $Sc_3C_2@C_{80}$, $Y_2@C_{79}N$, and many other EMF radicals. In addition, the magnetic properties of $Gd@C_{82}$, $Ln_3N@C_{80}$ (Ln = Ho, Tb), and $DySc_2N@C_{80}$ powders are illustrated.

5.5.1 KINDS AND LOCATIONS OF UNPAIRED SPIN

The magnetic properties of EMFs are originated from the unpaired electron spin. For EMFs, there are three parts where the electron spin can reside, that is, the fullerene cage, the endohedral metal, and the endohedral nonmetal unit. For example, the $La@C_{82}$ and $Y_2@C_{79}N$ [141] have their spin localizing on the fullerene cage and endohedral Y_2 dimer, respectively, but the spin in $Sc_3C_2@C_{80}$ distributes on the C_2 core unit. Sometimes, some EMFs have the spin situated both on the cage and on the endohedral metal. In $Sc@C_{82}-C_{2v}(9)$, for example, the unpaired electron mainly resides in the carbon cage and partially in the Sc(3d) orbital, which is different from those in $La@C_{82}-C_{2v}(9)$ and $Y@C_{82}-C_{2v}(9)$ with almost all spin on the cages [134]. In addition, the $TiM_2N@C_{80}-I_h(7)$ (M = Sc, Y) are new paramagnetic molecules, which have an unpaired electron on Ti^{3+} ion [81,82].

For many lanthanide ions in EMFs, they have one or more unpaired electron(s) embedded in f-orbitals. For example, the internal Gd^{3+} ion in Gd-based EMFs has seven unpaired f-electrons that also result in paramagnetic properties. For Gd@ C82-$C_{2v}(9)$, there are two kinds of spins: one is an unpaired electron delocalized on the C_{82} cage and the other is the seven unpaired f-electrons. Similarly, the $Er@C_{82}$, $Ho@C_{82}$, and $Eu@C_{82}$ all have two kinds of spins. On the other hand, the $Ln_3N@C_{80}$ (Ln = Gd, Ho, Tb) have closed-shell C_{80} cages, and thus, their magnetic properties are all originated from the embedded f-electrons.

5.5.2 COUPLINGS BETWEEN SPIN AND METAL NUCLEI

In order to detect the spin features of EMFs, ESR spectroscopy has been proved to be a powerful technique achieved by applying an oscillating magnetic field, which provides an energy splitting the spin states (Zeeman effect). Subsequently, the absorption of electromagnetic radiation induces the transitions of spin states and results in ESR signals. In ESR spectra of EMFs, there are two parameters that can generally reveal the electron spin characteristics, that is, g-factor and coupling constant (a). The g-factor can give information about the paramagnetic center. The hfcc is caused by the interaction between the unpaired electron and a metal nucleus (or metal nuclei) with a non-zero nuclear spin (I), splitting the ESR signal into multiline pattern.

For EMFs, most of the endohedral metals have non-zero nuclear spin, such as 1/2 for Y, 7/2 for Sc and La. The ESR lines can be counted by the following equation: $2nI + 1$, where n is the number of equivalent metal nuclei. For example, the $Sc@C_{82}$

has eight lines in its ESR spectrum. In $Sc_3C_2@C_{80}$, 22 lines can be observed resulted by three equivalent Sc nuclei. However, when the molecule has nonequivalent metal nuclei, the ESR pattern would be more complex, such as in $Sc_3C_2@C_{80}$ derivatives the number of ESR line is hard to count because of the three different Sc nuclei and multi-couplings [136].

Interestingly, the $TiY_2N@C_{80}$-I_h and $TiSc_2N@$ C_{80}-I_h both exhibit broad ESR signals detected at room temperature [81,82]. No hyperfine couplings were observed, this may be caused by the deeply embedded electron in Ti^{3+} ion, and the couplings between spin and Sc are largely lost. For EMFs encaging lanthanide ions (Gd, Ho, Tb, etc.), their several unpaired f-electrons also result in a broad ESR signal without hyperfine couplings [9]. That is because the serious spin-spin couplings dominate the ESR property.

5.5.3 MODULATION OF PARAMAGNETISM

Modulations of spin state and corresponding molecular paramagnetism are of great importance as they are useful in quantum information progressing, memory devices, and spintronics. The electron spin manipulation based on these paramagnetic species has been realized by changing temperatures and exohedral modification. For instance, in $Sc@C_{82}$-$C_{2v}(9)$ (Figure 5.13a), its ESR property exhibits obvious changes under low temperatures, that is, at low temperature the intensity of the ESR signals at higher magnetic field appears reduced and paramagnetic anisotropy emerges [134]. This kind of anisotropy could be caused by the insufficient rotational averaging of the g and hyperfine tensors. The $Y_2@C_{79}N$ has an open-shell electronic structure, where an unpaired electron resides on the endohedral Y_2 dimer (Figure 5.13b). Interestingly, the highly impressionable electron spin of $Y_2@C_{79}N$ was observed when decreasing the temperature [142]. Under low temperature, the intensity of ESR signals at higher magnetic field obviously increases. This phenomenon can be ascribed to the paramagnetic anisotropy and insufficient averaging of the paramagnetic tensors.

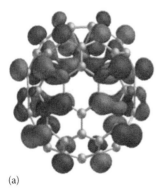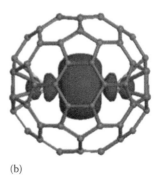

(a)   (b)

FIGURE 5.13 Calculated spin density distributions of (a) $Sc@C_{2v}(9)$-C_{82} and (b) $Y_2@C_{79}N$. (a) Reprinted with permission from Hachiya, M. et al., *J. Am. Chem. Soc.* 134, 15550–15555, 2012. Copyright 2012 American Chemical Society.)

The exohedral adducts modified by chemical reactions not only change the structures of EMFs, but also influence their spin states. For a $Sc_3C_2@I_h$-C_{80} derivative with adamantylidene carbene ($Sc_3C_2@C_{80}$-Ad), it shows an ESR spectrum with the hfcc of 7.39 G (two nuclei) and 1.99 G (one nucleus) [143]. However, the $Sc_3C_2@C_{80}$ fulleropyrrolidine exhibits different ESR spectrum with hfcc of 8.602 G (one nucleus) and 4.822 G (two nuclei) [136]. These ESR spectra are all different from that in pristine $Sc_3C_2@C_{80}$-I_h, indicating a considerable influence of the exohedral modification on the paramagnetism. The influence of adduct was also observed in $Sc@C_{82}$-Ad with four isomers, and moreover, each has a different hfcc, which are different from that of pristine $Sc@C_{82}$-$C_{2v}(9)$ [144]. In addition, in the ESR spectra of $Y_2@C_{79}N$ fulleropyrrolidines, two inhomogeneous Y nuclei were observed, and moreover, the ESR line in high magnetic field region shows high intensity, resulting in a paramagnetic anisotropy [142].

Due to their ball-like shape, the EMFs have advantages in involving supramolecular system, which is a platform to construct ordered clustering. For spin-active EMFs, this kind of host/guest system makes them controllable and also promising to be paramagnetic or ferromagnetic materials. An example is the host/guest system of unsaturated thiacrown ethers and $La@C_{82}$, in which obvious electron transfer was detected by ESR spectroscopy, resulting two spin states: spin-on and spin-off [145]. Another excellent example is the host/guest system of $La@C_{82}$ and a cyclodimeric copper porphyrin {cyclo-$[P_{Cu}]_2$} [146]. In cyclo-$[P_{Cu}]_2$⊃$La@C_{82}$, a ferromagnetically coupled property was found with a quartet spin ground state ($S = 3/2$) (Figure 5.14a). Interestingly, the cyclo-$[P_{Cu}]_2$⊃$La@C_{82}$ was transformed into its caged analogue though linking other two alkyl side chains of cyclo-$[P_{Cu}]_2$, resulting in a cage-$[P_{Cu}]_2$⊃$La@C_{82}$ that has ferromagnetic character with doublet spin ground state ($S = 3/2$) (Figure 5.14b).

5.5.4 Magnetic Properties

For many lanthanide ions in EMFs, such as Gd, Eu, Ho, Tb, and Dy, they have unpaired f-electrons, which make the molecule paramagnetic. The powders of these the f-electron-based paramagnetic species were widely studied by SQUID and

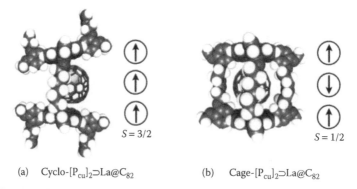

(a)   Cyclo-$[P_{Cu}]_2$⊃$La@C_{82}$ (b)   Cage-$[P_{Cu}]_2$⊃$La@C_{82}$

FIGURE 5.14 Optimized molecular models of (a) cyclo-$[P_{Cu}]_2$⊃$La@C_{82}$ and (b) cage-$[P_{Cu}]_2$⊃$La@C_{82}$. (Reprinted with permission from Hajjaj, F. et al., *J. Am. Chem. Soc.*, 133, 9290–9292, 2011. Copyright 2011 American Chemical Society.)

XMCD measurements. The magnetic properties of EMFs have prospective applications in such fields as quantum information processing, single molecule magnets, and metal magnetic memory (MMM).

The pioneering work in Ln@C_{82} (Ln = lanthanides) was fulfilled in 1995, which revealed that Gd@C_{82} features paramagnetism at the temperature of 3 K and shows antiferromagnetic property under room temperature [140]. It should be noted that the Ln@C_{82} (Ln = Gd, Tb, Dy, Ho, and Er) exhibit reducing magnetic moment compared to their corresponding free metal ions revealed by SQUID, which may be ascribed to a quenched orbital moment caused by antiferromagnetic coupling between the *f*-electrons of the entrapped metal and remaining unpaired electrons in the hybridized molecular orbital [147].

However, when it comes to Ln$_3$N@C_{80} (Ln = Tb, Ho, Er, Gd, etc.), which have a cage with closed shell, the magnetic behavior is governed only by the internal metal nuclei. In C_{80}-based metal NCFs, the magnetization curves recorded on Ln$_3$N@C_{80} (Ln = Tb, Ho, Tm, Er, Gd) above 1.8 K showed paramagnetism. In Ho$_3$N@C_{80} and Tb$_3$N@C_{80}, the magnetic anisotropy was found to be induced by the ligand field of the centered N atom [148].

Recently, Greber and Dunsch discovered the presence of magnetic hysteresis below 6 K with long relaxation time in DySc$_2$N@C_{80} measured by XMCD combined with SQUID magnetometry, in which the magnetic moment is merely attributed to the encaged Dy^{3+} ion (Figure 5.15) [149]. These studies on the magnetic properties of EMFs have substantially extended the potential application of EMFs to many typical fields like SMM; however, there are many problems to be solved.

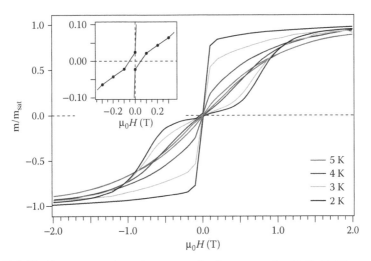

FIGURE 5.15 Temperature-dependent magnetization curves for DySc$_2$N@C_{80} recorded using SQUID magnetometry. (Reprinted with permission from Westerstrom, R. et al., *J. Am. Chem. Soc.*, 134, 9840–9843, 2012. Copyright 2012 American Chemical Society.)

5.6 CONCLUSIONS AND PROSPECTS

Thanks to recent advances in synthesis and isolation techniques, numerous EMFs with versatile encapsulated species and carbon cages have been synthesized and isolated during the past two decades, and their intrinsic properties have been extensively studied, including their electronic, electrochemical, optical, and magnetic properties. Resulting from the electron transfer from the encapsulated species (metal ions or clusters) to the carbon cage, EMFs exhibit fascinating electronic, electrochemical, optical, and magnetic properties as reviewed here. The electronic properties of EMFs typically studied by UV-Vis-NIR absorption spectroscopy provide valuable structural information, and the facile absorption spectroscopic characterization may in some cases be used for structural determination of new EMFs. The electrochemical properties of EMFs appear quite intriguing, especially the endohedral electrochemistry unveiled very recently for cluster fullerenes, and electrochemical study offers a direct access to the energetics and mechanisms of the charge transfer within EMFs. Limited studies on the luminescence properties of Er-based EMFs enable a better understanding to their luminescence mechanisms and show potential applications of Er-EMFs in quantum information processing and molecular memories. The magnetic properties of EMFs are originated from the unpaired electron as well as the endohedral lanthanides with unpaired f-electrons. The electron spin manipulation of the paramagnetic EMFs has been now realized by changing temperatures and exohedral modification, and this promises the prospect of EMF applications in quantum information processing and single-molecule magnets.

Despite the comprehensive studies on the properties of EMFs, there are still many open questions to be addressed particularly toward the potential applications of EMFs. For instance, so far endohedral electrochemistry has been revealed for only limited number of cluster fullerenes, which is, however, quite crucial for understanding their peculiar redox behaviors and more importantly their potential applications in charge storage inside a fullerene cage. Besides, considering the potential applications of EMFs based on their intriguing magnetic properties, more accurate manipulation of paramagnetic EMFs should be improved, and the ordered arrays of EMFs are waiting to be prepared in order to realize the information processing. Furthermore, complex systems composed by EMFs and organic or inorganic frameworks with higher stability and variable magnetic property should be constructed.

ACKNOWLEDGMENTS

We thank the present and former members of our groups for the contributions as well as our collaborators on the published works cited in this review. SY thanks the National Basic Research Program of China (2010CB923300, 2011CB921401) and National Natural Science Foundation of China (21132007, 21371164). CW and TW thank the National Basic Research Program of China (2012CB932900), National Natural Science Foundation of China (21121063, 21203205), NSAF (11179006), and China Postdoctoral Science Foundation (201104153).

REFERENCES

1. Popov, A. A.; Yang, S. F.; Dunsch, L. *Chem. Rev.* **2013**, 113, 5989–6113.
2. Yang, S. F.; Liu, F. P.; Chen, C. B.; Jiao, M. Z.; Wei, T. *Chem. Commun.* **2011**, 47, 11822–11839.
3. Lu, X., Feng, L.; Akasaka, T.; Nagase, S. *Chem. Soc. Rev.* **2012**, 41, 7723–7760.
4. Chaur, M. N.; Melin, F.; Ortiz, A. L.; Echegoyen, L. *Angew. Chem. Int. Ed.* **2009**, 48, 7514–7538.
5. Lu, X.; Akasaka, T.; Nagase, S. *Acc. Chem. Res.* **2013**, 46, 1627–1635.
6. Lu, X.; Akasaka, T.; Nagase, S. *Chem. Commun.* **2011**, 47, 5942–5957.
7. Popov, A. A.; Dunsch, L. *J. Phys. Chem. Lett.* **2011**, 2, 786–794.
8. Yang, S. F.; Dunsch, L. *J. Phys. Chem. B* **2005**, 109, 12320–12328.
9. Shinohara, H. *Rep. Prog. Phys.* **2000**, 63, 843–892.
10. Ding, J. Q.; Yang, S. H. *Angew. Chem. Int. Ed. Engl.* **1996**, 35, 2234–2235.
11. Ding, J. Q.; Yang, S. H. *J. Am. Chem. Soc.* **1996**, 118, 11254–11257.
12. Yamada, M.; Wakahara, T.; Nakahodo, T.; Tsuchiya, T.; Maeda, Y.; Akasaka, T.; Yoza, K.; Horn, E.; Mizorogi, N.; Nagase, S. *J. Am. Chem. Soc.* **2006**, 128, 1402–1403.
13. Inoue, T.; Tomiyama, T.; Sugai, T.; Okazaki, T.; Suematsu, T.; Fujii, N.; Utsumi, H.; Nojima, K.; Shinohara, H. *J. Phys. Chem. B* **2004**, 108, 7573–7579.
14. Inakuma, M.; Yamamoto, E.; Kai, T.; Wang, C. R.; Tomiyama, T.; Shinohara, H. *J. Phys. Chem. B* **2000**, 104, 5072–5077.
15. Tagmatarchis, N.; Shinohara, H. *Chem. Mater.* **2000**, 12, 3222–3226.
16. Tagmatarchis, N.; Aslanis, E.; Shinohara, H.; Prassides, K. *J. Phys. Chem. B* **2000**, 104, 11010–11012.
17. Liu, F. P.; Wei, T.; Wang, S.; Guan, J.; Lu, X.; Yang, S. F. *Fuller. Nanotub. Carbon Nanostruct.* **2014**, 22, 215–226.
18. Ito, Y.; Okazaki, T.; Okubo, S.; Akachi, M.; Ohno, Y.; Mizutani, T.; Nakamura, T.; Kitaura, R.; Sugai, T.; Shinohara, H. *ACS Nano* **2007**, *1*, 456.
19. Dunsch, L.; Yang, S. F. *Small* **2007**, 3, 1298–1320.
20. Krause, M.; Dunsch, L. *ChemPhysChem.* **2004**, 5, 1445–1449.
21. Chen, N.; Fan, L. Z; Tan, K.; Wu, Y. Q.; Shu, C. Y.; Lu, X.; Wang, C. R. *J. Phys. Chem. C* **2007**, 111, 11823–11828.
22. Dunsch, L.; Krause, M.; Noack, J.; Georgi, P. *J. Phys. Chem. Solid.* **2004**, 65, 309–315.
23. Yang, S. F.; Popov, A. A.; Dunsch, L. *Chem. Commun.* **2008**, 2885–2887.
24. Yang, S. F.; Popov, A. A.; Chen, C. B.; Dunsch, L. *J. Phys. Chem. C* **2009**, 113, 7616–7623.
25. Olmstead, M. M.; Bettencourt, A. D. D; Duchamp, J. C.; Stevenson, S.; Marciu, D.; Dorn, H. C.; Balch, A. L. *Angew. Chem. Int. Ed.* **2001**, 40, 1223–1225.
26. Cao, B. P.; Wakahara, T.; Tsuchiya, T.; Kondo, M.; Maeda, Y.; Rahman, G. M. A. Akasaka, T.; Kobayashi, K.; Nagase, S.; Yamamoto, K. *J. Am. Chem. Soc.* **2004**, 126, 9164–9165.
27. Cao, B. P.; Hasegawa, M.; Okada, K.; Tomiyama, T.; Okazaki, T.; Suenaga, K.; Shinohara, H. *J. Am. Chem. Soc.* **2001**, 123, 9679–9680.
28. Dunsch, L.; Yang, S. F.; Zhang, L.; Svitova, A.; Oswald, S.; Popov, A. A. *J. Am. Chem. Soc.* **2010**, *132*, 5413–5421.
29. Mercado, B. Q.; Chen, N.; Rodriguez-Fortea, A.; Mackey, M. A.; Stevenson, S.; Echegoyen, L.; Poblet, J. M.; Olmstead, M. M.; Balch, A. L. *J. Am. Chem. Soc.* **2011**, 133, 6752.
30. Hino, S.; Umishita, K.; Iwasaki, K.; Miyamae, T.; Inakuma, M.; Shinohara, H. *Chem. Phys. Lett.* **1999**, 300, 145.
31. Morley, G. W.; Herbert, B. J.; Lee, S. M.; Porfyrakis, K.; Dennis, T. J. S.; Nguyen-Manh, D.; Scipioni, R. et al. *Nanotechnology* **2005**, 16, 2469.
32. Takahashi, T.; Ito, A.; Inakuma, M.; Shinohara, H. *Phys. Rev. B* **1995**, 52, 13812.

33. Pichler, T.; Hu, Z.; Grazioli, C.; Legner, S.; Knupfer, M.; Golden, M. S.; Fink, J. et al. *Phys. Rev. B* **2000**, 62, 13196.

34. Alvarez, L.; Pichler, T.; Georgi, P.; Schwieger, T.; Peisert, H.; Dunsch, L.; Hu, Z. et al. *Phys. Rev. B* **2002**, 66, 035107.

35. Hino, S.; Zenki, M.; Zaima, T.; Aoki, Y.; Okita, S.; Ohta, T.; Yagi, H. et al. *J. Phys. Chem. C* **2012**, 116, 165.

36. Hino, S.; Ogasawara, N.; Ohta, T.; Yagi, H.; Miyazaki, T.; Nishi, T.; Shinohara, H. *Chem. Phys.* **2013**, 421, 39–43.

37. Hino, S.; Takahashi, H.; Iwasaki, K.; Matsumoto, K.; Miyazaki, T.; Hasegawa, S.; Kikuchi, K.; Achiba, Y. *Phys. Rev. Lett.* **1993**, 71, 4261.

38. Ton-That, C.; Shard, A. G.; Dhanak, V. R.; Shinohara, H.; Bendall, J. S.; Welland, M. E. *Phys. Rev. B* **2006**, 73, 205406.

39. Hino, S.; Wanita, N.; Iwasaki, K.; Yoshimura, D.; Akachi, T.; Inoue, T.; Ito, Y.; Sugai, T.; Shinohara, H. *Phys. Rev. B* **2005**, 72, 195424.

40. Chai, Y.; Guo, T.; Jin, C. M.; Haufler, R. E.; Chibante, L. P. F.; Fure, J.; Wang, L. H.; Alford, J. M.; Smalley, R. E. *J. Phys. Chem.* **1991**, 95, 7564–7568.

41. Kubozono, Y.; Takabayashi, Y.; Kashino, S.; Kondo, M.; Wakahara, T.; Akasaka, T.; Kobayashi, K.; Nagase, S.; Emura, S.; Yamamoto, K. *Chem. Phys. Lett.* **2001**, 335, 163.

42. Ding, J. Q.; Weng, L. T.; Yang, S. H. *J. Phys. Chem.* **1996**, 100, 11120.

43. Shibata, K.; Rikiishi, Y.; Hosokawa, T.; Haruyama, Y.; Kubozono, Y.; Kashino, S.; Uruga, T. et al. *Phys. Rev. B* **2003**, 68, 094104.

44. Wang, X. L.; Zuo, T. M.; Olmstead, M. M.; Duchamp, J. C.; Glass, T. E.; Cromer, F.; Balch, A. L.; Dorn, H. C. *J. Am. Chem. Soc.* **2006**, 128, 8884.

45. Hino, S.; Umishita, K.; Iwasaki, K.; Miyazaki, T.; Miyamae, T.; Kikuchi, K.; Achiba, Y. *Chem. Phys. Lett.* **1997**, 281, 115.

46. Giefers, H.; Nessel, F.; Gyory, S. I.; Strecker, M.; Wortmann, G.; Grushko, Y. S.; Alekseev, E. G.; Kozlov, V. S. *Carbon* **1999**, 37, 721.

47. Suenaga, K.; Iijima, S.; Kato, H.; Shinohara, H. *Phys. Rev. B* **2000**, 62, 1627.

48. Pagliara, S.; Sangaletti, L.; Cepek, C.; Bondino, F.; Larciprete, R.; Goldoni, A. *Phys. Rev. B* **2004**, 70, 035420.

49. Tang, J.; Xing, G. M.; Yuan, H.; Cao, W. B.; Jing, L.; Gao, X. F.; Qu, L. et al. *J. Phys. Chem. B* **2005**, 109, 8779.

50. Sabirianov, R. F.; Mei, W. N.; Lu, J.; Gao, Y.; Zeng, X. C.; Bolskar, R. D.; Jeppson, P.; Wu, N.; Caruso, A. N.; Dowben, P. A. *J. Phys.-Condens. Matter* **2007**, 19, 6.

51. Huang, H. J.; Yang, S. H. *J. Organomet. Chem.* **2000**, 599, 42.

52. Iida, S.; Kubozono, Y.; Slovokhotov, Y.; Takabayashi, Y.; Kanbara, T.; Fukunaga, T.; Fujiki, S.; Emura, S.; Kashino, S. *Chem. Phys. Lett.* **2001**, 338, 21.

53. Kubozono, Y.; Takabayashi, Y.; Shibata, K.; Kanbara, T.; Fujiki, S.; Kashino, S.; Fujiwara, A.; Emura, S. *Phys. Rev. B* **2003**, 67, 115410.

54. Kanbara, T.; Kubozono, Y.; Takabayashi, Y.; Fujiki, S.; Iida, S.; Haruyama, Y.; Kashino, S.; Emura, S.; Akasaka, T. *Phys. Rev. B* **2001**, 6411, 113403.

55. Shiozawa, H.; Rauf, H.; Pichler, T.; Grimm, D.; Liu, X.; Knupfer, M.; Kalbac, M. et al. *Phys. Rev. B* **2005**, 72, 195409.

56. Shiozawa, H.; Rauf, H.; Pichler, T.; Knupfer, M.; Kalbac, M.; Yang, S.; Dunsch, L.; Buechner, B.; Batchelor, D.; Kataura, H. *Phys. Rev. B* **2006**, 73, 205411.

57. Shiozawa, H.; Rauf, H.; Grimm, D.; Knupfer, M.; Kalbac, M.; Yang, S.; Dunsch, L.; Buchner, B.; Pichler, T. *Phys. Status Solidi B* **2006**, 243, 3004.

58. Pichler, T.; Golden, M. S.; Knupfer, M.; Fink, J.; Kirbach, U.; Kuran, P.; Dunsch, L. *Phys. Rev. Lett.* **1997**, 79, 3026.

59. Pichler, T.; Knupfer, M.; Golden, M. S.; Boske, T.; Fink, J.; Kirbach, U.; Kuran, P.; Dunsch, L.; Jung, C. *Appl. Phys. A: Mater. Sci. Process.* **1998**, 66, 281.

60. Pichler, T.; Winter, J.; Grazioli, C.; Golden, M. S.; Knupfer, M.; Kuran, P.; Dunsch, L.; Fink, J. *Synth. Met.* **1999**, 103, 2470.

61. Krause, M.; Liu, X. J.; Wong, J.; Pichler, T.; Knupfer, M.; Dunsch, L. *J. Phys. Chem. A* **2005**, 109, 7088.

62. Liu, X.; Krause, M.; Wong, J.; Pichler, T.; Dunsch, L.; Knupfer, M. *Phys. Rev. B* **2005**, 72, 085407.

63. Hino, S.; Umishita, K.; Iwasaki, K.; Aoki, M.; Kobayashi, K.; Nagase, S.; Dennis, T. J. S.; Nakane, T.; Shinohara, H. *Chem. Phys. Lett.* **2001**, 337, 65.

64. Cao, B. P.; Hasegawa, M.; Okada, K.; Tomiyama, T.; Okazaki, T.; Suenaga, K.; Shinohara, H. *J. Am. Chem. Soc.* **2001**, 123, 9679.

65. Suzuki, T.; Kikuchi, K.; Oguri, F.; Nakao, Y.; Suzuki, S.; Achiba, Y.; Yamamoto, K.; Funasaka, H.; Takahashi, T. *Tetrahedron* **1996**, 52, 4973.

66. Lu, X.; Slanina, Z.; Akasaka, T.; Tsuchiya, T.; Mizorogi, N.; Nagase, S. *J. Am. Chem. Soc.* **2010**, *132*, 5896–5905.

67. Suzuki, T.; Maruyama, Y.; Kato, T.; Kikuchi, K.; Nakao, Y.; Achiba, Y.; Kobayashi, K.; Nagase, S. *Angew. Chem. Int. Ed. Engl.* **1995**, 34, 1094.

68. Yamada, M.; Nakahodo, T.; Wakahara, T.; Tsuchiya, T.; Maeda, Y.; Akasaka, T.; Kako, M. et al. *J. Am. Chem. Soc.* **2005**, 127, 14570.

69. Yamada, M.; Mizorogi, N.; Tsuchiya, T.; Akasaka, T.; Nagase, S. *Chem. Eur. J.* **2009**, 15, 9486.

70. Cao, B. P.; Wakahara, T.; Tsuchiya, T.; Kondo, M.; Maeda, Y.; Rahman, G. M. A.; Akasaka, T.; Kobayashi, K.; Nagase, S.; Yamamoto, K. *J. Am. Chem. Soc.* **2004**, 126, 9164.

71. Lu, X.; Nikawa, H.; Nakahodo, T.; Tsuchiya, T.; Ishitsuka, M. O.; Maeda, Y.; Akasaka, T. et al. *J. Am. Chem. Soc.* **2008**, 130, 9129.

72. Fu, W.; Zhang, J.; Fuhrer, T.; Champion, H.; Furukawa, K.; Kato, T.; Mahaney, J. E. et al. *J. Am. Chem. Soc.* **2011**, 133, 9741.

73. Xu, W.; Feng, L.; Calvaresi, M.; Liu, J.; Liu, Y.; Niu, B.; Shi, Z. J.; Lian, Y. F.; Zerbetto, F. *J. Am. Chem. Soc.* **2013**, 135, 4187–4190.

74. Feng, L.; Wakahara, T.; Nakahodo, T.; Tsuchiya, T.; Piao, Q.; Maeda, Y.; Lian, Y. et al. *Chem. Eur. J.* **2006**, 12, 5578.

75. Wakahara, T.; Yamada, M.; Takahashi, S.; Nakahodo, T.; Tsuchiya, T.; Maeda, Y.; Akasaka, T. et al. *Chem. Commun.* **2007**, 2680.

76. Yamada, M.; Nakahodo, T.; Wakahara, T.; Tsuchiya, T.; Maeda, Y.; Akasaka, T.; Kako, M. et al. *J. Am. Chem. Soc.* **2005**, 127, 14570.

77. Ishitsuka, M. O.; Sano, S.; Enoki, H.; Sato, S.; Nikawa, H.; Tsuchiya, T.; Slanina, Z. et al. *J. Am. Chem. Soc.* **2011**, 133, 7128.

78. Elliott, B.; Yu, L.; Echegoyen, L. *J. Am. Chem. Soc.* **2005**, 127, 10885.

79. Cardona, C. M.; Elliott, B.; Echegoyen, L. *J. Am. Chem. Soc.* **2006**, 128, 6480.

80. Yang, S. F.; Zalibera, M.; Rapta, P.; Dunsch, L. *Chem.Eur. J.* **2006**, 12, 7848.

81. Popov, A. A.; Chen, C.; Yang, S.; Lipps, F.; Dunsch, L. *ACS Nano* **2010**, 4, 4857.

82. Chen, C.; Liu, F.; Li, S.; Wang, N.; Popov, A. A.; Jiao, M.; Wei, T.; Li, Q.; Dunsch, L.; Yang, S. *Inorg. Chem.* **2012**, 51, 3039.

83. Zhang, L.; Popov, A. A.; Yang, S. F.; Klod, S.; Raptaad, P.; Dunsch, L. *Phys. Chem. Chem. Phys.* **2010**, 12, 7840–7847.

84. Zhang, Y.; Schiemenz, S.; Popov, A. A.; Dunsch, L. *J. Phys. Chem. Lett.* **2013**, 4, 2404–2409.

85. Manuel, N. C.; Andreas, J. A.; Echegoyen, L. *Tetrahedron* **2008**, 64, 11387–11393.

86. Rivera-Nazarioa, D. M.; Pinzónc, J. R.; Stevenson, S.; Echegoyen, L. A. *J. Phys. Org. Chem.* **2013**, 26, 194–205.

87. Chen, N.; Pinzón, J. R.; Echegoyen, L. *ChemPhysChem* **2011**, 12, 1422.

88. Wakahara, T.; Sakuraba, A.; Iiduka, Y.; Okamura, M.; Tsuchiya, T.; Maeda, Y.; Akasaka, T. et al. *Chem. Phys. Lett.* **2004**, 398, 553.

89. Xu, W.; Wang, T.-S.; Wu, J.-Y.; Ma, Y.-H.; Zheng, J.-P.; Li, H.; Wang, B.; Jiang, L.; Shu, C.-Y.; Wang, C.-R. *J. Phys. Chem. C* **2011**, 115, 402.

90. Iiduka, Y.; Wakahara, T.; Nakajima, K.; Nakahodo, T.; Tsuchiya, T.; Maeda, Y.; Akasaka, T. et al. *Angew. Chem. Int. Ed.* **2007**, 46, 5562.

91. Wang, T. S.; Chen, N. Xiang, J. F.; Li, B.; Wu, J. Y.; Xu, W.; Jiang, L. et al. *J. Am. Chem. Soc.* **2009**, 131, 16646–16647.

92. Lu, X.; Nakajima, K.; Iiduka, Y.; Nikawa, H.; Mizorogi, N.; Slanina, Z.; Tsuchiya, T.; Nagase, S.; Akasaka, T. *J. Am. Chem. Soc.* **2011**, 133, 19553.

93. Lu, X.; Nikawa, H.; Kikuchi, T.; Mizorogi, N.; Slanina, Z.; Tsuchiya, T.; Nagase, S.; Akasaka, T. *Angew. Chem. Int. Ed. Engl.* **2011**, 50, 6356.

94. Lu, X.; Nakajima, K.; Iiduka, Y.; Nikawa, H.; Tsuchiya, T.; Mizorogi, N.; Slanina, Z.; Nagase, S.; Akasaka, T. *Angew. Chem. Int. Ed. Engl.* **2012**, 51, 5889.

95. Chen, N.; Chaur, M. N.; Moore, C.; Pinzon, J. R.; Valencia, R.; Rodrıguez-Fortea, A.; Poblet, J. M.; Echegoyen, L. *Chem. Commun.*, **2010**, 46, 4818–4820.

96. Chen, N.; Beavers, C. M.; Mulet-Gas, M.; Rodriguez-Fortea, A.; Munoz, E. J.; Li, Y.-Y.; Olmstead, M. M.; Balch, A. L.; Poblet, J. M.; Echegoyen, L. *J. Am. Chem. Soc.* **2012**, 134, 7851.

97. Chen, N.; Mulet-Gas, M.; Li, Y.-Y.; Stene, R. E.; Atherton, C. W.; Rodriguez-Fortea, A.; Poblet, J. M.; Echegoyen, L. *Chem. Sci.* **2013**, 4, 180.

98. Li, F. F.; Chen, N.; Mulet-Gas, M.; Triana, V.; Murillo, J.; Rodr´ıguez-Fortea, A.; Poblet, J. M.; Echegoyen, L. *Chem. Sci.* **2013**, 4, 3404–3410.

99. Popov, A. A.; Chen, N.; Pinzón, J. R.; Stevenson, S.; Echegoyen, L. A.; Dunsch, L. *J. Am. Chem. Soc.* **2012**, 134, 19607.

100. Wang, T.-S.; Feng, L.; Wu, J.-Y.; Xu, W.; Xiang, J.-F.; Tan, K.; Ma, Y.-H. et al. *J. Am. Chem. Soc.* **2010**, 132, 16362.

101. Yang, S. F.; Chen, C. B.; Liu, F. P.; Xie, Y. P.; Li, F. Y.; Jiao, M. Z.; Suzuki, M. et al. *Sci. Rep.* **2013**, 3, 1487.

102. Hoffman, K. R.; DeLapp, K.; Andrews, H.; Sprinkle, P.; Nickels, M.; Norris, B. *J. Lumin.* **1996**, 66–7, 244.

103. Hoffman, K. R.; Norris, B. J.; Merle, R. B.; Alford, M. *Chem. Phys. Lett.* **1998**, 284, 171.

104. Hoffman, K. R.; Conley, W. G. *J. Lumin.* **2001**, 94, 187.

105. Ding, X. Y.; Alford, J. M.; Wright, J. C. *Chem. Phys. Lett.* **1997**, 269, 72.

106. Macfarlane, R. M.; Wittmann, G.; vanLoosdrecht, P. H. M.; deVries, M.; Bethune, D. S.; Stevenson, S.; Dorn, H. C. *Phys. Rev. Lett.* **1997**, 79, 1397.

107. Plant, S. R.; Dantelle, G.; Ito, Y.; Ng, T. C.; Ardavan, A.; Shinohara, H.; Taylor, R. A.; Briggs, A. D.; Porfyrakis, K. *Chem. Phys. Lett.* **2009**, 476, 41.

108. Macfarlane, R. M.; Bethune, D. S.; Stevenson, S.; Dorn, H. C. *Chem. Phys. Lett.* **2001**, 343, 229.

109. Jones, M. A. G.; Taylor, R. A.; Ardavan, A.; Porfyrakis, K.; Briggs, G. A. D. *Chem. Phys. Lett.* **2006**, 428, 303.

110. Jones, M. A. G.; Morton, J. J. L.; Taylor, R. A.; Ardavan, A.; Briggs, G. A. D. *Phys. Status Solidi B* **2006**, 243, 3037.

111. Tiwari, A.; Dantelle, G.; Porfyrakis, K.; Taylor, R. A.; Watt, A. A. R.; Ardavan, A.; Briggs, G. A. D. *J. Chem. Phys.* **2007**, 127, 194504.

112. Tiwari, A.; Dantelle, G.; Porfyrakis, K.; Ardavan, A.; Briggs, G. A. D. *Phys. Status Solidi B* **2008**, 245, 1998.

113. Morton, J. G. M.; Tiwari, A.; Dantelle, G.; Porfyrakis, K.; Ardavan, A.; Briggs, G. A. D. *Phys. Rev. Lett.* **2008**, 101, 013002.

114. Dantelle, G.; Tiwari, A.; Rahman, R.; Plant, S. R.; Porfyrakis, K.; Mortier, M.; Taylor, R. A.; Briggs, A. D. *Opt. Mater.* **2009**, 32, 251.

115. Gimenez-Lopez, M. D. C.; Gardener, J. A.; Shaw, A. Q.; Iwasiewicz-Wabnig, A.; Porfyrakis, K.; Balmer, C.; Dantelle, G. et al. *Phys. Chem. Chem. Phys.* **2010**, 12, 123.

116. Gu, G.; Huang, H. J.; Yang, S. H.; Yu, P.; Fu, J. S.; Wong, G. K.; Wan, X. G.; Dong, J. M.; Du, Y. W. *Chem. Phys. Lett.* **1998**, 289, 167.

117. Heflin, J. R.; Marciu, D.; Figura, C.; Wang, S.; Burbank, P.; Stevenson, S.; Dorn, H. C. *Appl. Phys. Lett.* **1998**, 72, 2788.

118. Yaglioglu, G.; Pino, R.; Dorsinville, R.; Liu, J. Z. *Appl. Phys. Lett.* **2001**, 78, 898.

119. Xenogiannopoulou, E.; Couris, S.; Koudoumas, E.; Tagmatarchis, N.; Inoue, T.; Shinohara, H. *Chem. Phys. Lett.* **2004**, 394, 14.

120. Hu, H.; Cheng, W. D.; Huang, S. P.; Xie, Z.; Zhang, H. J. *Theor. Comput. Chem.* **2008**, 7, 737.

121. Fujitsuka, M.; Ito, O.; Kobayashi, K.; Nagase, S.; Yamamoto, K.; Kato, T.; Wakahara, T.; Akasaka, T. *Chem. Lett.* **2000**, 902.

122. Takano, Y.; Obuchi, S.; Mizorogi, N.; García, R.; Herranz, M. Á.; Rudolf, M.; Wolfrum, S. et al. *J. Am. Chem. Soc.* **2012**, 134, 16103.

123. Tsuchiya, T.; Rudolf, M.; Wolfrum, S.; Radhakrishnan, S. G.; Aoyama, R.; Yokosawa, Y.; Oshima, A.; Akasaka, T.; Nagase, S.; Guldi, D. M. *Chem. Eur. J.* **2013**, 19, 558.

124. Takano, Y.; Herranz, M. A.; Martin, N.; Radhakrishnan, S. G.; Guldi, D. M.; Tsuchiya, T.; Nagase, S.; Akasaka, T. *J. Am. Chem. Soc.* **2010**, 132, 8048.

125. Yaglioglu, G.; Pino, R.; Dorsinville, R.; Liu, J. Z. *Appl. Phys. Lett.* **2001**, 78, 898.

126. Kawashima, Y.; Ohkubo, K.; Fukuzumi, S. *J. Phys. Chem. A* **2012**, 116, 8942.

127. Bharadwaj, L.; Novotny, L. *J. Phys. Chem. C* **2010**, 114, 7444.

128. Toth, K.; Molloy, J. K.; Matta, M.; Heinrich, B.; Guillon, D.; Bergamini, G.; Zerbetto, F.; Donnio, B.; Ceroni, P.; Felder-Flesch, D. *Angew. Chem. Int. Ed.* **2013**, 52, 12303–12307.

129. Tagmatarchis, N.; Kato, H.; Shinohara, H. *Phys. Chem. Chem. Phys.* **2001**, 3, 3200.

130. McCluskey, D. M.; Smith, T. N.; Madasu, P. K.; Coumbe, C. E.; Mackey, M. A.; Fulmer, P. A.; Wynne, J. H.; Stevenson, S.; Phillips, J. P. *ACS Appl. Mater. Interfaces* **2009**, 1, 882.

131. Yanagi, K.; Okubo, S.; Okazaki, T.; Kataura, H. *Chem. Phys. Lett.* **2007**, 435, 306.

132. Alidzhanov, E. K.; Lantukh, Y. D.; Letuta, S. N.; Pashkevich, S. N.; Kareev, I. E.; Bubnov, V. P.; Yagubskii, E. B. *Opt. Spectrosc.* **2010**, 109, 578.

133. Kato, T.; Suzuki, S.; Kikuchi, K.; Achiba, Y. *J. Phys. Chem.* **1993**, 97, 13425–13428.

134. Inakuma, M.; Shinohara, H. *J. Phys. Chem. B* **2000**, 104, 7595–7599.

135. Kikuchi, K.; Nakao, Y.; Suzuki, S.; Achiba, Y.; Suzuki, T.; Maruyama, Y. *J. Am. Chem. Soc.* **1994**, 116, 9367–9368.

136. Wang, T. S.; Wu, J. Y.; Xu, W.; Xiang, J. F.; Lu, X.; Li, B.; Jiang, L.; Shu, C. Y.; Wang, C. R. *Angew. Chem. Int. Ed.* **2010**, 49, 1786–1789.

137. Tan, K.; Lu, X. *J. Phys. Chem. A* **2006**, 110, 1171–1176.

138. Shu, C. Y.; Zhang, E. Y.; Xiang, J. F.; Zhu, C. F.; Wang, C. R.; Pei, X. L.; Han, H. B. *J. Phys. Chem. B* **2006**, 110, 15597–15601.

139. Shu, C. Y.; Slebodnick, C.; Xu, L. S.; Champion, H.; Fuhrer, T.; Cai, T.; Reid, J. E. et al. *J. Am. Chem. Soc.* **2008**, 130, 17755–17760.

140. Funasaka, H.; Sakurai, K.; Oda, Y.; Yamamoto, K.; Takahashi, T. *Chem. Phys. Lett.* **1995**, 232, 273–277.

141. Zuo, T. M.; Xu, L. S.; Beavers, C. M.; Olmstead, M. M.; Fu, W. J.; Crawford, D.; Balch, A. L.; Dorn, H. C. *J. Am. Chem. Soc.* **2008**, 130, 12992–12997.

142. Ma, Y. H.; Wang, T. S.; Wu, J. Y.; Feng, Y. Q.; Jiang, L.; Shu, C. Y.; Wang, C. R. *Chem. Commun.* **2012**, 48, 11570–11572.

143. Iiduka, Y.; Wakahara, T.; Nakahodo, T.; Tsuchiya, T.; Sakuraba, A.; Maeda, Y.; Akasaka, T. et al. *J. Am. Chem. Soc.* **2005**, 127, 12500–12501.

144. Hachiya, M.; Nikawa, H.; Mizorogi, N.; Tsuchiya, T.; Lu, X.; Akasaka, T. *J. Am. Chem. Soc.* **2012**, 134, 15550–15555.

145. Tsuchiya, T.; Kurihara, H.; Sato, K.; Wakahara, T.; Akasaka, T.; Shimizu, T.; Kamigata, N.; Mizorogi, N.; Nagase, S. *Chem. Commun.* **2006**, 3585–3587.

146. Hajjaj, F.; Tashiro, K.; Nikawa, H.; Mizorogi, N.; Akasaka, T.; Nagase, S.; Furukawa, K.; Kato, T.; Aida, T. *J. Am. Chem. Soc.* **2011**, 133, 9290–9292.

147. Huang, H. J.; Yang, S. H.; Zhang, X. X. *J. Phys. Chem.* B **2000**, 104, 1473–1482.

148. Wolf, M.; Muller, K. H.; Skourski, Y.; Eckert, D.; Georgi, P.; Krause, M.; Dunsch, L. *Angew. Chem. Int. Ed.* **2005**, 44, 3306–3309.

149. Westerstrom, R.; Dreiser, J.; Piamonteze, C.; Muntwiler, M.; Weyeneth, S.; Brune, H.; Rusponi, S. et al. *J. Am. Chem. Soc.* **2012**, 134, 9840–9843.

6 Chemistry of Conventional Endohedral Metallofullerenes and Cluster Endohedral Fullerenes

CONTENTS

6.1   CHEMISTRY OF CONVENTIONAL EMFs

Muqing Chen and Xing Lu

Conventional endohedral metallofullerenes (EMFs) refer to these species discovered at the early stage of research, mainly pointing to mono-EMF and di-EMFs. Novel EMFs are these compounds containing a metallic cluster of various kinds which were discovered within the past 10 to 12 years. Because the internal metallic species exert fundamental influences on the cage carbons, the inherent properties of conventional EMFs differ significantly from those of cluster EMFs, especially in terms of chemical properties. Accordingly, we address the chemistry of EMFs in accordance with the type and composition of the internal metallic species.

6.1.1   CHEMISTRY OF MONO-EMFs

Conventional EMFs include such species containing merely metal atoms. In the following section, we will briefly discuss classic chemical functionalization studies of EMFs, focusing on the chemical properties of mono-EMFs, especially $M@C_{82}$.[1] The reactions that have been successfully performed on mono-EMFs are summarized in Table 6.1.

6.1.1.1   Disilylation of mono-EMFs

The first exohedral functionalization of EMFs was undertaken by Akasaka et al. in 1995 via photochemical and thermal reactions of $La@C_{82}$ with disilirane [1,1,2,2,-tetrakis(2,4,6-trimethylphenyl)-1,2-disilarane], leading to the formation of a mono-addition product (Figure 6.1, reaction a).[2] An interesting result was that the high thermal reactivity of $La@C_{82}$ with disilirane was attributable to the stronger electron donor and acceptor of the endohedral fullerene relative to the empty fullerenes (C_{60} and C_{70}).[2] Similar photochemical and thermal reactions of $La@C_{82}$ with digermirane were reported by Akasaka in 1996.[3] The finding was that even when the reaction temperature went down from 80°C to 20°C, the thermal addition of digermirane onto $La@C_{82}$ was observed, unlike the disilirane case.[3] It also indicated that the degree of the electron transfer from the electropositive metal to the fullerene cage played a crucial role in modifying the electronic properties of the fullerene cage, unlike the novel gas case ($He@C_{60}$ and $He@C_{70}$).[3] The chemical reactivity of $Gd@C_{82}$ toward functionalization with disilirane was almost identical to that of $La@C_{82}$.[4] Kato et al. chose diphenyldiazomethane (Ph_2CN_2) as a reactant to functionalize $La@C_{82}$. The reaction was carried out at 60°C, and the electron paramagnetic resonance (EPR) spectra were recorded every 30 min at room temperature. The EPR results showed that the EMF can be functionalized effectively, retaining its unique electronic properties. Further studies revealed that different reactivities of the two $La@C_{82}$ isomers with disilirane were speculated to the different nature of LUMO of the endohedral fullerenes.[5] The silylated products of mono-EMFs ($Y@C_{82}$ and $La@C_{82}$) were synthesized and isolated by Akasaka et al. and the redox properties of

TABLE 6.1

Summary of the Reactions That Have Been Performed on Conventional EMFs

EMFs	Radical Reaction	Carbene Addition	Silylation	1,3-Dipolar Reaction	Diels–Alder Cycloaddition	[2 + 2] Cycloaddition	Bingel–Hirsch
La@$C_{72,74,80,82}$	✓						✓
M@$C_{2v}(9)$-C_{82} (M = Sc, Y, La, Ce, Pr, Gd)	✓	✓	✓	✓	✓	✓	✓
Gd@C_{60}							✓
La@$C_s(6)$-C_{82}	✓	✓	✓				
Yb@$C_{2v}(3)$-C_{80}		✓					
Yb@$C_2(13)$-C_{84}		✓	✓				
La$_2$@$D_2(10611)$-C_{72}		✓	✓	✓			
La$_2$@C_{78}		✓	✓				
La$_2$@C_{80}		✓	✓	✓			
Ce$_2$@C_{80}			✓				
Ce$_2$@C_{78}			✓				
Sc$_2$C$_2$@C_{82}			✓	✓			

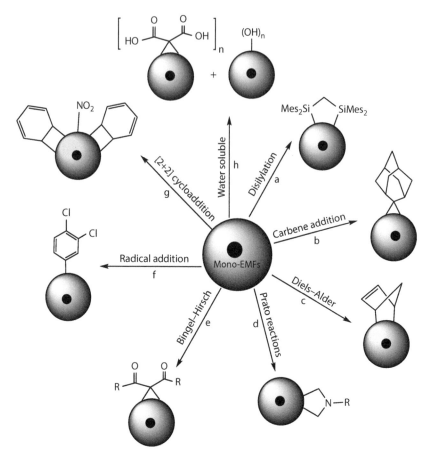

FIGURE 6.1 Chemical reactions that have been performed on mono-EMF M@C_{2n}.

silylated M@C_{82} (M = Y and La) were investigated.[6] The cyclic voltammetry (CV) and differential pulse voltammetry (DPV) results indicated that the bis-silylated mono-EMFs had lower oxidation and higher reduction potentials than the parent ones. They were the first examples to show that electronic properties of EMFs can be adjusted by exohedral addition.[6] Later on, the same group also reported the chemical reactivities of the cation and anion of M@C_{82} (M = Y, La, and Ce) with disilirane derivatives.[7] The interesting correlation was obtained between the reactivity toward disilirane derivative and the redox potentials, as well as their HOMO/LUMO levels of the EMFs. Electronically positive fullerenes, such as [M@C_{82}]$^+$, reacted readily with the nucleophillic disilirane reagent. However, the anions such as [M@C_{82}]$^-$ did not react with disilirane. The conclusion was that the reactivity of M@C_{82} could be tuned by ionization.[7] Two isomers of Pr@C_{82} were isolated and reacted with disilirane derivatives in 1,2,4-trichlorobenzene by photoirradiation and thermal methods, but the adducts were very unstable under photochemical and thermal conditions.[8] Different stabilities of the adducts of Pr@C_{82} and La@C_{82} might be ascribed to the nature of the encapsulated metal atom.[8]

6.1.1.2 Carbene Addition to mono-EMFs

Akasaka's group successfully isolated the first isomers of regiospecific addition reaction of La@C_{2v}(9)-C_{82} with 2-adamantane-2,3-[3H]-diazirine (Ad) in 2004 (Figure 6.1, reaction b).[9] The electrochemical behavior of the La@C_{82} derivative showed that the introduction of an Ad group results in decreasing the electron-accepting property, and the encapsulated metal was useful for controlling the reactivity and selectivity of fullerenes.[9] Later in 2008, the same group first reported the regioselective reaction of La@$(C_s)C_{82}$ with Ad. The obtained derivatives of La@$(C_s)C_{82}$, were characterized by the absorption and EPR spectroscopic analyses.[10] At the same time, Gd@C_{82}-Ad derivatives were also synthesized by irradiating high-pressure mercury-arc lamp (cutoff 300 nm).[11] The single-crystal X-ray crystallographic analysis of Gd@C_{82}-Ad suggested that Gd@C_{82} had a normal endohedral structure in which the Gd atom was located at an off-center position near a hexagonal ring (not near the C–C double bond) along the C_2 axis of the C_{2v}-C_{82} cage. The results indicated that the maximum entropy method (MEM)/Rietveld analysis was not always reliable for metal positions as well as cage structures. The highly selective derivatization of Gd@C_{82} suggested that an encapsulated metal played an important role in controlling the reactivity and selectivity of fullerenes.[11] The adamantyl derivatives of Ce@C_{82} and Y@C_{82} were synthesized in high yields by a carbene addition reaction in 2009.[12,13] The chemical reactivity of Ce@C_{82} was similar to that of La@C_{82}, and the observations of paramagnetic shifts helped to reveal the magnetic properties of metal-encapsulating carbon clusters by structural and substituent effects on Ce@C_{82} derivatives.[11] The first single crystallographic result of the crucial EMF Y@C_{82} was obtained by measuring the adamantyl derivatives of Y@C_{82}. The Y atom in Y@C_{82} was confirmed to be located under a hexagonal ring along the C_2 axis, which makes only one of the cage carbons sufficiently reactive toward the electrophile Ad. Furthermore, the electron-donating ability of the Ad group to EMFs was confirmed in all systems.[13] In 2012, Hachiya et al. found that Sc@C_{2v}(9)-C_{82} had an exceptional chemical reactivity toward the electrophile diazirine adamantylidene, affording four monoadduct isomers. This was unexpected because the reactions of M@C_{2v}(9)-C_{82} (M = Y, La, Ce, Gd) with diazirine adamantylidene gave rise to only two monoadduct isomers, indicating that the cage reactivity of mono-EMFs was not dependent on the type of the internal metal ion. The exceptional chemical property of Sc@C_{2v}(9)–C_{82} was associated with the small ionic radius of Sc^{3+}, which allows stronger metal–cage interactions and makes back-donation of charge from the cage to the metal more pronounced.[14]

In 2013, Lu group reported the corresponding adamantyl derivatives of Yb@C_{2v}(3)-C_{80} and Yb@C_2(13)-C_{84}, which were synthesized by the irradiation of the mixture of appropriate EMFs and 2-adamantane-2,3-[3H]-diazirine (AdN$_2$).[15,16] These derivatives represented the first exohedral functionalized adducts of divalent EMFs. The results suggested that the chemical reactivity of divalent EMFs differed greatly from that of the trivalent analogues, but that it was similar to that of empty fullerenes.[14] Yb@C_2(13)-C_{84}-Ad was the first X-ray structure of any EMF-derivatives bearing a cage larger than C_{82}. Both derivatives show [5,6]-open structures (Figure 6.2). The interesting finding was that the dynamic motion of the metal ion in Yb@C_2(13)-C_{84} was completely stopped in the derivative, although the metal is not trapped inside the cavity provided by bond cleavage. In contrast, the metal atoms tend to hop inside the functionalized Yb@C_{80} cage.[16]

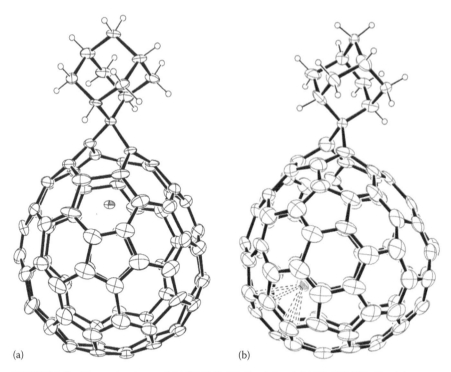

(a)   (b)

FIGURE 6.2   X-ray structures of (a) Yb@C_{2v}(3)-C_{80}-Ad and (b) Yb@C_2(13)-C_{84}-Ad.

6.1.1.3   Diels–Alder Reactions of mono-EMFs

In 2005, Akasaka et al. reported the Diels–Alder addition reaction of cyclopentadiene (Cp) with La@C_{82}, affording La@C_{82}Cp in 44% yield with high regioselectivity (Figure 6.1, reaction c).[17] This reaction of La@C_{82} with Cp was reversible. The La@C_{82}Cp could be decomposed to La@C_{82} and Cp even at 298 K in toluene. The retro-reaction of La@C_{82}Cp was much faster than that of C_{60}Cp, which was attributed to the lower activation energy of La@C_{82}Cp than that of the retro-reaction of C_{60}Cp.[17] In 2010, Akasaka et al. modified the cyclopentadiene by replacing the hydrogen with methyl. The novel cyclopentadiene derivative (Cp*) reacted with La@C_{82} forming a stable monoadduct (La@C_{82}Cp*) via the Diels–Alder cycloaddition reaction.[18] X-ray analysis results showed that La@C_{82}Cp* takes two orientations in the crystal structure, and 60% of the monoadduct forms a dimer in the solid state. The length of the C–C bond connecting two C_{82} cages in the La@ C_{82}Cp* dimer (1.606 Å) was only about 0.1 Å longer than the typical C–C single bond length.[18] The mechanism of the reaction between La@C_{82} and Cp* was also investigated both theoretically and experimentally by the same authors in 2013. The results confirmed a concerted mechanism that includes formation of a stable intermediate.[19]

6.1.1.4   Prato Reactions of mono-EMFs

La@C_{82} was first reported by Akasaka et al. to react with azomethine ylides, the so-called Prato reaction (Figure 6.1, reaction d), which afforded multiple derivatives.[20] Gu group then reported the same reaction with Gd@C_{82}.[21] Two kinds of pyrrolidine ring-fused EMF derivatives via different 1,3-dipolar reactions were synthesized. The results showed that Gd@C$_{82}$ was more reactive than C$_{60}$ when reacting with azomethine ylides, because up to eight pyrrolidine rings could be added to the Gd@C$_{82}$ cage within 30 min whereas only a monoadduct was formed for C$_{60}$ after 2 hr.[21] Later, the same group studied the Prato reaction of M@C$_{82}$ (M = Gd, Y) by sarcosine and different aldehydes. By changing the aldehydes with different substituted groups, the numbers of the rings on the EMFs M@C$_{82}$ (M = Gd, Y) could be tuned.[22] The La@C$_{82}$-pyrrolidine derivatives with different functional groups were synthesized for further photophysical studies.[23,24] The novel electron donor–acceptor conjugate consisting of a paramagnetic La@C$_{82}$ and an electron-donating exTTF was also reported. The ground states of La@C$_{82}$ derivatives lacked appreciable electronic interactions between La@C$_{82}$ and exTTF, whereas the electronically excited state undergoes electron transfer deactivation of the photoexcited La@C$_{82}$. The unique properties of La@C$_{82}$ and open-shell electronic structure assisted in stabilizing the radical ion pair state for more than 2 ns.[20] Another La@C$_{82}$ derivative via the Prato reaction with an open-shell electronic configuration was synthesized, which coordinated noncovalently with zinc tetraphenylporphyrin (ZnP) forming electron donor–acceptor hybrid.[21]

6.1.1.5 Bingel–Hirsch Reactions of mono-EMFs

The Bingel–Hirsch reaction (Figure 6.1, reaction e) was applied to synthesize novel endohedral fullerene derivatives. The highly soluble, air-stable Bingel adduct of EMFs (Gd@C$_{60}$[C(COOC$_2$H$_5$)]$_{10}$) was reported by Bolskar et al. in 2003 via cycloaddition of bromomalonates to Gd@C$_{60}$.[25] Gd@C$_{60}$[C(COOC$_2$H$_5$)]$_{10}$ was further hydrolyzed to the water-soluble Gd@C$_{60}$[C(COOH)$_2$]$_{10}$, as a new MRI contrast agent that has also been evaluated by *in vivo* MRI biodistribution measurements, relaxometry, and dynamic light scattering. This carboxylated EMF derivative was proved to be the first fullerene compound with favorable biodistribution (non-RES localizing).[25] Gd@C$_{60}$[C(COOH)$_2$]$_{10}$ tended to aggregate in aqueous solution under different pH, temperature, and concentration.[26] A singly bonded derivative of La@C$_{82}$CBr(COOC$_2$H$_5$)$_2$ was reported by Akasaka's group in 2005 based on the Bingel–Hirsch reaction of La@C$_{82}$ with diethyl bromomalonate in the presence of 1,8-diazabicyclo[5.4.0]-undec-7-ene (DBU).[27] One monoadduct was most abundantly produced and isolated in the yield of 40% based on consumed La@C$_{82}$. The structure of the La@C$_{82}$CBr(COOC$_2$H$_5$)$_2$ was fully determined by NMR spectroscopy and X-ray crystallographic analyses, which was very different from the conventional Bingel–Hirsch adduct of empty fullerenes (Figure 6.3a).[27] In the next year, they reported an unconventional derivative of La@C$_{82}$, that is, La@C$_{82}$CBr(COOC$_2$H$_5$)$_2$.[28] Five monoadducts (mono-A, mono-B, mono-C, mono-D, and mono-E) of the La@C$_{82}$CBr(COOC$_2$H$_5$)$_2$ were isolated from the reaction mixture by multistage high-performance liquid chromatography (HPLC) (5PYE column, Buckyprep M column and 5PBB column). Mono-A, mono-B, mono-C, and mono-D were revealed to be regioisomers of singly bonded monoadducts, whereas mono-E was suggested to be a cycloadduct that was similar to conventional Bingel adducts. This research indicated that the reactivity of an EMF was distinctly different from those of empty fullerenes in some reactions. Mono-A, mono-B, and mono-C exhibited nearly identical physicochemical properties, whereas mono-E possessed different properties,

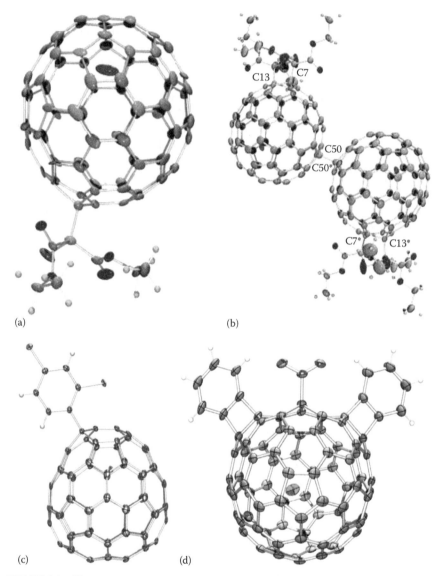

FIGURE 6.3 X-ray structures of (a) La@C_{82}CBr(COOC$_2$H$_5$)$_2$ and (b) La@C_{82}[CH(COOC$_2$H$_5$)$_2$]$_2$, which is a dimeric structure. (Reprinted with permission from Feng, L. et al. *J. Am. Chem. Soc.* 127(49), 17136–17137, 2005. Copyright 2005 American Chemical Society; and Feng, L. et al., *J. Am. Chem. Soc.* 128(18), 5990–5991, 2006, Copyright 2006 American Chemical Society, respectively.) X-ray structures of (c) La@D_{3h}-C$_{74}$-C$_6$H$_3$C$_{12}$ and (d) La@C_{2v}(9)-C$_{82}$-(C$_6$H$_4$)$_2$NO$_2$.

which was in good agreement with their different structures.[28,29] At the same year, 7,13-bisadduct of La@C_{2v}(9)-C$_{82}$ was synthesized by a Bingel–Hirsch reaction with a good yield. This was the first unambiguous example of the dimer formation of EMFs, which was verified by X-ray crystallographic analysis (Figure 6.3b). The addition of bromomalonate occurred at the C7 and C13 atoms of La@(C_{2v})C$_{82}$, which were located at the apex of two hexagons and one pentagon.[30] The monoadduct malonic ester derivative of paramagnetic gadofullerene ($Gd@C_{82}[C(COOCH_2CH_3)_2]$) was synthesized by the Bingel–Hirsch reaction and was purified by HPLC, which was further hydrolyzed to water-soluble $Gd@C_{82}[C(COONa)_2]$. The size distribution of $Gd@C_{82}[C(COONa)_2]$ was 167 nm in aqueous solution by number-weighted pattern with dynamic light scattering analysis.[31]

6.1.1.6 Radical Addition Reactions of mono-EMFs

The radical addition reaction of EMFs was first reported by Suzuki and Kato et al. in 1995 (Figure 6.1, reaction f). An excess of diphenyldiazomethane was added to a toluene solution of $La@C_{82}$ in an ESR tube at 60°C, affording the methanofullerene derivatives $La@C_{82}$-$(CPh_2)_n$ ($n = 0$–3), which showed that the EMF can be functionalized but the derivative retained its unique electronic properties.[32]

Insoluble EMFs can also undergo radical reactions and the derivatives can be isolated in solution. Soot-containing lanthanum EMFs was refluxed in the 1,2,4-trichlorobenzene (TCB). The soluble fraction was separated and mass spectrum showed a new molecular ion peak at m/z 1179–1172 attributed to the corresponding derivative of $La@C_{74}$ (m/z 1027) with a dichlorophenyl group ($C_6H_3Cl_2$, mass m/z = 145). $La@C_{74}(D_{3h})$ had a radical character and could be trapped with a dichlorophenyl radical produced during the extraction.[33] X-ray structure of $La@D_{3h}$-C_{74}-$C_6H_3Cl_2$ (Figure 6.3c) reveals that the dichlorophenyl group adds to one of the cage carbons that have high radical characters. Then the same group reported $La@C_{72}(C_6H_3Cl_2)$ with a non-IPR cage, which were formed during the extraction process of soot with 1,2,4-trichlorobenzene.[34] A dichlorophenyl radical was added to the insoluble $La@C_{2v}$-C_{80} with high radical character, resulting in the stable $La@C_{80}(C_6H_3Cl_2)$ with a closed-shell structure by refluxing the mixture solution of 1,2,4-trichlorobenzene and raw soot containing lanthanum EMFs.[35] The structure of $La@C_{80}(C_6H_3Cl_2)$ was determined unambiguously using single-crystal X-ray crystallographic analysis. $La@C_{2v}(7)$-$C_{82}(C_6H_3Cl_2)$ was also isolated and determined by X-ray crystallography.[36] The dichlorophenyl group was singly bonded to a triple-hexagon junction (THJ) carbon atom on the C_3 axis based on the X-ray results.[36]

In 2008, Akasaka's group reported that benzyl monoadducts $La@C_{2v}(9)$-$C_{82}(CH_2C_6H_5)$ was found by the thermal reaction of $La@C_{2v}(9)$-C_{82} with 3-triphenylmethyl-5-oxazolidinone in toluene. The same monoadducts were also obtained by the photoirradiation of $La@C_{2v}(9)$-C_{82} in toluene without the existence of 3-triphenylmethyl-5-oxazolidinone.[37] Photoirradiation of $La@C_{2v}(9)$-C_{82} in 1,2-dichlorobenzene in the presence of tetrachlorotoluene also affords the monoadducts $La@C_{2v}(9)$-$C_{82}(CHClC_6H_3Cl_2)$ on the basis of the X-ray structural analysis. The conclusion is that this radical addition reaction can be widely used to the reaction of paramagnetic EMFs [$La@C_{82}(C_s)$, $Ce@C_{2v}(9)$-C_{82}, and substituted benzenes (toluene, p-xylene, o-xylene, p-tert-butyltoluene) under photoirradiation.[37] Later in 2010, the same group found that retro-reaction of radical monoadducts of a paramagnetic EMF ($La@C_{2v}$-C_{82}) proceeded thermally in the presence of a radical-trapping reagent, affording pristine $La@C_{2v}$-C_{82} in high yield (96%). Using this retro-reaction, pristine missing EMFs are possibly obtainable from their derivatives, such as $La@C_nC_6H_3Cl_2$ ($n = 72, 74, 80$).[38]

Shinohara et al. synthesized the first perfluoroalkylated paramagnetic EMF deriva-tive La@$C_{82}(C_8F_{17})_2$ by the fluorous-phase partitioning method (or liquid–liquid extrac-tion).[39] The authors concluded that the perfluoroalkyl addends should be located in the 1,2- rather than in the 1,4- positions because of the attack at the [6,6] double bond adja-cent to La. With the former structural arrangement, all double bonds of the exohedrally functionalized EMFs were located exocyclic to all pentagons. The favorable π electron system of the endohedral fullerene was preserved to a large extent and steric strain in the carbon cage is released, and the solubility of EMFs could be drastically increased in perfluorocarbon solvents by fluorous-phase approach with long and heavily fluorinated alkyl ponytails.[39] Another fluoroalkylated derivative of EMFs was prepared by Kareev et al. in 2005 via a high temperature reaction of the Y@C_{82}-enriched extract and silver(I) trifluoroacetate(AgCF$_3$CO$_2$).[40] Trifluoromethylation was chosen to study the derivatiza-tion of EMFs because CF$_3$ derivatives were less prone to hydrolysis than fluorofullerenes and were more soluble in common organic solvents.[19] F NMR spectroscopic analysis and DFT calculations elucidated three possible structures for the two isomers of Y@$C_{82}(CF_3)_5$.[40] The high-temperature (300°C–400°C) solid-phase reaction of M@C_{82} (M = Gd, Ce) with silver trifluoroacetate CF$_3$COOAg lead to the formation of two iso-mers M@$C_{82}(CF_3)_5$ (M = Gd, Ce) and Ce@C_{82} derivatives, similar to Y@$C_{82}(CF_3)_5$.[41]

6.1.1.7   [2 + 2] Cycloaddition to mono-EMFs

Lu et al. first reported the preferential addition of benzyne to the [5,6]-bond of La@C_{82} to form closed cyclobutene rings between the substituents and the cage via [2 + 2] cycloaddition of benzyne in 2010 (Figure 6.1, reaction g).[42] An unexpected NO$_2$ group was also found in the benzyne adducts that has an oxidation effect on the electro-chemical properties of La@C_{82}. The three addends apparently preferred a 1,4-addition pattern, which implies a positional directing effect of the NO$_2$ group (Figure 6.3d).[42]

6.1.1.8   Water-Soluble Derivatives of mono-EMFs

Modification of mono-EMFs with water-soluble functional groups was the prerequisite for their potential biological and medical applications (Figure 6.1, reaction h). The first water-soluble EMF derivative Ho@$C_{82}(OH)_x$ was synthesized by Wilson et al. through the electrophilic addition of nitronium tetrafluoroborate in the presence of arenecar-boxylic acid in a nonaqueous medium.[43] Later, the water-soluble holmium EMFs were neutron irradiated and employed in a radiotracer biodistribution study.[44] In 1999, water-soluble multihydroxyl endohedral metallofullerol, Pr@$C_{82}O_m$-(OH)$_n$($m \approx 10$ and $n \approx 10$), was successfully synthesized by treatment with nitric acid, followed by hydrolysis.[45] NO$_2$ radical addition in the reaction was considered to be a crucial step for the formation of metallofullerols via the aqueous chemistry.[45] The next year, Shinohara et al. synthesized a new type of water-soluble metallofullerenol [Gd@$C_{82}(OH)_n$] for use as magnetic resonance imaging (MRI) contrast agents. Gd $M_{4,5}$-edge EELS spectra showed that the encapsulated Gd atom has a trivalent Gd^{3+} state, which was maintained after hydroxylation.[46] In 2001, Shinohara and coworkers reported that the paramag-netic Gd@$C_{82}(OH)_n$ can serve as novel MRI contrast agents for next generation. The *in vitro* water proton relaxivity R_1 of Gd-fullerenols was significantly higher (20-folds) than that of the commercial MRI contrast agent Magnevist (gadolinium diethylenetri-aminepentaacetic acid, Gd-DTPA) at 1.0 T close to the common field of clinical

MRI.[47] Two years later, the same group synthesized a series of multi-hydroxyl lanthanoid metallofullerenols, $M@C_{82}(OH)_n$ (M = La, Ce, Gd, Dy, and Er) and studied systematically their performances as MRI contrast agents.[48] The La-, Ce-, Gd-, Dy-, and Er-metallofullerenols have R_1 values in the range of 0.8–73 (sec^{-1}mM^{-1}) at 0.47 T. In addition, Gd-, Dy-, and Er-metallofullerenols can be used as T_2-enhancing MRI contrast agents at higher magnetic fields.[48] The water-soluble carboxylated EMF derivative $Gd@C_{60}[C-(COOH)_2]_{10}$ was prepared by Bolskar et al. through the hydrolysis of the $Gd@C_{60}[C(COOCH_2CH_3)_2]_{10}$ synthesized via Bingel–Hirsch reaction.[25] The $Gd@C_{60}[C-(COOH)_2]_{10}$ as a new MRI contrast agent was studied by both *in vivo* and *in vitro* MRI measurements.[25,26] Several other different types of water-soluble derivatives of $Gd@C_{82}$ have also been synthesized by many groups. In 2004, Gu reported the nucleophilic addition of glycine esters to $Gd@C_{82}$, affording the singly bonded products with glycine methyl esters and hydroxyl.[49] In 2006, Wang group prepared an amino acid derivatives $Gd@C_{82}Om(OH)_n$ (NHCH$_2$CH$_2$COOH)$_l$ ($m \approx 6$, $n \approx 16$, and $l \approx 8$) by a direct reaction of pure $Gd@C_{82}$ with an excess of an alkaline solution of β-alanine. Both water proton relaxivity and MRI phantom indicated the efficiency of the $Gd@C_{82}$ derivative as MRI contrast agents.[50] In 2008, Wang et al. synthesized a water-soluble gadofulleride $Gd@C_{82}O_6(OH)_{16}(NHCH_2CH_2COOH)_8$, which exhibited high efficiency as MRI contrast agents by a two-step reaction under ambient conditions.[47] Later on, the same group synthesized $Gd@C_{82}O_2(OH)_{16}(C(PO_3Et_2)_2)_{10}$ as an organophosphonate functionalized derivative of $Gd@C_{82}$ by a simple synthesis procedure, just stirring the toluene mixture of NaH and tetraethyl methylenediphosphonate at room temperature.[51]

6.1.2 CHEMICAL REACTIONS OF DI-EMFs

The chemical reactions of mono-EMFs spread to a large range of different reactions, but the reactions of the di-EMFs are relatively limited. As shown in Figure 6.4, only silylation, carbine addition, and Prato reaction have been performed on di-EMFs.

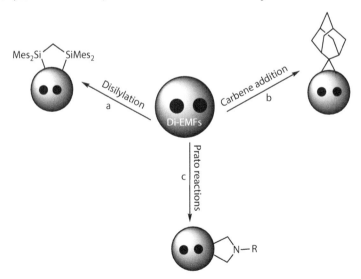

FIGURE 6.4 Chemical reactions that have been performed on di-EMFs $M_2@C_{2n}$.

6.1.2.1 Disilylation Reactions of di-EMFs

The reaction of $La_2@C_{80}$ and $Sc_2@C_{84}$ with 1,1,2,2-tetramesityl-1,2-disilirane was reported by Akasaka et al. under irradiation ($\lambda > 400$ nm) at 20°C to obtain 1:1 adducts, whereas thermal addition of disilirane to $Sc_2@C_{82}$ was restrained (Figure 6.4a).[52] The silylated $Ce_2@C_{80}$ adducts were synthesized in toluene solution of $Ce_2@C_{80}$ and 1,1,2,2-tetrakis(2,4,6-trimethylphenyl)-1,2-disilirane at 80°C for 4 hr.[53] X-ray crystal structure of the silylated $Ce_2@C_{80}$ adduct revealed that the free random motion of two Ce atoms was fixed at specific positions by exohedral chemical functionalization. This was the first experimental evidence for the control of the motion of encapsulated atoms inside the fullerene cage.[53] Another new $Ce_2@C_{78}$ bis-silylated derivative was successfully prepared at 80°C for 85 hr. The X-ray crystallographic analysis clearly confirmed that regioselective 1,4-addition of disilirane to $Ce_2@C_{78}$ takes place and the two Ce atoms stand still by facing toward the hexagonal rings at the equator.[54]

6.1.2.2 Carbene Addition Reactions of di-EMFs

The study of photochemical carbene addition to the class of di-EMFs (Figure 6.4b) was also reported by Akasaka group. The reactions of $La_2@(D_{3h})C_{78}$ and adamantylidene were carried out under two different conditions, in 2008.[54] One was that the blend solution refluxed for 30 min, whereas another was photochemical reaction ($\lambda > 300$ nm, 60 s) in a quartz tube. Three isolated isomers of $La_2@C_{78}$-Ad had C_s symmetry; one of the isomers had an open structure with two La atoms on the C_3 axis of $La_2@C_{78}$ (D_{3h}). The addition occurred at both [5,6]- and [6,6]-junctions around the pole and the equator of $La_2@C_{78}$.[55] The photochemical reaction of $M_2@C_{80}$ (M = La and Ce) with adamantylidene afforded the corresponding adducts by carbene addition. Crystallographic data for the adduct $La_2@C_{80}$-Ad revealed that the two La atoms are collinear with the spiro carbon of the [6,6]-open adduct (Figure 6.5a), whereas $Ce_2@C_{80}$-Ad indicated that the two Ce atoms were also collinear with the spiro carbon at room temperature in solution.[56] The photochemical reaction was also studied on a non-IPR di-EMF $La_2@D_2(10611)$-C_{72}, giving six isomers of monoadducts in 2008.[57] Interestingly, the X-ray analyses of the major monoadduct isomers revealed that the fused-pentagon site are very reactive toward carbene addition. However, the carbon atoms forming the [5,5] junctions were not reactive because of the bonding to the encaged metal.[57] X-ray structure of the bisadduct revealed that the fused-pentagon parts (Figure 6.5b) are still very reactive so that the Ad groups tend to add to the two poles of the molecule. In 2011, Akasaka et al. described bis-functionalization of EMF $La_2@C_{80}$ by carbene addition using different carbene reagents to afford *heterobisadduct* with high yield and high regioselectivity at each stage. Both carbene additions took place at the [6,6]-bond junction, along with bond cleavages on the C_{80} cage to afford the open-cage structures.[58] Guldi et al. reported for the first time the synthesis of a novel electron donor–acceptor conjugate $Ce_2@I_h$-C_{80}-ZnP by carbene addition reactions, and the results shed new light on the nature of $Ce_2@I_h$-C_{80} in charge transfer chemistry.[59]

6.1.2.3 Prato Reaction of di-EMFs

Akasaka et al. reported the Prato reaction of a di-EMF $La_2@C_{80}$ with 3-triphenylmethyl-5-oxazoli-inone (NTrt) yielding both [5,6]- and [6,6]-adducts in 2006 (Figure 6.4c).[60] The structure of the [6,6]-adduct was fully determined by means of X-ray crystallographic

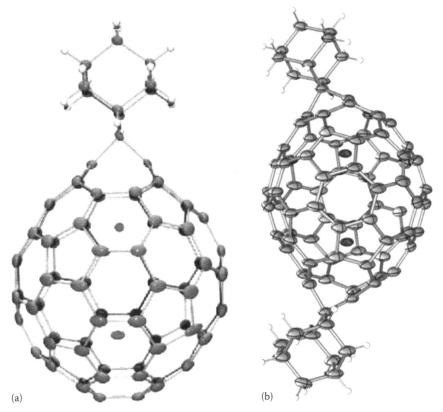

(a)

(b)

FIGURE 6.5 X-ray structures of (a) $La_2@C_{80}$-Ad and (b) $La_2@C_{72}$-Ad$_2$. (Reprinted with permission from Yamada, M. et al. *J. Am. Chem. Soc.* 130(4), 1171–1176, 2008. Copyright 2008 American Chemical Society.)

analysis, revealing that the two originally moving La atoms are fixed in the [6,6]-adduct.[60] The same reaction of $Ce_2@C_{80}$ was reported in 2009.[61] The structure of the [6,6]-adducts was fully determined by means of X-ray crystallographic analysis, and the two metal atoms were fixed at the slantwise positions on the mirror plane in the [6,6]-pyrrolidino-EMF.[61] Akasaka et al. synthesized a donor–acceptor conjugate of lanthanum EMF and π-extended tetrathiafulvalene by highly regioselective 1,3-dipolar cycloadditions of exTTF-containing azomethine ylides to the endofullerene.[62]

6.1.3 THE CHEMICAL REACTIONS OF CARBIDE CLUSTER EMFS

Actually, many carbide cluster EMFs have been erroneously assigned as conventional EMFs in earlier studies.[63] $Sc_3@C_{82}$ has been confirmed to be $Sc_3C_2@C_{80}$. To make single crystals of this EMF suitable for X-ray crystallographic study, $Sc_3C_2@C_{80}$ was functionalized with adamantylidene carbene (Ad) and the X-ray results of the isolated adduct unambiguously identified the correct structure.[64] Similarly, $Sc_2C_2@C_{3v}(8)$-C_{82}, which was previously proposed as $Sc_2@C_{84}$ isomer, was functionalized with Ad and the isolated derivative was characterized with X-ray crystallography.[65] Lu et al. reported the chemical property of $Sc_2C_2@C_s(6)$-C_{82} using 3-triphenylmethyl-5-oxazolidinone which provided a 1,3-dipolar reagent under heating (Figure 6.6). It was surprising to find that only one monoadduct isomer was formed in the reaction, regarding the low cage symmetry of this endohedral, which contained 44 types of nonequivalent cage carbons.[66] Single-crystal X-ray results of $Sc_2C_2@C_s(6)$-$C_{82}N(CH_2)_2Trt$ showed that the addition location was at a [6,6]-bond junction, which was far from either of the two Sc atoms (Figure 6.7a).[66] The important

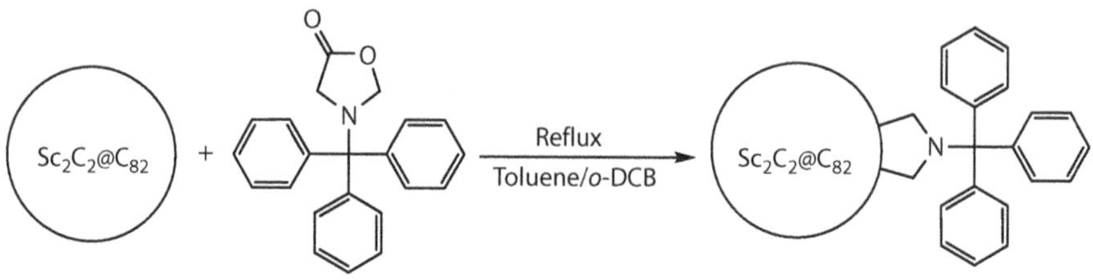

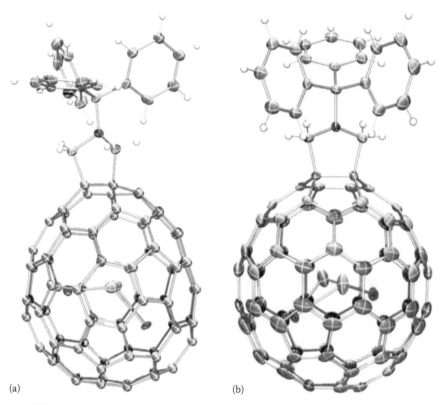

FIGURE 6.6   The 1,3-dipolar cycloaddition of $Sc_2C_2@C_{82}$.

(a)

(b)

FIGURE 6.7   X-ray structures of (a) $Sc_2C_2@Cs(6)$-$C_{82}N(CH_2)_2Trt$ and (b) $Sc_2C_2@$ $C_{2v}(9)$-$C_{82}N(CH_2)_2Trt$.

point was that this $Sc_2@C_{84}$ isomer was unambiguously identified as a new carbide cluster EMF $Sc_2C_2@C_s(6)$-C_{82} previously.[66] The next year, another $Sc_2@C_{84}$ isomer $Sc_2@C_{2v}(17)$-C_{84} was identified as new carbide cluster EMF $Sc_2C_2@C_{2v}(9)$-C_{82} and was functionalized by the 1,3-dipolar reaction, affording three monoadduct isomers. X-ray data of the most abundant adduct was confirmed to be a [5,6]-closed structure, whereas the other two minor adducts were seemingly [6,6]-isomers (Figure 6.7b).[67]

6.1.4 SUPRAMOLECULAR COMPLEXES OF CONVENTIONAL EMFS

The electronic features of EMFs are more complicated because of the electron transfer from the internal metallic species to the fullerene cage. In this regard, the bonding character between the encapsulated metal ions and the cage carbons was mainly ionic. Accordingly, the EMF molecules were considered as a kind of salt which were not dissociated upon solvation.

Supramolecular complexes of EMFs with macrocyclic compounds such as calixarenes, crown ethers, and porphyrins were also synthesized by several groups. The first example of host–guest supramolecular complexes of EMFs is the one prepared by Yang and Yang in 2002.[68] When they mixed $Dy@C_{82}$ and p-tert-butylcalix[8]arene (C8A), a black precipitate ($Dy@C_{82}$-C8A) was formed with a yield of 33% after stirring for 72 h at room temperature, but the yield was much lower than that of the (C_{60}-C8A) complex (100%). This was explained by the larger size of $Dy@C_{82}$ (8.0 Å) than that of C_{60} (7.1 Å) and the entropy effect which was associated with the loss of the rotational freedom of EMF upon complexation with the calixarene.[68] In 2006, Akasaka et al. studied the complexation of $La@C_{82}$-$C_{2v}(9)$ with other macrocyclic compounds, such as 1,4,7,10,13,16-hexaazacyclooctadecane, 1,4,7,10,13,16-hexam-ethyl-1,4,7,10,13,16-hexaaza-cyclooctadecane, monoaza-18-crown-6 ether, 18-crown-6 ether, and p-tert-butylcalix[n]arenes ($n = 4$–8).[69] In solution, $La@C_{82}$ formed complexes only with azacrown ethers, and the complex formation was accompanied by the electron transfer to the $La@C_{82}$ as the result of the low reduction potentials of the EMF. The authors successfully applied such a complexation approach to the selective separation of EMFs from the soot extract.[69] In the same year, this group also reported that the 15-, 18-, 21- and 24-membered unsaturated thiacrown ethers form 1:1 complexes with $La@C_{2v}(9)$-C_{82} in solution via an electron transfer process. The 21-membered unsaturated thiacrown ethers had the best ring-size for the complexation with this metallofullerene.[70] Another organic donor molecule, N,N,N',N'-tetramethyl-p-phenylenediamine (TMPD) with $La@C_{82}$ in nitrobenzene was studied by the standard titration technique using Vis-NIR spectroscopy, which formed a stable complex and behaved as the reversible intermolecular spin-site exchange systems in solution.[71] In 2010, Shinohara and coworkers reported the formation of the 1:1 noncovalent complex of $La@C_{82}$ with isophthaloyl-bridged porphyrin dimer.[72] The association ability of porphyrin dimer and $La@C_{82}$ was based exclusively on cooperative supramolecular forces, namely, structural noncovalent interactions.[72] Paramagnetic cyclo-[PCu]$_2$ was an interesting host molecule that could be ferromagnetically coupled with paramagnetic $La@C_{82}$ (Figure 6.8). This ferromagnetic nature turned ferrimagnetic when cyclo-[PCu]$_2$⊃$La@C_{82}$ was transformed into its caged analogue (cage-[PCu]$_2$⊃$La@C_{82}$).[73] The same year, Fukuzumi et al. studied supramolecular complexes of $Li^+@C_{60}$ with tetrathiafulvalene calix[4]pyrroles.[74] Binding of chloride anion to

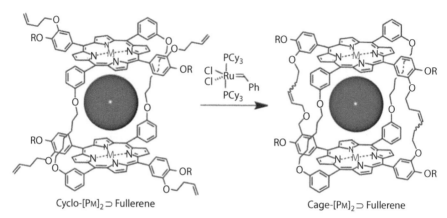

FIGURE 6.8 The structures of cyclo-[PM]$_2$⊃Fullerene and cage-[PM]$_2$⊃Fullerene [R = C$_{12}$H$_{25}$; M = 2H, Cu, Zn; Fullerene (blue-colored sphere) = La@C$_{82}$] and their transformation by Ring-Closing Metathesis. (Reprinted with permission from Lu, X. et al., *Angew. Chem. Int. Ed. Engl.*, 51(24), 5889–5892, 2012. Copyright 2011 American Chemical Society.)

a tetrathiafulvalene calix[4]pyrrole (TTF-C$_4$P) donor resulted in electron transfer to Li$^+$@C$_{60}$ to produce the radical pair (TTF-C$_4$P•$^+$/Li$^+$@C$_{60}$•-), which were characterized by X-ray crystallographic analysis.[74] Two different lanthanum metallofullerene derivatives La@C$_{82}$Py with respective open-shell electronic and closed-shell electronic configuration were synthesized by Akasaka et al. in 2013 (Figure 6.9, see further in the text). The La@C$_{82}$Py and zinc tetraphenylporphyrin (ZnP) were formed as the noncovalent electron donor–acceptor coordinative hybrid.[24]

6.2 CHEMICAL PROPERTIES OF CLUSTER ENDOHEDRAL FULLERENES

Maira R. Cerón, Marta Izquierdo, and Luis Echegoyen

Metals, metal clusters, and small molecules can be encapsulated inside fullerenes. In this chapter, we will focus on cluster EMFs, which have attracted increasing attention since the discovery of the third most abundant fullerene (after C$_{60}$ and C$_{70}$), Sc$_3$N@I$_h$-C$_{80}$ in 1999.[75]

Exohedral functionalization is crucial to improve the solubilities of endohedral fullerenes in order to explore their properties and potential applications. In this chapter, different types of functionalization reactions on cluster endohedral fullerenes, such as Diels–Alder reactions, [2 + 2] cycloadditions, sylilations, carbene additions, 1,3-dipolar cycloadditions, Bingel–Hirsch reactions, and electrosynthesis, among others, are reviewed.

6.2.1 Pericyclic Reactions: [4 + 2] and [2 + 2] Cycloadditions

6.2.1.1 [4 + 2] Cycloadditions

After the production of trimetallic nitride EMFs in high yields in 1999,[75] the first exohedral functionalization was reported in 2002 by Iezzi et al. (Scheme 6.1).[76]

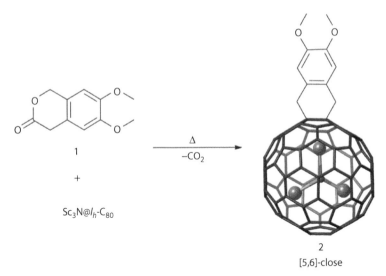

2

[5,6]-close

SCHEME 6.1 Diels-Alder reaction of 6,7-dimethoxyisochroman-3-one and $Sc_3N@I_h$-C_{80} to give rise compound **2**.

A solution of 6,7-dimethoxyisochroman-3-one (99% ^{13}C labeled) **1** and $Sc_3N@$ I_h-C_{80} in 1,2,4-trichlorobenzene (TCB) was heated at reflux for 24 hr (Scheme 6.1) to obtain the monoadduct **2**. The ^{13}C NMR spectrum exhibited a single signal for the equivalent methylene carbons, indicating the presence of a plane of symmetry. Therefore, the Diels–Alder reaction occurs at a [5,6] double bond ring junction. Crystallographic characterization was reported later, which confirmed the suggested structure.[77] The C—C bond at the addition site is elongated and pulled out from the fullerene cage and the Sc_3N cluster is positioned away from the addition site.

The Diels–Alder reaction was also used as a separation technique to isolate the I_h isomer of both $Lu_3N@C_{80}$ and $Sc_3N@C_{80}$ from their corresponding D_{5h} isomer.[78]

6.2.1.2 [2 + 2] Cycloaddition

In 2011, Li et al. reported the [2 + 2] cycloaddition of $Sc_3N@I_h$-C_{80} with benzyne to afford both [5,6]- and [6,6]- monoadducts.[79] A solution of $Sc_3N@I_h$-C_{80}, isoamyl nitrite **3**, and anthranilic acid **4** in ortho-dichlorobenzene (o-DCB) and toluene was stirred at room temperature under argon for 3 hr to obtain compounds **5** and **6** at 70% and 15% yields, respectively, based on consumed $Sc_3N@I_hC_{80}$ (Scheme 6.2).

NMR experiments showed a symmetric addition pattern for compound **5** and an unsymmetrical one for compound **6**, which suggest [5,6] and [6,6] addition patterns respectively. The crystal structure of both regioisomers confirmed the assignment based on spectroscopic data and showed that for both compounds the C—C bond at the addition site was elongated from 1.43 ± 0.01 Å to 1.65 ± 0.01 Å, but not broken. The Sc atoms are located away from the addition sites, although the Sc_3N cluster position differs for both isomers. The electrochemical behavior of the [6,6] isomer was reported for the first time, which exhibited an unexpected electrochemically reversible cathodic behavior.[79]

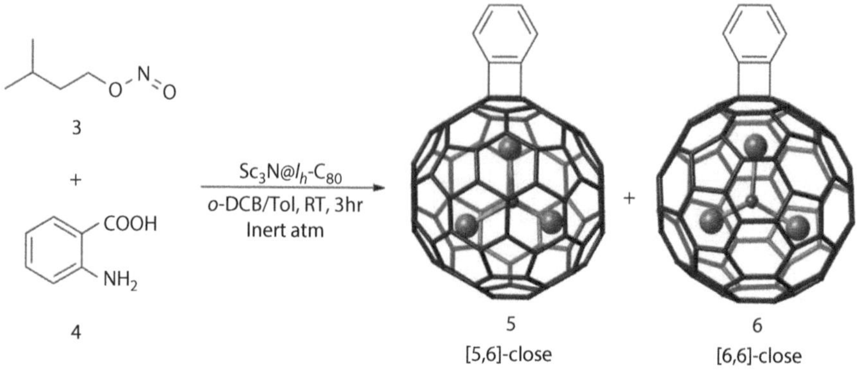

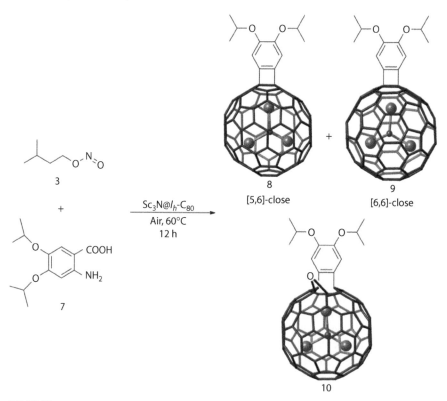

SCHEME 6.2 Anaerobic [2 + 2] cycloaddition reaction on $Sc_3N@I_h$-C_{80}.

The [2 + 2] cycloaddition on $Sc_3N@I_h$-C_{80} was also explored by Wang et al. using 4,5 diisopropoxybenzyne generated *in situ*.[80] When the reaction was carried out under inert atmosphere, the expected benzyne adducts **8** ([5,6] regioisomer) and **9** ([6,6] regioisomer) were obtained. However, when the reaction was conducted under aerobic conditions and 25 equivalents of water, a new open-cage compound was formed **10** (Scheme 6.3).

SCHEME 6.3 Aerobic [2 + 2] cycloaddition reaction on $Sc_3N@I_h$-C_{80}.

Compound **8** was completely transformed to **10** when more than 8 equivalents of isoamyl nitrite **3** were used in the reaction.

6.2.2 1,3-DIPOLAR CYCLOADDTION

In 2005, Cardona et al. reported the first and most studied fulleropyrrolidine derivative of a cluster endohedral fullerene, resulting from the addition of the N-ethylazomethine ylide **13** to $Sc_3N@I_h$-C_{80} to yield regioselectively a [5,6]-pyrrolidine derivative **14a** (Scheme 6.4),[81,82] exactly in the same position as that described previously for a Diels–Alder adduct of the same cluster fullerene.[76] Later Cai et al. also reported the [5,6]-fulleropyrrolidine of $Sc_3N@I_h$-C_{80} and $Er_3N@I_h$-C_{80} **14a**.[83]

Interestingly, the same reaction conditions initially led to a [6,6]-pyrrolidine addition pattern on the analogous Y-based compound, $Y_3N@I_h$-C_{80} (Scheme 6.4).[84] Further studies by Cardona et al. showed that the [6,6]-adduct of $Y_3N@I_h$-C_{80} corresponded to the kinetic product, which eventually isomerizes to the thermodynamically preferred [5,6] isomer derivative **14b** (Scheme 6.4).[85,86] The process is rather slow, with an unimolecular rate constant $k = 0.01$ min^{-1} at 145°C.[85] Similar behavior was reported by Cai et al. for the isomerization of the [6,6]-kinetic adduct to the [5,6]-thermodynamic adduct of $Sc_3N@I_h$-C_{80}.[87] Unlike I_h, the D_{5h} isomer of $Sc_3N@C_{80}$ yields two adducts, a thermodynamically stable [6,6]-symmetric product **14c** (Scheme 6.4) and a [6,6]-unsymmetric one that is partially converted to other unidentified monoadduct isomers.[78]

An interesting example of the effect of the size of the encapsulated cluster on the addition pattern was reported by Chen et al. who showed that when the number of Gd atoms increases in the series $Sc_xGd_{3-x}N@I_h$-C_{80} ($x = 0$–3), the relative amount of the [6,6]-isomer also increases; going from exclusively the [5,6]-isomer in $Sc_3N@I_h$-C_{80} (**14a**) to only the [6,6]-isomer in $Gd_3N@I_h$-C_{80} (**14c**).[88]

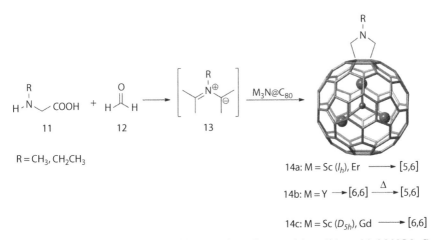

$R = CH_3, CH_2CH_3$

14a: M = Sc (I_h), Er ⟶ [5,6]

14b: M = Y ⟶ [6,6] $\xrightarrow{\Delta}$ [5,6]

14c: M = Sc (D_{5h}), Gd ⟶ [6,6]

SCHEME 6.4 1,3-dipolar cycloaddition reaction of azomethine ylides with $M_3N@I_h$-C_{80} (M = Sc (I_h and D_{5h}), Er, Y, Gd).

Recently. Aroua et al. reported the study of the isomerization kinetics of the conversion from [6,6] to the [5,6] isomer, as a function of the size of the trimetallic nitride inside of the I_h-C_{80} cage (M = Gd^{3+} > Y^{3+} > Lu^{3+} > Sc^{3+}). This article showed that larger cluster sizes lead to slower isomerization rates as was the case of Gd^{3+} and Y^{3+}. The reason for this behavior could be attributed to more stabilization of the [6,6]-adduct via the pyramidalization of the sp^3 carbons at the [6,6]-junction observed for the larger metal clusters. The kinetic study indicated that the initial 1,3-dipolar cycloaddition occurred at a [6,6]-junction followed by a thermal sigmatropic rearrangement to a [5,6]-junction. In the case of Gd^{3+} and Y^{3+}, the isomerization seems to be reversible and equilibria are attained, where both [6,6] and [5,6]-adducts are observed simultaneously.[89]

Other examples of 1,3-dipolar cycloadditions have been reported for $Sc_3N@C_{78}$,[90] $Sc_{3-x}Y_xN@C_{80}$[91], and $Sc_2C_2@C_{82}$.[66] Several examples of donor–acceptor systems using 1,3-dipolar cycloaddition reactions of azomethine ylides were also reported for $Sc_3N@I_h$-C_{80} and $Y_3N@I_h$-C_{80} as acceptors and ferrocene,[92] phorphyrin,[93] or triphenylamine as donors.[94]

6.2.3 METHANO DERIVATIVES

6.2.3.1 Bingel–Hirsch Reaction

In 2005, Cardona et al. reported the first Bingel–Hirsch reaction on a cluster endohedral fullerene.[84a] The reaction of $Y_3N@I_h$-C_{80} with an excess of ethyl bromomalonate **15** with 1,8-Diazabicyclo[5.4.0]undec-7-ene (DBU) in o-DCB afforded compound **16a** (Scheme 6.5). If the reaction takes place at a [6,6] junction, the ethyl groups would be symmetric by ^1H NMR; however, if the addition occurs at a [5,6] junction, the ethyl groups would be nonequivalent (Figure 6.9).

After 10 min of reaction at room temperature, the monoadduct was formed together with traces of bisadducts. The ^1H NMR showed only one set of signals for the ethyl groups, indicating selectivity for the [6,6]-addition pattern. This was later confirmed by X-ray crystallography, which showed an open-cage fulleroid structure.[95] When the same conditions were used with $Sc_3N@I_h$-C_{80}, no products were detected, clearly showing drastically different reactivities for these two endohedral fullerenes.

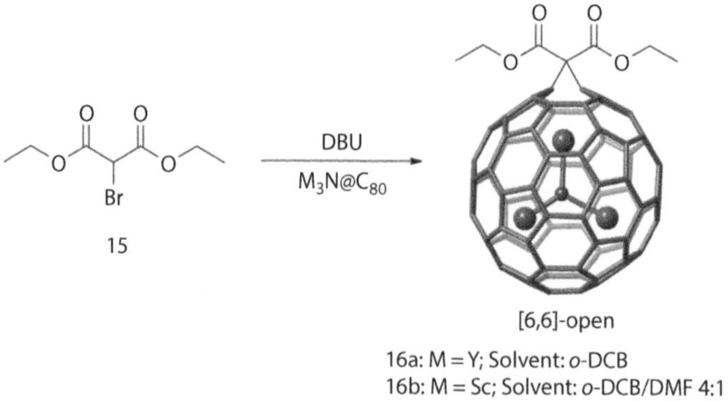

[6,6]-open

16a: M = Y; Solvent: o-DCB
16b: M = Sc; Solvent: o-DCB/DMF 4:1

SCHEME 6.5   Bingel-Hirsch reaction on $Y_3N@I_h$-C_{80} and $Sc_3N@I_h$-C_{80}.

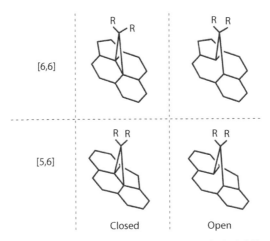

FIGURE 6.9 Possible cycloaddition products of cluster endohedral fullerene.

Cai et al. explored the same reaction with $Sc_3N@C_{78}$, which produced one monoadduct, presumably attached at a closed-[6,6] bond.[96] When the reaction was run with $Sc_3N@C_{68}$ (non-IPR endofullerene), the addition occurred at the adjacent bond of the [5,5] junction.[97] This indicates that the [5,5]-ring junctions of non-IPR metallofullerenes are very stable, probably because of a strong electrostatic interaction with the clusters inside.

The cage size effect on the Bingel–Hirsch reaction was studied by Chaur et al. with samples of $Gd_3N@C_n$ (n = 80, 84, and 88).[98] The reaction with $Gd_3N@I_h$-C_{80} under standard conditions afforded both monoadduct and bisadducts almost instantaneously on open-[6,6] bonds. For $Gd_3N@C_{84}$ and $Gd_3N@C_{82}$ under identical conditions as those described for $Gd_3N@I_h$-C_{80}, after 20 min of reaction yielded only monoadducts, no further addition was observed demonstrating their lower reactivity compared with that for $Gd_3N@I_h$-C_{80}. In the case of the largest cage, $Gd_3N@C_{88}$, the reaction was unsuccessful, and no products were obtained, which indicates that the reactivity increases as the cage size decreases, due to the higher degree of pyramidalization at the C atoms and the consequent increase in cage strain.[99]

A successful Bingel reaction with $Sc_3N@I_h$-C_{80} was not accomplished until 2010 by Pinzon et al., who reported a modification of the reaction conditions to obtain the methano derivative in relatively good yields. The presence of dimethylformamide (DMF) in the solvent mixture allowed the isolation of compound **16b** (Scheme 6.5). The reason why this modified reaction condition works is that the DMF stabilizes the transition state during the elimination step.[100,101]

6.2.3.2 Phenyl Butyric Acid Esters Type Functionalization

The thermolysis of tosylhydrazones in the presence of strong bases produces diazocompounds *in situ*, a protocol known as the Bamford–Stevens reaction. Apparently, both carbene as well as [3 + 2] cycloadditions followed by N_2 extrusion can occur with this reaction. An important example is phenyl-C_{81}-butyric acid esters of $M_3N@$ I_h-C_{80} (M = Lu, Sc, and Y) (Scheme 6.6).[102]

17

[6,6]-open

18a: M = Lu, R = Hexyl
18b: M = Sc, R = Me
18c: M = Y, R = Me

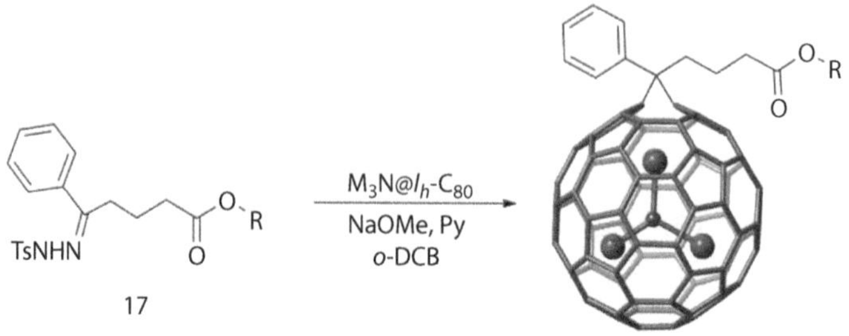

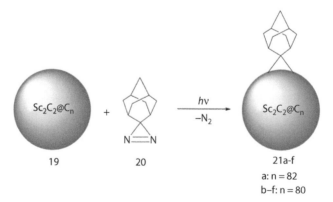

SCHEME 6.6 Synthesis of PCBR derivatives of $M_3N@I_h$-C_{80} (M = Sc, Lu, Y).

This reaction uses the general procedure implemented for C_{60} or C_{70}, giving rise in all cases to a [6,6]-open addition pattern. An example is PCBH-$Lu_3N@C_{80}$ (**18a**), which is the first and currently the only endohedral derivative used as an acceptor material in an organic photovoltaic (OPV) device.[102b]

6.2.3.3 Carbene Additions

Diazo compounds are known to be effective carbene precursors and carbenes are one of the most useful intermediates used to functionalize fullerenes. In 2005, Iiduka et al. confirmed that $Sc_2@C_{84}$(III) is actually a scandium carbide metallofullerene ($Sc_2C_2@C_{82}$(III)), **19**.[103] Given that spectroscopic data was somewhat controversial, X-ray single-crystal analysis was essential to confirm the endohedral carbide structure. Derivatization of $Sc_2C_2@C_{82}$(III) was achieved by photoreaction with 2-adamantane-2,3-[3H]-diazirine **20** (50 equiv) at room temperature using a high-pressure mercury-arc lamp (cutoff < 350 nm) for 35 s which afforded compound **21a** ([5,6]-open regioisomer) (Scheme 6.7). The crystal structure of compound **21a** showed C_1 symmetry, which results from a C_{3v} carbon cage C_{82} (not C_{84}).

19 20 21a-f

a: n = 82
b-f: n = 80

SCHEME 6.7 Synthesis of compounds **21a-f** $Sc_2C_2@C_n$(Ad) (n = 80,82).

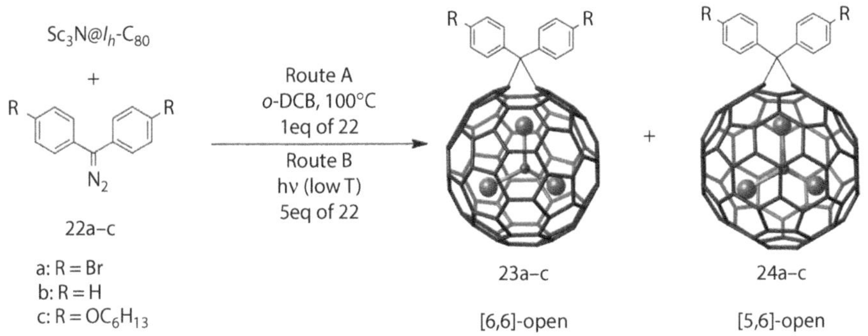

SCHEME 6.8 Synthesis of DPM derivatives of $Sc_3N@I_h-C_{80}$ (**23** and **24**) under photoirradation and thermal conditions.

The same reaction was used with $Sc_2C_2@C_{2v}(5)-C_{80}$, and in this case, five regioisomers were detected (**21b–f**); four of them are [6,6]-open regioisomers and only one is an unstable [5,6]-regioisomer which isomerizes upon heating to a [6,6]-open regioisomer.[104] The preferred addition occurs at a cage region on top of the cluster, and the Sc atoms are mobile but the C2-unit is practically motionless.

The [6,6]-junction is favored for methano-bridge formation on I_h-C_{80},[105] but recently, Izquierdo et al. reported the first [5,6]-methanoadduct of $Sc_3N@I_h-C_{80}$, resulting from the reaction with diaryl-diazocompound reagent, **22** (Scheme 6.8).

Photoirradiation of the carbene precursor **22** (5 equiv) in an o-DCB solution containing $Sc_3N@I_h-C_{80}$ for 10 min at 0°C (Scheme 6.8, route B) yielded compounds **23** and **24**. ^1H NMR showed that compound **23** has a [6,6]-addition pattern, whereas compound **24** corresponds to the [5,6] isomer. The ratio of the [5,6]:[6,6] isomers depended significantly on the nature of the *para* substituent on the aromatic ring.[106]

6.2.4 SILYLATION

The addition of heteroatoms such as electropositive silicon to EMFs has helped to study the changes in the electronic characteristics, such as the increase in the negative charge on the cage in silylated fullerene derivatives and the movement of the encapsulated metal atoms as a function of the number of silicon atoms on the surface.[107] Sato et al. reported the reaction of 9,9-bis(2,6-diethylphenyl)-9-silabicyclo[6.1.0]nonane, **25** (which can efficiently generate a silylene under both thermal and photochemical conditions) with $Lu_3N@I_h-C_{80}$ which led to two isomers of $Lu_3N@I_h-C_{80}(SiDep_2)$, **26** and **27** (Scheme 6.9). Unexpectedly, the [6,6]-closed $Lu_3N@I_h-C_{80}(SiDep_2)$ **26** derivative isomerizes under ambient light to a new [5,6]-open-silylene-bridged silafulleroid structure **27**. The formation of the [5,6]-open $Lu_3N@I_h-C_{80}(SiDep_2)$ **27** can be avoided if the reaction is conducted in the dark under thermal conditions.[108] As predicted, monosilylation is effective to fine-tune the electronic properties of endohedral metallofullerenes as well as empty fullerenes.

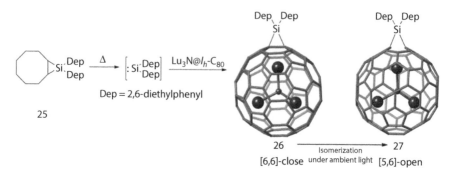

SCHEME 6.9 Synthesis of Lu$_3$N@I_h-C$_{80}$(SiDep$_2$) **27** using photochemical or thermal conditions.

Photochemical reactions of cluster endohedral fullerenes with disiliranes yield several isomeric adducts as reported by Sato et al. for the reaction of Lu$_3$N@I_h-C$_{80}$ with disilirane **28** (Scheme 6.10).[109]

There are two kinds of nonequivalent carbon atoms on the I_h-C$_{80}$ cage; therefore, the possible addition sites for 1,2- and 1,4-addition of disiliranes are 1,2(AA); 1,2(AB); 1,4(AA); and 1,4(BB) (Figure 6.10a). The disilirane addend can also adopt three conformations (Figure 6.10b); consequently, the number of possible isomers is 18 for an I_h-C$_{80}$ cage. Analysis of the two Lu$_3$N@I_h-C$_{80}$ bis-silylated monoadducts showed that one of the products isomerized under ambient light to a more stable 1,4(AA)-adduct **29a**, which consists of paired twist conformers at room temperature.

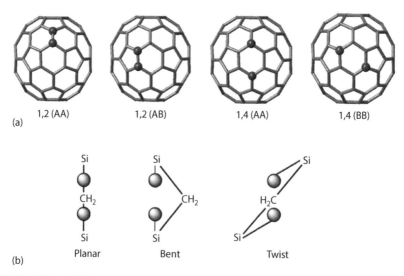

FIGURE 6.10 Addition patterns of disilirane on cluster endohedral fullerenes. (a) Four possible addition sites for 1,2-addition and 1,4-addition. (b) Three possible conformations of the disilirane moiety.

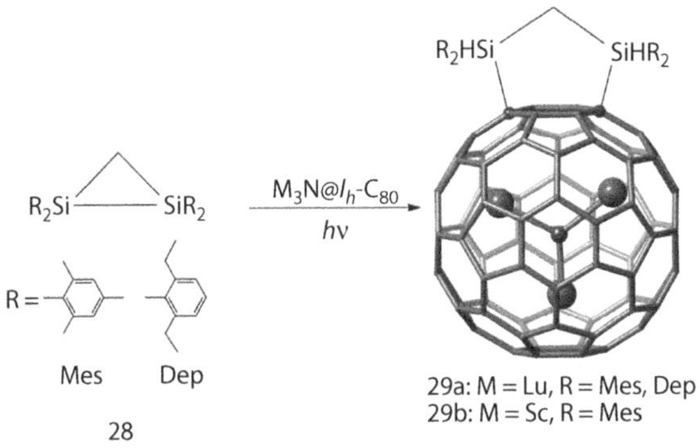

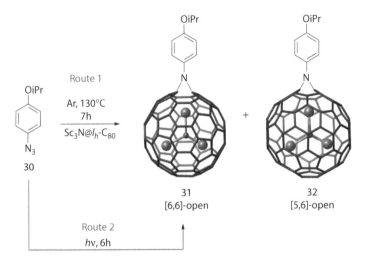

SCHEME 6.10 Synthesis of 1,4(AA)-$M_3N@I_h$-$C_{80}(R_2Si)_2CH_2$ derivatives.

The same reaction with $Sc_3N@I_h$-C_{80} yielded two isomers, ^{13}C NMR spectrum revealed that one of the isomers results from a 1,2-addition of the disilirane addend to obtain a closed structure (1,2(AA)-closed), and the second isomer results from a 1,4-addition (1,4(AA)-adduct). As was observed with the $Lu_3N@I_h$-C_{80}, one of the isomers thermally isomerizes to a mixture of 1,4(AA)-conformers **29b**, which means that the 1,2(AA)-adduct is thermodynamically less stable than the 1,4(AA)-adduct but is kinetically more favorable.

6.2.5 AZIDE ADDITIONS

The synthesis of aza-bridged metallofullerenes was first reported by Liu et al. in 2012.[110] The reaction of 4-isopropoxyphenyl azide **30** (60 equivalents) with $Sc_3N@$ I_h-C_{80} in o-DCB at 130°C for 7 hr gave rise to compound **31** ([6,6]-open regioisomer)

SCHEME 6.11 Synthesis of aza-bridged metallofullerenes.

and compound **32** ([5,6]-open regioisomer) in 13% and 7% yields, respectively (Scheme 6.11). When the reaction occurs at room temperature under irradiation ($\lambda > 290$ nm), compound **31** is the only product observed.

The isomerization of compounds 31 and 32 was also explored. Heating pure 31 (130°C, 7 hr) gave rise to a mixture of 31 and 32 in a ratio of 44:56. In contrast, when pure 32 was heated (130°C, 14 hr), a mixture of 31 and 32 was found in a 16:84 ratio. Consequently, regioisomer 31 is the kinetic product and regioisomer 32 is the thermodynamic product.

6.2.6 ELECTROSYNTHESIS

Electrochemical synthesis using cluster endohedral fullerenes is a specialized method used for the selective synthesis of endohedral fullerene derivatives that are otherwise not accessible by typical synthetic procedures. This type of reaction involves anionic fullerene species generated using electrochemical reduction which act as nucleophiles to produce methano derivatives with high regioselectivity. Li et al. recently reported that the reactivities of anionic cluster endohedral fullerene species depend strongly on the nature of the cluster inside of the carbon cage. They used the electroreductive synthetic method with $M_3N@I_h$-C_{80} (M = Sc, Lu) and found very different reactivities under identical conditions. $[Sc_3N@I_h$-$C_{80}]^{2-}$ **33a** was completely unreactive with benzal bromide **34**, whereas $[Lu_3N@I_h$-$C_{80}]^{2-}$ **33b** readily reacted to form the methano derivative $Lu_3N@I_h$-$C_{80}(CHC_6H_5)$ **35** (Scheme 6.12).[111]

Calculations showed that $[Sc_3N@I_h$-$C_{80}]^{2-}$ **33a** has the charge density mainly localized on the encapsulated cluster and not on the cage, and thus, the dianion exhibits low nucleophilicity, whereas the opposite is true for $[Lu_3N@I_h$-$C_{80}]^{2-}$, **33b**.[112] Computational studies suggested that $[Sc_3N@I_h$-$C_{80}]^{3-}$, **36**, should exhibit higher reactivity with electrophiles because of a much higher charge density localized on the cage. In agreement with the theoretical prediction, the nucleophilic reaction of $[Sc_3N@I_h$-$C_{80}]^{3-}$ **36** with benzal bromide (PhCHBr$_2$), **34**, produced the methano

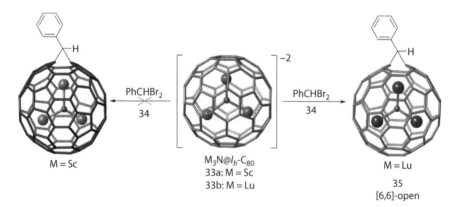

SCHEME 6.12   Nucleophilic reaction of $[M_3N@I_h$-$C_{80}]^{3-}$ with benzal bromide (PhCHBr$_2$), **34**.

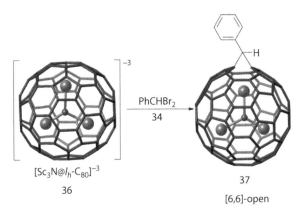

SCHEME 6.13 Nucleophilic reaction of $[Sc_3N@I_h\text{-}C_{80}]^{3-}$ with benzal bromide (PhCHBr$_2$).

derivative $Sc_3N@I_h\text{-}C_{80}(CHC_6H_5)$, **37**, in which the addend was selectively attached to a [6,6]- ring junction (Scheme 6.13).

6.2.7 OTHER ADDITION REACTIONS

In 2007, Shu et al. reported the first radical reaction to $Sc_3N@I_h\text{-}C_{80}$ using carbon radicals generated from diethyl malonate **15** catalyzed by manganese(III) acetate **38** (Scheme 6.14).[113] Two different [6,6]-ring-bridged methano derivatives were obtained, $Sc_3N@C_{80}[^{13}C(COOC_2H_5)_2]$ **39**, which corresponds to the addition of the diethyl malonate to a [6,6]-bond and $Sc_3N@C_{80}[^{13}CHCOOC_2H_5]$ **40**, which results from loss of a caboxylate group from the monoadduct **39** (Scheme 6.14). The same conditions were used with $Lu_3N@I_h\text{-}C_{80}$ to produce similar analogues. Multiadducts were also obtained for both endohedral compounds, octa-adducts in the case of $Sc_3N@I_h\text{-}C_{80}$ and deca-adducts for $Lu_3N@I_h\text{-}C_{80}$.

A different procedure to generate radicals to functionalize cluster endohedral fullerenes was later reported by Shustove et al. as well as by Yang et al. The carbon radicals were generated by two methods, either using gaseous CF$_3$I **41** or $Ag(CF_3CO_2)$ **42** at high temperatures. The CF$_3$ radicals generated reacted with the two isomers of $Sc_3N@C_{80}$ (I_h and D_{5h}) to produce a family of compounds, $Sc_3N@C_{80}(CF_3)_n$ ($n = 2–16$) **43** (Scheme 6.15a).[114] Based on the analysis of a large number of single-crystal X-ray structures of $Sc_3N@C_{80}(CF_3)_n$, it was possible to generalize that CF$_3$ groups tend to form *ribbon* addition patterns due to their preferred addition to adjacent hexagons (rarely pentagons) on the endohedral fullerene cage (Scheme 6.15b). Because of the steric congestion of bulky CF$_3$ groups, they rarely occupy vicinal (*ortho*) positions. It was also observed that the exohedral attachment influences the position of the Sc$_3$N cluster inside the C$_{80}$ cage.[115]

In 2008, Shu et al. reported a method to functionalize the less reactive cluster endohedral fullerenes such as the trimetallic nitrides, using a photochemical reaction between benzyl bromide **44** and $Sc_3N@I_h\text{-}C_{80}$, which yielded highly selectively

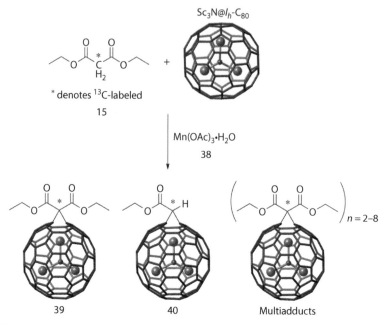

SCHEME 6.14   Radical reaction of diethyl malonate **15** and Sc$_3$N@I_h-C$_{80}$, catalyzed by manganese(III) acetate **38**.

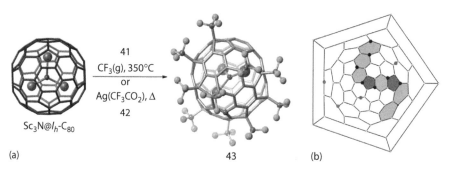

SCHEME 6.15   (a) Synthesis of Sc$_3$N@C$_{80}$(CF$_3$)$_n$ ($n = 2$–16) via radical reactions. (b) Schlegel diagrams of Sc$_3$N@I_h-C$_{80}$(CF$_3$)$_{10}$ with a *ribbon* addition pattern. Black dots denote CF$_3$-bearing carbon atoms, blue dots denote the CF$_3$-bearing triple-hexagon junctions. Green dots denote the positions of Sc atoms.

a 1,4(AA)-adduct **45**, (Figure 6.10a), that was unambiguously assigned by X-ray crystallographic analysis (Scheme 6.16). Under the same conditions, Lu$_3$N@ I_h-C$_{80}$ yielded an adduct with similar properties to those of Sc$_3$N@I_h-C$_{80}$. Analysis of the UV-Vis spectra showed that the 1,4-dibenzyl adduct of Sc$_3$N@I_h-C$_{80}$ had similar electronic properties to those reported for disilirane and trifluoromethyl derivatives.[116]

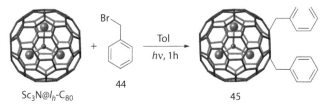

SCHEME 6.16 Photochemical synthesis of 1,4-dibenzyl adduct of $Sc_3N@I_h$-C_{80}.

6.2.8 CONCLUSIONS AND OUTLOOK

To summarize, all the known reactions of cluster endohedral fullerenes with different reagents were reviewed in the present chapter. Chemical functionalization has shown that cluster fullerenes react very differently than the corresponding empty cages, indicating a very strong influence of the encapsulated clusters, which make certain bonds more reactive toward exohedral functionalization.

In the case of [4 + 2] cycloadditions, it was observed that these occur only at [5,6] double bond ring junctions, unlike the [2 + 2] cycloadditions for which mixtures of [5,6]- and [6,6]-monoadducts are obtained. Very interesting behavior was observed for the 1,3-dipolar cycloaddition reactions, for which the ratios of [5,6]- to [6,6]-regioisomers depend on the cluster size inside the fullerene. The [5,6] isomer predominates in the case of $M_3N@I_h$-C_{80} (M = Sc, Er and Y), whereas the [6,6] isomer predominates for $Sc_3N@D_{5h}$-C_{80} and $Gd_3N@I_h$-C_{80}.

For reactions that yield methano derivatives, [6,6]-regioisomers are formed in most of the cluster endohedral fullerenes. Specifically, in I_h-C_{80}, there is only one exception reported of a [5,6]-regioisomer, which was formed when diaryl-diazocompounds were reacted. The addition of heteroatoms such as silicon and nitrogen to the endohedral metallofullerene surfaces was also discussed in this chapter, which yield silylene-bridged and aza-bridged derivatives.

Exohedral functionalization of endohedral cluster fullerenes have expanded their potential uses, such as in organic photovoltaic solar cells,[102b] as MRI contrast agents,[117] radiotherapeutic drugs[118] and single molecule magnets,[119] to name a few. However, there is still a lot to be explored, for example, the multiple additions and the effect that the clusters have in influencing the regiochemistry of the additions. There are also many potential uses of cluster fullerene derivatives that remain unexplored in biomedicine, nanoscience, and others.[120]

REFERENCES

1. Maeda, Y.; Tsuchiya, T.; Lu, X.; Takano, Y.; Akasaka, T.; Nagase, S. Current progress on the chemical functionalization and supramolecular chemistry of M@C_{82}. *Nanoscale* 2011; 3(6): 2421–2429. Epub April 13, 2011.
2. Akasaka, T.; Kato, T.; Kobayashi, K.; Nagase, S.; Yamamoto, K.; Funasaka, H.; Takahashi, T. Exohedral adducts of La@C_{82}. *Nature* 1995; 374(6523): 600–601.
3. Akasaka, T. Chemical derivatization of endohedral metallfullerene La@C_{82} with digermirane. *Tetrahedron* 1996; 52(14): 5015–5020.

4. Takeshi Akasaka, S. N.; Kobayashi, K.; Suzuki, T.; Kato, T. Exohedral derivatization of an endohedral metallofullerene Gd@C_{82}. *Journal of the Chemical Society, Chemical Communications* 1995; 1343–1344.

5. Kato, T.; Akasaka, T.; Kobayashi, K.; Nagase, S.; Yamamoto, K.; Funasaka, H. et al. ESR study on the reactivity of two isomers of LaC$_{82}$ with disilirane. *Applied Magnetic Resonance* 1996; 11(2): 293–300.

6. Yamada, M.; Feng, L.; Wakahara, T.; Tsuchiya, T.; Maeda, Y.; Lian, Y. F. et al. Synthesis and characterization of exohedrally silylated M@C_{82} (M = Y and La). *Journal of Physical Chemistry B* 2005; 109(13): 6049–6051.

7. Maeda, Y.; Miyashita, J.; Hasegawa, T.; Wakahara, T.; Tsuchiya, T.; Feng, L. et al. Chemical reactivities of the cation and anion of M@C_{82} (M = Y, La, and Ce). *Journal of the American Chemical Society* 2005; 127(7): 2143–2146.

8. Akasaka, T.; Okubo, S.; Kondo, M.; Maeda, Y.; Wakahara, T.; Kato, T. et al. Isolation and characterization of two Pr@C_{82} isomers. *Chemical Physics Letters* 2000; 319(1–2): 153–156.

9. Maeda, Y.; Matsunaga, Y.; Wakahara, T.; Takahashi, S.; Tsuchiya, T.; Ishitsuka, M. O. et al. Isolation and characterization of a carbene derivative of La@C_{82}. *Journal of the American Chemical Society* 2004; 126(22): 6858–6859.

10. Akasaka, T.; Kono, T.; Matsunaga, Y.; Wakahara, T.; Nakahodo, T.; Ishitsuka, M. O. et al. Isolation and characterization of carbene derivatives of La@C_{82}(Cs). *Journal of Physical Chemistry A* 2008; 112(6): 1294–1297.

11. Akasaka, T.; Kono, T.; Takematsu, Y.; Nikawa, H.; Nakahodo, T.; Wakahara, T. et al. Does Gd@C_{82} have an anomalous endohedral structure? Synthesis and single crystal X-ray structure of the carbene adduct. *Journal of the American Chemical Society* 2008; 130(39): 12840–12841.

12. Lu, X.; Nikawa, H.; Feng, L.; Tsuchiya, T.; Maeda, Y.; Akasaka, T. et al. Location of the yttrium atom in Y@C_{82} and its influence on the reactivity of cage carbons. *Journal of the American Chemical Society* 2009; 131(34): 12066–12067.

13. Takano, Y.; Aoyagi, M.; Yamada, M.; Nikawa, H.; Slanina, Z.; Mizorogi, N. et al. Anisotropic magnetic behavior of anionic Ce@C_{82} carbene adducts. *Journal of the American Chemical Society* 2009; 131(26): 9340–9346.

14. Hachiya, M.; Nikawa, H.; Mizorogi, N.; Tsuchiya, T.; Lum, X.; Akasaka, T. Exceptional chemical properties of Sc@C_{2v}(9)-C_{82} probed with adamantylidene carbene. *Journal of the American Chemical Society* 2012; 134: 15550–15555.

15. Xie, Y.; Suzuki, M.; Cai, W.; Mizorogi, N.; Nagase, S.; Akasaka, T. et al. Highly regioselective addition of adamantylidene carbene to Yb@C_{2v}(3)-C_{80} to afford the first derivative of divalent metallofullerenes. *Angewandte Chemie International Edition in English* 2013; 52(19): 5142–5145.

16. Zhang, W.; Suzuki, M.; Xie, Y. P.; Bao, L. P.; Cai, W. T.; Slanina, Z. et al. Molecular structure and chemical property of a divalent metallofullerene Yb@C2(13)-C84. *Journal of the American Chemical Society* 2013; 135(34): 12730–12735.

17. Maeda, Y.; Miyashita, J.; Hasegawa, T.; Wakahara, T.; Tsuchiya, T.; Nakahodo, T. et al. Reversible and regioselective reaction of La@C_{82} with cyclopentadiene. *Journal of the American Chemical Society* 2005; 127(35): 12190–12191.

18. Maeda, Y.; Sato, S.; Inada, K.; Nikawa, H.; Yamada, M.; Mizorogi, N. et al. Regioselective exohedral functionalization of La@C-82 and its 1,2,3,4,5-Pentamethylcyclopentadiene and adamantylidene adducts. *Chemistry A European Journal* 2010; 16(7): 2193–2197.

19. Sato, S.; Maeda, Y.; Guo, J. D.; Yamada, M.; Mizorogi, N.; Nagase, S. et al. Mechanistic study of the Diels-Alder reaction of paramagnetic endohedral metallofullerene: Reaction of La@C_{82} with 1,2,3,4,5-pentamethylcyclopentadiene. *Journal of the American Chemical Society* 2013; 135(15): 5582–5587. Epub March 23, 2013.

20. Cao, B. P.; Wakahara, T.; Maeda, Y.; Han, A. H.; Akasaka, T.; Kato, T. et al. Lanthanum endohedral metallofulleropyrrolidines: Synthesis, isolation, and EPR characterization. *Chemistry A European Journal* 2004; 10(3): 716–720.

21. Lu, X.; He, X. R.; Feng, L.; Shi, Z. J.; Gu, Z. N. Synthesis of pyrrolidine ring-fused metallofullerene derivatives. *Tetrahedron* 2004; 60(16): 3713–3716.

22. Feng, L.; Lu, X.; He, X. R.; Shi, Z. J.; Gu, Z. N. Reactions of endohedral metallofullerenes with azomethine ylides: An efficient route toward metallofullerene-pyrrolidines. *Inorganic Chemistry Communications* 2004; 7(9): 1010–1013.

23. Takano, Y.; Obuchi, S.; Mizorogi, N.; Garcia, R.; Herranz, M. A.; Rudolf, M. et al. Stabilizing ion and radical ion pair states in a paramagnetic endohedral metallofullerene/pi-extended tetrathiafulvalene conjugate. *Journal of the American Chemical Society* 2012; 134(39): 16103–16106. Epub September 14, 2012.

24. Tsuchiya, T.; Rudolf, M.; Wolfrum, S.; Radhakrishnan, S. G.; Aoyama, R.; Yokosawa, Y. et al. Coordinative interactions between Porphyrins and C-60, La@C-82, and La-2@C-80. *Chemistry A European Journal* 2013; 19(2): 557–564.

25. Bolskar, R. D. B.; Husebo, A. F.; Price, L. O.; Jackson, R. E.; Wallace, E. F.; Wilson, S.; Alford, L. J. First soluble M@C_{60} derivatives provide enhanced access to metallofullerenes and permit in vivo evaluation of Gd@C_{60}[C(COOH)$_2$]$_{(10)}$ as a MRI contrast agent. *Journal of the American Chemical Society* 2003; 125(18): 5471–5478.

26. Sitharaman, B.; Bolskar, R. D.; Rusakova, I.; Wilson, L. J. Gd@C60[C(COOH)(2)](10) and Gd@C60(OH)(x): Nanoscale aggregation studies of two metallofullerene MRI contrast agents in aqueous solution. *Nano Letters* 2004; 4(12): 2373–2378.

27. Feng, L.; Nakahodo, T.; Wakahara, T.; Tsuchiya, T.; Maeda, Y.; Akasaka, T. et al. A singly bonded derivative of endohedral metallofullerene: La@C_{82}CBr(COOC2H5) (2). *Journal of the American Chemical Society* 2005; 127(49): 17136–17137.

28. Feng, L.; Wakahara, T.; Nakahodo, T.; Tsuchiya, T.; Piao, Q.; Maeda, Y. et al. The Bingel monoadducts of La@C_{82}: Synthesis, characterization, and electrochemistry. *Chemistry A European Journal* 2006; 12(21): 5578–5586.

29. Hirsch, A.; Brettreich, M. *Fullerenes-Chemistry and Reactions.* Weinheim, Germany: Wiley-VCH; 2005.

30. Feng, L.; Tsuchiya, T.; Wakahara, T.; Nakahodo, T.; Piao, Q.; Maeda, Y.; et al. Synthesis and characterization of a bisadduct of La@C82. *Journal of the American Chemical Society* 2006; 128(18): 5990–5991.

31. He, R.; Zhao, H.; Liu, J. H.; Jiao, Y. H.; Liang, Y. Y.; Li, X. et al. Synthesis and aggregation studies of Bingel-Hirsch monoadducts of gadofullerene. *Fullerenes Nanotubes and Carbon Nanostructures* 2013; 21(6): 549–559.

32. Suzuki, T.; Maruyama, Y.; Kato, T.; Akasaka, T.; Kobayashi, K.; Nagase, S. et al. Chemical reactivity of a metallofullerene—EPR study of diphenylmethano-La@C_{82} radicals. *Journal of the American Chemical Society* 1995; 117(37): 9606–9607.

33. Nikawa, H.; Kikuchi, T.; Wakahara, T.; Nakahodo, T.; Tsuchiya, T.; Rahman, G. M. A. et al. Missing metallofullerene La@C_{74}. *Journal of the American Chemical Society* 2005; 127(27): 9684–9685.

34. Wakahara, T.; Nikawa, H.; Kikuchi, T.; Nakahodo, T.; Rahman, G. M. A.; Tsuchiya, T. et al. La@C_{72} having a non-IPR carbon cage. *Journal of the American Chemical Society* 2006; 128(44): 14228–14229.

35. Nikawa, H.; Yamada, T.; Cao, B. P.; Mizorogi, N.; Slanina, Z.; Tsuchiya, T. et al. Missing metallofullerene with C80 cage. *Journal of the American Chemical Society* 2009; 131(31): 10950–10954.

36. Akasaka, T.; Lu, X.; Kuga, H.; Nikawa, H.; Mizorogi, N.; Slanina, Z. et al. Dichlorophenyl derivatives of La@C_{3v}(7)-C_{82}: Endohedral metal induced localization of pyramidalization and spin on a Triple-Hexagon Junction. *Angewandte Chemie International Edition* 2010; 49(50): 9715–9719.

37. Takano, Y.; Yomogida, A.; Nikawa, H.; Yamada, M.; Wakahara, T.; Tsuchiya, T. et al. Radical coupling reaction of paramagnetic endohedral metallofullerene La@C_{82}. *Journal of the American Chemical Society* 2008; 130: 16224–16230.

38. Takano, Y.; Ishitsuka, M. O.; Tsuchiya, T.; Akasaka, T.; Kato, T.; Nagase, S. Retro-reaction of singly bonded La@C-82 derivatives. *Chemical Communications* 2010; 46(42): 8035–8036.

39. Tagmatarchis, N.; Taninaka, A.; Shinohara, H. Production and EPR characterization of exo-hedrally perfluoroalkylated paramagnetic lanthanum metallofullerenes: La@C82(C8F17)2. *Chemical Physics Letters* 2002; 355(3–4): 226–232.

40. Kareev, I. E.; Lebedkin, S. F.; Bubnov, V. P.; Yagubskii, E. B.; Ioffe, I. N.; Khavrel, P. A. et al. Trifluoromethylated endohedral metallofullerenes: Synthesis and characterization of Y@C82(CF3)(5). *Angewandte Chemie International Edition* 2005; 44(12): 1846–1849.

41. Kareev, I. E.; Yagubskii, E. B. Endohedral gadolinium containing metallofullerenes in the trifluoromethylation reaction. *Russian Chemical Bulletin, International Edition* 2008; 57(7): 1486–1491.

42. Lu, X.; Nikawa, H.; Tsuchiya, T.; Akasaka, T.; Toki, M.; Sawa, H.; et al. Nitrated Benzyne derivatives of La@C-82: Addition of NO2 and its positional directing effect on the subsequent addition of Benzynes. *Angewandte Chemie International Edition* 2010; 49(3): 594–597.

43. Popov, A.; Yang, S.; Dunsch, L. Endohedral fullerenes. *Chemical Reviews* 2013; 113: 5989–6113.

44. Wilson, L. J.; Cagle, D. W.; Thrash, T. P.; Kennel, S. J.; Mirzadeh, S.; Alford, J. M.; Ehrhard G. J. Metallofullerene Drug Design, *Coordination Chemistry Reviews* 1999; 190–192: 199–207.

45. Sun, D. Y.; Huang, H. J.; Yang, S. H.; Liu, Z. Y.; Liu, S. Y. Synthesis and characterization of a water-soluble endohedral metallofullerol. *Chemistry of Materials* 1999; 11(4): 1003–1006.

46. Kato, H.; Suenaga, K.; Mikawa, W.; Okumura, M.; Miwa, N.; Yashiro, A. et al. Syntheses and EELS characterization of water-soluble multi-hydroxyl Gd@C82 fullere-nols. *Chemical Physics Letters* 2000; 324(4): 255–259.

47. Mikawa, M. K.; Okumura, H.; Narazaki, M.; Kanazawa, M.; Miwa, Y.; Shinohara, N. Paramagnetic water-soluble metallofullerenes having the highest relaxivity for MRI contrast agents. *Bioconjugate Chemistry* 2001; 12(4): 510–514.

48. Kato, H.; Kanazawa, Y.; Okumura, M.; Taninaka, A.; Yokawa, T.; Shinohara, H. Lanthanoid endohedral metallofullerenols for MRI contrast agents. *Journal of the American Chemical Society* 2003; 125(14): 4391–4397.

49. Lu, X.; Zhou, X. H.; Shi, Z. J.; Gu, Z. N. Nucleophilic addition of glycine esters to Gd@C82. *Inorganica Chimica Acta* 2004; 357(8): 2397–2400.

50. Shu, C. Y.; Gan, L. H.; Wang, C. R.; Pei, X. L.; Han, H. B. Synthesis and character-ization of a new water-soluble endohedral metallofullerene for MRI contrast agents. *Carbon* 2006; 44(3): 496–500.

51. Shu, C. Y. W.; Zhang, C. R.; Gibson, J. F.; Dorn, H. W.; Corwin, H. C.; Fatouros, F. D.; Dennis, P. P. Organophosphonate functionalized Gd@C82 as a magnetic resonance imaging contrast agent. *Chemistry of Materials* 2008; 20(6): 2106–2109.

52. Akasaka, T.; Nagase, S.; Kobayashi, K.; Suzuki, T.; Kato, T.; Kikuchi, K. et al. Syntheis of the first adducts of the dimetallofullerenes La2C80 and Sc2C84 by addition of a disilirane. *Angewandte Chemie International Edition in English* 1995; 34(19): 2139–2141.

53. Yamada, M.; Nakahodo, T.; Wakahara, T.; Tsuchiya, T.; Maeda, Y.; Akasaka, T. et al. Positional control of encapsulated atoms inside a fullerene cage by exohedral addition. *Journal of the American Chemical Society* 2005; 127(42): 14570–14571.

54. Yamada, M.; Wakahara, T.; Tsuchiya, T.; Maeda, Y.; Kako, M.; Akasaka, T. et al. Location of the metal atoms in Ce2@C78 and its bis-silylated derivative. *Chemical Communications (Cambridge)* 2008(5): 558–560. Epub Januay 23, 2008.

55. Cao, B.; Nikawa, H.; Nakahodo, T.; Tsuchiya, T.; Maeda, Y.; Akasaka, T. et al. Addition of adamantylidene to $La_2@C_{78}$: Isolation and single-crystal X-ray structural determination of the monoadducts. *Journal of the American Chemical Society* 2007; 130(3): 983–989.

56. Yamada, M.; Someya, C.; Wakahara, T.; Tsuchiya, T.; Maeda, Y.; Akasaka, T. et al. Metal atoms collinear with the spiro carbon of 6,6-open adducts, $M_2@C_{80}$(Ad) (M = La and Ce, Ad = adamantylidene). *Journal of the American Chemical Society* 2008; 130(4): 1171–1176.

57. Lu, X.; Nikawa, H.; Nakahodo, T.; Tsuchiya, T.; Ishitsuka, M. O.; Maeda, Y. et al. Chemical understanding of a non-IPR metallofullerene: Stabilization of encaged metals on fused-pentagon bonds in $La_2@C_{72}$. *Journal of the American Chemical Society* 2008; 130(28): 9129–9136.

58. Ishitsuka, M. O.; Sano, S.; Enoki, H.; Sato, S.; Nikawa, H.; Tsuchiya, T. et al. Regioselective bis-functionalization of endohedral dimetallofullerene, La-2@C-80: Extremal La-La distance. *Journal of the American Chemical Society* 2011; 133(18): 7128–7134.

59. Guldi, D. M.; Feng, L.; Radhakrishnan, S. G.; Nikawa, H.; Yamada, M.; Mizorogi, N. et al. A molecular Ce-2@I-h-C-80 switch-unprecedented oxidative pathway in photoinduced charge transfer reactivity. *Journal of the American Chemical Society* 2010; 132(26): 9078–9086.

60. Yamada, M.; Wakahara, T.; Nakahodo, T.; Tsuchiya, T.; Maeda, Y.; Akasaka, T. et al. Synthesis and structural characterization of endohedral pyrrolidinodimetallofullerene: $La_2@C_{80}$(CH2)(2)NTrt. *Journal of the American Chemical Society* 2006; 128(5): 1402–1403.

61. Yamada, M.; Okamura, M.; Sato, S.; Someya, C. I.; Mizorogi, N.; Tsuchiya, T. et al. Two regioisomers of endohedral pyrrolidinodimetallofullerenes M_2@I-h-C-80(CH2)(2)NTrt (M = La, Ce; Trt = trityl): Control of metal atom positions by addition positions. *Chemistry A European Journal* 2009; 15(40): 10533–10542.

62. Yuta Takano, M. A. N. H.; Martín, N. Donor acceptor conjugates of lanthanum endohedral metallofullerene and π-extended tetrathiafulvalene. *Journal of the American Chemical Society* 2010; 132: 132, 8048–8055.

63. Lu, X.; Akasaka, T.; Nagase, S. Carbide cluster metallofullerenes: Structure, properties, and possible origin. *Accounts of Chemical Research* 2013; 47(7): 1627–1635.

64. Iiduka, Y.; Wakahara, T.; Nakahodo, T.; Tsuchiya, T.; Sakuraba, A.; Maeda, Y. et al. Structural determination of metallofullerene Sc_3C_{82} revisited: A surprising finding. *Journal of the American Chemical Society* 2005; 127(36): 12500–12501.

65. Iiduka, Y.; Wakahara, T.; Nakajima, K.; Nakahodo, T.; Tsuchiya, T.; Maeda, Y. et al. Experimental and theoretical studies of the scandium carbide endohedral metallofullerene $Sc_2C_2@C_{82}$ and its carbene derivative. *Angewandte Chemie International Edition* 2007; 46(29): 5562–5564.

66. Lu, X.; Nakajima, K.; Iiduka, Y.; Nikawa, H.; Mizorogi, N.; Slanina, Z. et al. Structural elucidation and regioselective functionalization of an unexplored carbide cluster metallofullerene $Sc_2C_2@C_s$(6)-C_{82}. *Journal of the American Chemical Society* 2011; 133(48): 19553–19558.

67. Lu, X.; Nakajima, K.; Iiduka, Y.; Nikawa, H.; Tsuchiya, T.; Mizorogi, N. et al. The long-believed $Sc_2@C_{2v}$(17)-C_{84} is actually $Sc_2C_2@C_{2v}$(9)-C_{82}: Unambiguous structure assignment and chemical functionalization. *Angewandte Chemie International Edition in English* 2012; 51(24): 5889–5892.

68. Yang, S. F.; Yang, S. H. Preparation and film formation behavior of the supramolecular complex of the endohedral metallofullerene $Dy@C_{82}$ with calix[8]arene. *Langmuir* 2002; 18(22): 8488–8495.

69. Tsuchiya, T.; Sato, K.; Kurihara, H.; Wakahara, T.; Nakahodo, T.; Maeda, Y. et al. Host-guest complexation of endohedral metallofullerene with azacrown ether and its application. *Journal of the American Chemical Society* 2006; 128(20): 6699–6703.

70. Tsuchiya, T.; Kurihara, H.; Sato, K.; Wakahara, T.; Akasaka, T.; Shimizu, T. et al. Supramolecular complexes of La@C$_{82}$ with unsaturated thiacrown ethers. *Chemical Communications* 2006(34): 3585–3587.

71. Tsuchiya, T.; Sato, K.; Kurihara, H.; Wakahara, T.; Maeda, Y.; Akasaka, T. et al. Spin-site exchange system constructed from endohedral metallofullerenes and organic donors. *Journal of the American Chemical Society* 2006; 128(45): 14418–14419.

72. Pagona, G.; Economopoulos, S. P.; Aono, T.; Miyata, Y.; Shinohara, H.; Tagmatarchis, N. Molecular recognition of La@C-82 endohedral metallofullerene by an isophthaloyl-bridged porphyrin dimer. *Tetrahedron Letters* 2010; 51(45): 5896–5899.

73. Hajjaj, F.; Tashiro, K.; Nikawa, H.; Mizorogi, N.; Akasaka, T.; Nagase, S. et al. Ferromagnetic spin coupling between endohedral metallofullerene La@C(82) and a cyclodimeric copper porphyrin upon inclusion. *Journal of the American Chemical Society* 2011; 133(24): 9290–9292.

74. Fukuzumi, S.; Ohkubo, K.; Kawashima, Y.; Kim, D. S.; Park, J. S.; Jana, A. et al. Ion-controlled on-off switch of electron transfer from tetrathiafulvalene Calix 4 pyrroles to Li$^+$@C-60. Journal of the American Chemical Society 2011; 133(40): 15938–15941.

75. Stevenson, S.; Rice, G.; Glass, T.; Harlch, K.; Cromer, F.; Jordan, M. R.; Craft, J.; Hadju, E.; Bible, R.; Olmstead, M. M. et al. Small-bandgap endohedral metallofullerenes in high yield and purity. *Nature* 1999; 401: 55–57.

76. Iezzi, E. B.; Duchamp, J. C.; Harich, K.; T. E.; Lee, H. M.; Olmstead, M. M.; Balch, A. L.; Dorn, H. C. A symmetric derivative of the trimetallic nitride endohedral metallofuller-ene, Sc$_3$N@C$_{80}$. *Journal of the American Chemical Society* 2001; 124(4): 524–525.

77. Lee, H. M.; Olmstead, M. M.; Iezzi, E. B.; Duchamp, J. C.; Dorn, H. C.; Balch, A. L. Crystallographic characterization and structural analysis of the first organic function-alization product of the endohedral fullerene Sc$_3$N@C$_{80}$. *Journal of the American Chemical Society* 2002; 124(14): 3494–3495.

78. Cai, T.; Xu, L.; Anderson, M. R.; Ge, Z.; Zuo, T.; Wang, X.; Olmstead, M. M.; Balch, A. L.; Gibson, H. W.; Dorn, H. C. Structure and enhanced reactivity rates of the D_{5h}-Sc$_3$N@ C$_{80}$ and Lu$_3$N@C$_{80}$ metallofullerene isomers: The importance of the pyracylene motif. *Journal of the American Chemical Society* 2006; 128(26): 8581–8589.

79. Li, F.-F.; Pinzón, J. R.; Mercado, B. Q. Olmstead. M. M.; Balch, A. L.; Echegoyen, L. [2 + 2] Cycloaddition reaction to Sc$_3$N@I_h-C$_{80}$. The formation of very stable [5,6]- and [6,6]-adducts. *Journal of the American Chemical Society* 2011; 133: 1563–1571.

80. Wang, G.-W.; Liu, T.-X.; Jiao, M.; Wang, N.; Zhu, S.-E.; Chen, C.; Yang, S.; Bowles, F. L.; Beavers, C. M.; Olmstead, M. M. et al. The cycloaddition reaction of I_h-Sc$_3$N@C$_{80}$ with 2-amino-4,5-diisopropoxybenzoic acid and isoamyl nitrite to pro-duce an open-cage metallofullerene. *Angewandte Chemie International Edition* 2011; 50: 4658–4662.

81. Cardona, C. M.; Kitaygorodskiy, A.; Ortiz, A.; Herranz, M. A.; Echegoyen, L. The first fulleropyrrolidine derivative of Sc$_3$N@C$_{80}$: Pronounced chemical shift differences of the geminal protons on the pyrrolidine ring. *Journal of Organic Chemistry* 2005; 70: 5092–5097.

82. Elliott, B.; Pykhova, A. D.; Rivera, J.; Cardona, C. M.; Dunsch, L.; Popov, A. A.; Echegoyen, L. Spin density and cluster dynamics in Sc$_3$N@C$_{80}$ upon [5,6] exohedral functionalization: An ESR and DFT study. *Journal of Physical Chemistry C* 2013; 117(5): 2344–2348.

83. Cai, T.; Ge, Z.; Iezzi, E. B.; Glass, T. E.; Harich, K.; Gibson, H. W.; Dorn, H. C. Synthesis and characterization of the first trimetallic nitride templated pyrrolidino endo-hedral metallofullerenes. *Chemical Communications* 2005; (28): 3594–3596.

84. a) Cardona, C. M.; Kitaygorodskiy, A.; Echegoyen, L. Trimetallic nitride endohedral metallofullerenes: Reactivity dictated by the encapsulated metal cluster. *Journal of the American Chemical Society* 2005; 127: 10448–10453. b) Echegoyen, L.; Chancellor, C.

J.; Cardona, C. M.; Elliott, B.; Rivera, J.; Olmstead, M. M.; Balch, A. L. X-Ray crystallographic and EPR spectroscopic characterization of a pyrrolidine adduct of $Y_3N@C_{80}$. *Chemical Communications* 2006; (25): 2653–2655.

85. a) Rodríguez-Fortea, A.; Campanera, J. M.; Cardona, C. M.; Echegoyen, L.; Poblet, J. M. Dancing on a fullerene surface: Isomerization of $Y_3N@(N$-ethylpyrrolidino-$C_{80})$ from the 6,6 to the 5,6 regioisomer. *Angewandte Chemie International Edition* 2006; 45: 8176–8180. b) Cardona, C. M.; Elliott, B.; Echegoyen, L. Unexpected chemical and electrochemical properties of $M_3N@C_{80}$ (M = Sc, Y, Er). *Journal of the American Chemical Society* 2006; 128(19): 6480–6485.

86. Chen, N.; Pinzón, J. R.; Echegoyen, L. Influence of the encapsulated clusters on the electrochemical behavior of endohedral fullerene derivatives: Comparative study of N-tritylpyrrolidino derivatives of $Sc_3N@I_h$-C_{80} and $Lu_3N@I_h$-C_{80}. *ChemPhysChem* 2011; 12: 1422–1425.

87. Cai, T.; Slebodnick, C.; Xu, L.; Harich, K.; Glass, T. E.; Chancellor, C.; Fettinger, J. C.; Olmstead, M. M.; Balch, A. L.; Gibson, H. W. et al. A pirouette on a metallofullerene sphere: Interconversion of isomers of N-tritylpyrrolidino I_h-$Sc_3N@C_{80}$. *Journal of the American Chemical Society* 2006; 128(19): 6486–6492.

88. Chen, N.; Zhang, E.-Y.; Tan, K.; Wang, C.-R.; Lu, X. Size effect of encaged clusters on the exohedral chemistry of endohedral fullerenes: A case study on the pyrrolidino reaction of $Sc_xGd_{3-x}N@C_{80}$ (x = 0–3). *Organic Letters* 2007; 9(10): 2011–2013.

89. Aroua, S.; Yamakoshi, Y. Prato reaction of $M_3N@I_h$-C_{80} (M = Sc, Lu, Y, Gd) with reversible isomerization. *Journal of the American Chemical Society* 2012; 134(50): 20242–20245.

90. Cai, T.; Xu, L.; Gibson, H. W.; Dorn, H. C.; Chancellor, C. J.; Olmstead, M. M.; Balch, A. L. $Sc_3N@C_{78}$: Encapsulated cluster regiocontrol of adduct docking on an ellipsoidal metallofullerene sphere. *Journal of the American Chemical Society* 2007; 129(35): 10795–10800.

91. Chen, N.; Fan, L.-Z.; Tan, K.; Wu, Y.-Q.; Shu, C.-Y.; Lu, X.; Wang, C.-R. Comparative spectroscopic and reactivity studies of $Sc_{3-x}Y_xN@C_{80}$ (x = 0–3). *Journal of Physical Chemistry C.* 2007; 111(32): 11823–11828.

92. a) Pinzón, J. R.; Plonska-Brezinska, M. E.; Cardona, C. M.; Athans, A. J.; Gayathri, S. S.; Guldi, D. M.; Herranz, M. A.; Martin, N.; Torres, T.; Echegoyen, L. $Sc_3N@C_{80}$-ferrocene electron-donor/acceptor conjugates as promising materials for photovoltaic applications. *Angewandte Chemie International Edition* 2008; 47: 4173–4176. b) Pinzón, J. R.; Cardona, C. M.; Herránz, M. Á.; Plonska-Brezinska, M. E.; Palkar, A.; Athans, A. J.; Martin, N.; Ropdriguez-Fortea, A.; Poblet, J. M.; Bottari, G. Metal nitride cluster fullerene $M_3N@C_{80}$ (M = Y, Sc) based dyads: Synthesis, and electrochemical, theoretical and photophysical studies. *Chemistry A European Journal* 2009; 15: 864–877.

93. Wolfrum, S.; Pinzón, J. R.; Molina-Ontoria, A.; Gouloumis, A.; Martin, N.; Echegoyen, L.; Guldi, D. M. Utilization of $Sc_3N@C_{80}$ in long-range charge transfer reactions. *Chemical Communications* 2011; 47: 2270–2272.

94. Pinzón, J. R.; Gasca, D. C.; Sankaranarayanan, S. G.; Bottari, G.; Torres, T.; Guldi, D. M.; Echegoyen, L. Photoinduced charge transfer and electrochemical properties of triphenylamine I_h-$Sc_3N@C_{80}$ donor-acceptor conjugates. *Journal of the American Chemical Society* 2009; 131: 7727–7734.

95. Lukoyanova, O.; Cardona, C. M.; Rivera, J.; Lugo-Morales, L. Z.; Chancellor, C. J.; Olmstead, M. M.; Rodriguez-Fortea, A.; Poblet, J. M.; Balch, A. L.; Echegoyen, L. Open rather than closed malonate methano-fullerene derivatives. The formation of methanofulleroid adducts of $Y_3N@C_{80}$. *Journal of the American Chemical Society* 2007; 129: 10423–10430.

96. Cai, T.; Xu, L.; Shu, C.; Champion, H. A.; Reid, J. E.; Anklin, C.; Anderson, M. R.; Gibson, H. W.; Dorn, H. C. Selective formation of a symmetric $Sc_3N@C_{78}$ bisadduct: Adduct docking controlled by an internal trimetallic nitride cluster. *Journal of the American Chemical Society* 2008; 130(7): 2136–2137.

97. Cai, T.; Xu, L.; Shu, C.; Reid, J. E.; Gibson, H. W.; Dorn, H. C. Synthesis and characterization of a Non-IPR fullerene derivative: $Sc_3N@C_{68}[C(COOC_2H_5)_2]$. *Journal of the Physical Chemistry C* 2008; 112(49): 19203–19208.

98. Chaur, M. N.; Melin, F.; Athans, A. J.; Elliott, B.; Walker, K.; Holloway, B. C.; Echegoyen, L. The influence of cage size on the reactivity of trimetallic nitride metallofullerenes: A mono- and bis-methanoadduct of $Gd_3N@C_{80}$ and a monoadduct of $Gd_3N@C_{84}$. *Chemical Communications* 2008; (23): 2665–2667.

99. Alegret, N.; Chaur, M. N.; Santos, E.; Rodriguez-Fortea, A.; Echegoyen, L.; Poblet, J. M. Bingel-Hirsch reactions on non-IPR $Gd_3N@C_{2n}$ (2n = 82 and 84). *Journal of Organic Chemistry* 2010; 75: 8299–8302.

100. Gao, X.; Ishimura, K.; Nagase, S.; Chen, Z. Dichlorocarbene addition to C_{60} from the trichloromethyl anion: Carbene mechanism or Bingel mechanism? *Journal of Physical Chemistry A* 2009; 113(15): 3673–3676.

101. Pinzón, J. R.; Zuo, T.; Echegoyen. L. Synthesis and electrochemical studies of Bingel-Hirsch derivatives of $M_3N@I_h$-C_{80} (M = Sc, Lu). *Chemistry A European Journal* 2010; 16: 4864–4869.

102. a) Shu, C.; Xu, W.; Slebodnick, C.; Champion, H.; Fu, W.; Reid, J. E.; Azurmendi, H.; Wang, C.; Harich, K.; Dorn, H. C. et al. Syntheses and structures of phenyl-C81-butyric acid methyl esters (PCBMs) from $M_3N@C_{80}$. *Organic Letters* 2009; 11(8): 1753–1756. b) Ross, R. B.; Cardona, C. M.; Guldi, D. M.; Sankaranarayanan, S. G.; Reese, M. O.; Kopidakis, N.; Peet, J.; Walker, B.; Bazan, G. C.; Van Keuren, E. et al. Endohedral fullerenes for organic photovoltaic devices. *Nature Materials* 2009; 8(3): 208–212.

103. a) Iiduka, Y.; Wakahara, T.; Nakahodo, T.; Tsuchiya, T.; Sakuraba, A.; Maeda, Y.; Akasaka, T.; Yoza, K.; Horn, E.; Kato, T. et al. Structural determination of metallofullerene Sc_3C_{82} revisited: A surprising finding. *Journal of the American Chemical Society* 2005; 127(36): 12500–12501. b) Iiduka, Y.; Wakahara, T.; Nakajima, K.; Nakahodo, T.; Tsuchiya, T.; Maeda, Y.; Akasaka, T.; Yoza, K.; Liu, M. T. H.; Mizorogi, N. et al. Experimental and theoretical studies of the scandium carbide endohedral metallofullerene $Sc_2C_2@C_{82}$ and its Carbene derivative. *Angewandte Chemie International Edition* 2007; 46(29): 5562–5564.

104. Kurihara, H.; Lu, X.; Iiduka, Y.; Nikawa, H.; Mizorogi, N.; Slanina, Z.; Tsuchiya, T.; Nagase, S.; Akasaka, T. Chemical understanding of carbide cluster metallofullerenes: A case study on $Sc_2C_2@C_{2v}(5)$–C_{80} with complete X-ray crystallographic characterizations. *Journal of the American Chemical Society* 2012; 134(6): 3139–3144.

105. a) Yamada, M.; Akasaka, T.; Nagase. S. Carbene additions to fullerenes. *Chemical Reviews* 2013; 113(9): 7209–7264. b) Feng, L.; Gayathri Radhakrishnan, S.; Mizorogi, N.; Slanina, Z.; Nikawa, H.; Tsuchiya, T.; Akasaka, T.; Nagase, S.; Martín, N.; Guldi, D. M. Synthesis and charge-transfer chemistry of $La_2@I_h$-C_{80}/$Sc_3N@I_h$-C_{80}–Zinc porphyrin conjugates: Impact of endohedral cluster. *Journal of the American Chemical Society* 2011; 133(19): 7608–7618. c) Maeda, Y.; Matsunaga, Y.; Wakahara T.; Takahashi, S.; Tsuchiya, T.; Ishitsuka, M. O.; Hasegawa, T.; Akasaka, T.; Liu, M. T. H.; Kokura, K. et al. Isolation and characterization of a Carbene derivative of $La@C_{82}$. *Journal of the American Chemical Society* 2004; 126(22): 6858–6859.

106. Izquierdo, M.; Cerón, M. R.; Olmstead, M. M.; Balch, A. L.; Echegoyen, L. [5,6]-open methanofullerene derivatives of I_h-C_{80}. *Angewandte Chemie International Edition* 2013; 52(45): 11826–11830.

107. Yamada, M.; Nakahodo, T.; Wakahara, T.; Tsuchiya, T.; Maeda, Y.; Akasaka, T.; Kako, M.; Yoza, K.; Horn, E.; Mizorogi, N. et al. Positional control of encapsulated atoms inside a fullerene cage by exohedral addition. *Journal of the American Chemical Society* 2005; 127(42): 14570–14571.

108. Sato, K.; Kako, M.; Suzuki, M.; Mizorogi, N.; Tsuchiya, T.; Olmstead, M. M.; Balch, A. L.; Akasaka, T.; Nagase, S. Synthesis of Silylene-Bridged endohedral metallofullerene $Lu_3N@I_h$-C_{80}. *Journal of the American Chemical Society* 2012; 134(38): 16033–16039.

109. a) Sato, K.; Kako, M.; Mizorogi, N.; Tsuchiya, T.; Akasaka, T.; Nagase, S. Bis-Silylation of $Lu_3N@I_h$-C_{80}: Considerable variation in the electronic structures. *Organic Letters* 2012; 14(23): 5908–5911. b) Iiduka, Y.; Ikenaga, O.; Sakuraba, A.; Wakahara, T.; Tsuchiya, T.; Maeda, Y.; Nakahodo, T.; Akasaka, T.; Kako, M.; Mizorogi, N. et al. Chemical reactivity of $Sc_3N@C_{80}$ and $La_2@C_{80}$. *Journal of the American Chemical Society* 2005; 127(28): 9956–9957.

110. Liu, T.-X.; Wei, T.; Zhu, S.-E.; Wang, G.-W.; Jiao, M.; Yang, S.; Bowles, F. L.; Olmstead, M. M.; Balch, A. L. Azide addition to an endohedral metallofullerene: Formation of azaful-leroids of $Sc_3N@I_h$-C_{80}. *Journal of the American Chemical Society* 2012; 134(29): 11956–11959.

111. Li, F.-F.; Rodríguez-Fortea, A.; Poblet, J. M.; Echegoyen, L. Reactivity of metallic nitride endohedral metallofullerene anions: Electrochemical synthesis of a $Lu_3N@I_{h^{-}80}$ derivative. *Journal of the American Chemical Society* 2011; 133(8): 2760–2765.

112. Li, F.-F.; Rodríguez-Fortea, A.; Peng, P.; Campos, C. G. A.; Poblet, J. M.; Echegoyen, L. Electrosynthesis of a $Sc_3N@I_h$-C_{80} methano derivative from trianionic $Sc_3N@I_h$-C_{80}. *Journal of the American Chemical Society* 2012; 134: 7480–7487.

113. Shu, C.; Cai, T.; Xu, L.; Zuo, T.; Reid, J.; Harich, K.; Dorn, H. C.; Gibson, H. W. Manganese(III)-catalyzed free radical reactions on trimetallic nitride endohedral metallofullerenes. *Journal of the American Chemical Society* 2007; 129(50): 15710–15717.

114. a) Shustova, N. B.; Popov, A. A.; Mackey, M. A.; Coumbe, C. E.; Phillips, J. P.; Stevenson, S.; Strauss, S. H.; Boltalina, O. V. Radical trifluoromethylation of $Sc_3N@C_{80}$. *Journal of the American Chemical Society* 2007; 129(38): 11676–11677. b) Popov, A. A.; Shustova, N. B.; Svitova, A. L.; Mackey, M. A.; Coumbe, C. E.; Phillips, J. P.; Stevenson, S.; Strauss, S. H.; Boltalina, O. V.; Dunsch, L. Redox-Tuning endohedral fullerene spin states: From the dication to the trianion radical of $Sc_3N@C_{80}(CF_3)_2$ in five reversible single-electron steps. *Chemistry A European Journal* 2010; 16: 4721–4724. c) Yang, S.; Chen, C.; Lanskikh, M. A.; Tamm, N. B.; Kemnitz, E.; Troyanov, S. I. New isomers of trifluoromethylated derivatives of metal nitride cluster fullerene: $Sc_3N@C_{80}(CF_3)n$ (n = 14 and 16). *Chemistry An Asian Journal* 2011; 6: 505–509.

115. a) Shustova, N. B.; Peryshkov, D. V.; Kuvychko, I. V.; Chen, Y.-S.; MacKey, M. A.; Coumbe, C. E.; Heaps, D. T.; Confait, B. S.; Heine, T.; Phillips, J. P. et al. Poly(perfluoroalkylation) of metallic nitride fullerenes reveals addition-pattern guide-lines: Synthesis and characterization of a family of $Sc_3N@C_{80}(CF_3)n$ (n = 2–16) and their radical anions. *Journal of the American Chemical Society* 2011; 133: 2672–2690. b) Yang, S.; Chen, C.; Jiao, M.; Tamm, N. B.; Lanskikh, M. A.; Kemnitz, E.; Troyanov, S. I. Synthesis, isolation, and addition patterns of trifluoromethylated D_{5h} and I_h isomers of $Sc_3N@C_{80}$:$Sc_3N@D_{5h}$-$C_{80}(CF_3)_{18}$ and $Sc_3N@I_h$-$C_{80}(CF_3)_{14}$. *Inorganic Chemistry* 2011; 50: 3766–3771.

116. Shu, C.; Slebodnick, C.; Xu, L.; Champion, H.; Fuhrer, T.; Cai, T.; Reid, J. E.; Fu, W.; Harich, K.; Dorn, H. C. et al. Highly regioselective derivatization of trimetallic nitride templated endohedral metallofullerenes via a facile photochemical reaction. *Journal of the American Chemical Society* 2008; 130(52): 17755–17760.

117. Zhang, J.; Fatouros, P. P.; Shu, C.; Reid, J.; Owens, L. S.; Cai, T.; Gibson, H. W.; Long, G. L.; Corwin, F. D.; Chen, Z.-J. et al. High relaxivity trimetallic nitride (Gd_3N) metallofullerene MRI contrast agents with optimized functionality. *Bioconjugate Chemistry* 2010; 21(4): 610–615.

118. Shultz, M. D.; Duchamp, J. C.; Wilson, J. D.; Shu, C.-Y.; Ge, J.; Zhang, J.; Gibson, H. W.; Fillmore, H. L.; Hirsch, J. I.; Dorn, H. C. et al. Encapsulation of a radiolabeled cluster inside a fullerene cage, $^{177}Lu_xLu_{(3-x)}N@C_{80}$: An Interleukin-13-conjugated radiolabeled metallofullerene platform. *Journal of the American Chemical Society* 2010; 132(14): 4980–4981.

119. Westerström, R.; Dreiser, J.; Piamonteze, C.; Muntwiler, M.; Weyeneth, S.; Brune, H.; Rusponi, S.; Nolting, F.; Popov, A.; Yang, S. et al. An endohedral single-molecule magnet with long relaxation times: $DySc_2N@C_{80}$. *Journal of the American Chemical Society* 2012; 134(24): 9840–9843.

120. a) Cerón, M. R.; Li, F.-F.; Echegoyen, L. A. Endohedral fullerenes: the importance of electronic, size and shape complementarity between the carbon cages and the corresponding encapsulated clusters. *Journal of Physical Organic Chemistry* 2014; 27(4): 258–264. b) Rivera-Nazario, D. M.; Pinzón, J. R.; Stevenson, S.; Echegoyen, L. A. Buckyball maracas: exploring the inside and outside properties of endohedral fullerenes. *Journal of Physical Organic Chemistry* 2013; 26(2): 194–205. c) Popov, A. A.; Yang, S.; Dunsch, L. Endohedral fullerenes. *Chemical Reviews* 2013; 113(8): 5989–6113. d) Yang, S.; Liu, F.; Chen, C.; Jiao, M.; Wei, T. Fullerenes encaging metal clusters-clusterfullerenes. *Chemical Communications* 2011; 47(43): 11822–11839. e) Lu, X.; Akasaka, T.; Nagase, S. Chemistry of endohedral metallofullerenes: the role of metals. *Chemical Communications* 2011; 47(21): 5942–5957. f) Yamada, M.; Akasaka, T.; Nagase, S. Endohedral metal atoms in pristine and functionalized fullerene cages. *Accounts of Chemical Research* 2010; 43: 92–102. g) Chaur, M. N.; Melin, F.; Ortiz, A. L.; Echegoyen, L. Chemical, electrochemical, and structural properties of endohedral metallofullerenes. *Angewandte Chemie International Edition* 2009; 48(41): 7514–7538.

7 Oxidative Charge Transfer of EMFs, Biomedical Applications of EMFs, and EMF-Based Nanomaterials and Molecular Materials

CONTENTS

7.1 OXIDATIVE CHARGE TRANSFER IN FULLERENES AND ENDOHEDRAL METALLOFULLERENES

Marcus Lederer, Marc Rudolf, Maximilian Wolf, and Dirk M. Guldi

7.1.1 INTRODUCTION

Fullerenes have been under investigation in different fields of research for nearly 30 years. In September 1985, a group of scientists including Robert Curl, James Heath, Harry Kroto, Yuan Liu, Sean O'Brian, and Richard Smalley discovered fullerenes via mass spectrometry at Rice University in Houston, Texas [1]. This new class of carbon allotropes was termed *buckminsterfullerenes* because the geodesic domes designed by inventor and architect Buckminster Fuller provided a decisive clue about their structure (Figure 7.1) [2]. A decade later, namely, in 1996, Curl, Kroto, and Smalley were awarded with the Nobel Prize in chemistry for the discovery of fullerenes [1].

The stage for a broad platform of fullerene research was only set in 1990, when appreciable amounts of fullerenes were available due to the work of Wolfgang Krätschmer and Donald Huffman assisted by their coworkers Konstantinos Fostiropoulos and Lowell Lamb [4]. They were able to produce C_{60} by vaporizing graphite via electrical arc discharge and resistance heating of graphite electrodes, respectively. In 1991, higher fullerenes, that is, C_{70}, C_{76}, C_{78} and C_{84}, were isolated (Figure 7.2) [5].

Besides the aforementioned techniques, pyrolysis of hydrocarbons and partial burning of carbon sources such as benzene are the most promising production strategies [6]. Variation of parameters such as pressure, temperature, and carbon source offers a way of enriching the raw product with respect to one type of fullerenes [7]. One of the most challenging aspects in this regard is the isolation of the desired

FIGURE 7.1 Biosphère aerial geodesic dome, St. Helen's Island, Montreal, Quebec, Canada. (Data from Ehrhardt, R., Wikimedia Commons, licensed under Creative Commons Attribution 2.0 Generic lizence, http://creativecommons.org/licenses/by/2.0/deed.en.)

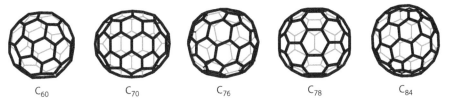

FIGURE 7.2   Structures of different fullerenes from left to right: C_{60} (I_h symmetry), C_{70} (D_{5h} symmetry), C_{76} (D_2 symmetry), C_{78} (C_{2v} symmetry), and C_{84} (D_{2d} symmetry).

FIGURE 7.3   Chocolate-covered buckyball as eye catcher on a Thanksgiving cake.

fullerene from other by-products in high purity, which is an essential criterion for the investigation of, for example, physical and chemical properties.

Consisting only of carbon atoms, fullerenes belong to the modifications of carbon, such as graphite and diamond. Additionally, they exhibit an esthetically pleasing three-dimensional structure (Figure 7.3). The presence of pentagons is crucial to endow spherical fullerenes with their curvature. The most prominent fullerene is the highly symmetric C_{60} with a point group of I_h consisting of 12 five-membered pentagons and 20 six-membered hexagons to afford a truncated icosahedron. Importantly, each of the pentagons is surrounded by five hexagons, which is the heart of the isolated pentagon rule (IPR) for stable fullerenes (Figure 7.4). For C_{60}, two different bond lengths are known—a shorter one of 135–140 pm (6,6 bond) and a longer one of 145–150 pm (5,6 bond) [8–11]. C_{60} reveals a delocalized π-system on its surface with almost an aromatic character. The true degree of hybridization for C_{60} is best described as sp^{2,278} [12].

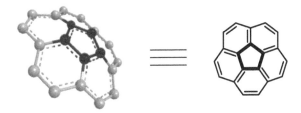

FIGURE 7.4   IPR demonstrated in corannulene.

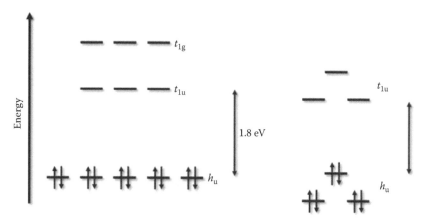

FIGURE 7.5 π-Orbital energy levels in C_{60} (left) and C_{70} (right).

Regarding molecular orbitals, all of the binding π-orbitals of C_{60} are completely filled with electrons. The highest occupied molecular orbital (HOMO) is fivefold degenerate and separated by 1.8 eV from the lowest unoccupied molecular orbital (LUMO) that is threefold degenerate (Figure 7.5) [13]. Fullerenes, in general, and C_{60}, in particular, are known to be excellent electron acceptors, possessing small reorganization energies in electron transfer reactions [13]. To this end, fullerenes are extensively studied in electron transfer applications such as solar energy conversion. Most notably, [60]PCBM [(6,6)-phenyl-C_{61}-butyric methyl ester] and [70]PCBM [(6,6)-phenyl-C_{71}-butyric methyl ester] are widely used in organic bulk heterojunction photovoltaic devices [14–16].

Regarding photophysical properties, C_{60} and C_{70} feature strong absorptions in the ultraviolet range of the solar spectrum. Their rather low fluorescence quantum yields are due to a dominating intersystem crossing. In turn, they exhibit high triplet quantum yields, which are close to unity, [17] and long-lived triplet excited lifetimes, whose values reach tens to hundreds of microseconds (Figure 7.6) [18]. To systematically fine-tune properties such as solubility, electrochemical potentials, and bonding motifs, a myriad of functionalization methodologies has been developed. Most prominent among the functionalization methods are cyclopropanations (Bingel–Hirsch adduct) [19], 1,3-dipolar cycloadditions (Prato adduct) [20], and heterofullerenes [21].

Ever since the early days of fullerene research, the encapsulation of atoms inside fullerenes has attracted enormous attention (Figure 7.7). Already in 1985, mass spectrometry indicated the existence of endohedral fullerenes [22]. In the early years, synthetic protocols and purification processes were under development and, as such, macroscopic amounts were rare. Since 1991 macroscopic amounts of endohedral fullerenes became accessible with the help of the arc-discharge method [23–26]. With steady progress in the synthesis and advanced separation techniques, the availability of endohedral metallofullerenes (EMFs) has been systematically improved and has resulted in more detailed studies [27–30]. The separation is mainly based on the different chemical and electrochemical reactivities of EMFs.

EMFs still retain the three-dimensional, spherical structure found for empty fullerenes, but host metals or metal cluster inside their carbon cage. Unlike nonmetallic

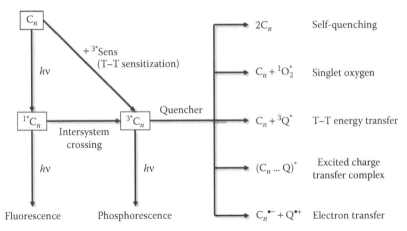

FIGURE 7.6 Different pathways of excited-state deactivation in fullerenes ($n = 60, 70, 76, 78, 84$, etc.).

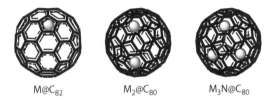

$$M@C_{82} \qquad M_2@C_{80} \qquad M_3N@C_{80}$$

FIGURE 7.7 Structures of different types of EMFs from left to right: $M@C_{82}$, $M_2@C_{80}$, and $M_3N@C_{80}$.

endohedral fullerenes such as $N@C_{60}$ [31], $He@C_{60}$ [32], and $H_2@C_{60}$ [33,34] in which the encapsulated species reveal very weak interactions with the fullerene, EMFs are positively charged cores surrounded by negatively charged fullerenes [35,36]. In this instance, substantial interactions between the encapsulated species and the fullerenes include strong orbital mixing and electronic reorganization, which lead to modified π-systems and novel physicochemical properties [35,37].

In general, monometallofullerenes that comprise only one metal atom are represented by $M^{3+}@(C_{2n})^{3-}$, $M^{2+}@(C_{2n})^{2-}$, or $M^{+}@(C_{2n})^{-}$, in which the encapsulated metals cover group 1–4 metals and most lanthanide metals [35,36]. Trivalent monometallofullerenes exhibit small band gaps on the order of 0.5 eV, which renders them highly amphoteric. Nevertheless, two [38], three [39], and also four metal atoms [36] have been trapped inside fullerenes. In the case of two metal atoms, the I_h isomer of C_{80} is the most favored fullerene—thermodynamically as well as kinetically—that accommodates two La (i.e., $La_2@C_{80}$), Ce (i.e., $Ce_2@C_{80}$), and so on [38]. The resulting electronic structure is best described as $M_2^{6+}@C_{80}^{6-}$. Dimetallofullerenes of the form $M_2@I_h$-C_{80} have similar first reduction potentials and about 0.5 V higher oxidation potentials than those found for monometallofullerene $La@C_{82}$. Still they are much lower than those of C_{60} [37].

In 1999, it was found that small amounts of nitrogen led to the formation of trimetallic nitride endohedral fullerenes $M_3N@C_{2n}$ (M = Sc, Y, Gd-Lu; $2n$ = 68–96) [40,41]. Importantly, the yields of trimetallic nitride endohedral fullerenes exceed those of most empty fullerenes rendering $Sc_3N@C_{80}$ the third most abundant fullerene under conventional arcing condition, after C_{60} and C_{70}. While their first reductions are shifted by 0.1–0.3 V to more negative values compared to C_{60}, their first oxidations are 0.5–0.6 V lower than those of C_{60} [37].

Besides pure metallic or pure nonmetallic species, methano- [42], oxide- [43], carbide- [44], cyano- [45], and sulfide-metal clusters [46] have also been stabilized inside fullerenes.

7.1.2 FULLERENES

7.1.2.1 Oxidation of Fullerenes—Ground-State Oxidation

In contrast to the versatile electron acceptor property of C_{60}, its electron donating property is rather poor. As a matter of fact, electrochemical studies have shown that C_{60} is easily reduced, but it renders very difficult to oxidize it [47–52]. C_{70} gives rise to a similar picture with slightly easier oxidation. In the ground state, the oxidation of, for example, C_{60} has been found to be limited to three irreversible one-electron processes resulting in the subsequent formation of fairly unstable $C_{60}^{\bullet+}$, C_{60}^{2+}, and C_{60}^{3+}, respectively [53]. The first oxidation potential of C_{60} is high. A value of +1.20 V versus ferrocene/ferrocinium, as, for example, measured in tetrachloroethylene [54], relates well to the fully occupied and, in turn, resonance-stabilized HOMO. The oxidation potential of C_{70} is +1.20 V versus ferrocene/ferrocinium [54]. Stabilization of $C_{60}^{\bullet+}$ requires the addition of trifluoromethanesulfonic acid (CF_3SO_3H) to 1,1',2,2'-tetrachloroethane (TCE), where electrochemically formed $C_{60}^{\bullet+}$ was found to be stable for several hours at 233 K—close to the freezing point of the solvent [55].

The electronic absorption spectrum of $C_{60}^{\bullet+}$ was first observed in γ-irradiated frozen polyatomic matrices at 77 K [56]. Later on, pulse radiolysis was introduced as a suitable tool for investigating elementary redox reactions of C_{60}, C_{70}, and so on in solution at room temperature [57]. Using pulse radiolysis techniques, $C_{60}^{\bullet+}$ has been generated for the first time via electron transfer to highly oxidizing radical cations of dichloroethane or CCl_4 with $k \geq 2 \times 10^{10}$ M^{-1} s^{-1} [58]. Similar to the one-electron reduced form of C_{60}, $C_{60}^{\bullet+}$ displays a characteristic band in the near infrared around 980 nm (Figure 7.8) [55–62].

A similar transient was found when C_{60} was codeposited with excess argon and argon resonance radiation was performed. Strong new absorptions at 973 nm ($C_{60}^{\bullet+}$) and 1068 nm ($C_{60}^{\bullet-}$) evolved in solid argon at 11 K. In the presence of CCl_4, which serves as an electron trap, the yield of the 1068 nm band was reduced, while little impact was noted on the 973 nm absorption [63]. In the case of $C_{70}^{\bullet+}$, however, various absorption bands have been reported, including 770–780 [61,62], 980 [60], and 1400 nm [60]. The overall low absorption coefficients impede reliable detection [64].

An interesting study includes the spontaneous oxidation of C_{60} by Au ions. When, for example, C_{60} was guided to make contacts with Au^{3+} ions in aqueous $HAuCl_4$ solution, electrons were spontaneously transferred from C_{60} to Au^{3+} resulting in $C_{60}^{\bullet+}$ and Au nanoparticles on a C_{60} layer. This spontaneous electron transfer occurs due to the

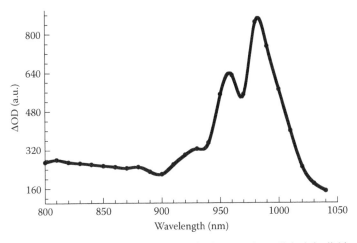

FIGURE 7.8   Near-infrared spectrum of $C_{60}^{\bullet+}$ obtained upon pulse radiolysis in dichloroethane.

galvanic displacement from C_{60} to Au^{3+} ions owing to the energy difference between the Fermi energy level of C_{60} and the standard reduction potential of Au^{3+} [65].

In light of the latter, stable solutions of $C_{60}^{\bullet+}$ and $C_{70}^{\bullet+}$ were prepared by using the conjugate anion of the superacid $H(CB_{11}H_6X_6)$ (where X = Cl or Br) and a strong chemical oxidant, that is, hexabrominated phenylcarbazole radical cation, concluding that the non-nucleophilic carborane counter anion was critical to their success [66].

7.1.2.2   Oxidation of Fullerenes—Excited-State Oxidation

In the excited state, oxidative quenching of triplet excited state of C_{60} has been probed with a series of electron acceptors ranging from tetracyano-*p*-quinodimethane (TCNQ) and tetracyanoethylene (TCNE) to chloraniline [58,60,61,67]. Although very fast quenching of the triplet excited states of C_{60} and C_{70} occurs in all of these cases, the underlying processes are surprisingly not accompanied by the separation of the free radical pairs. This behavior stems from the triplet exciplex formation, which prevents escaping from the cage. Only in the case of TCNE the exciplex has been considered as to feature a radical ion pair character [67]. In a complementary EPR investigation, it was shown that the stable radical anion $TCNE^{\bullet-}$ interacts with the triplet excited state of C_{60} [68]. Chemically induced dynamic electron polarization of the reduced acceptor, $TCNE^{\bullet-}$, has been demonstrated in these studies.

Importantly, in the presence of scandium triflate [$Sc(OTf)_3$], which binds to the product of an electron transfer, an efficient photoinduced electron transfer from the triplet excited state of C_{60} to *p*-chloranil and/or *p*-benzoquinone occurs to afford $C_{60}^{\bullet+}$ (Figure 7.9). No photoinduced electron transfer occurs, however, from the triplet excited state of C_{60} to *p*-chloranil in the absence of $Sc(OTf)_3$ in benzonitrile. Clearly, $Sc(OTf)_3$ is catalytically active in the photoinduced electron transfer from

$$^{3*}C_{60} + Cl_4Q \xrightarrow{+Sc^{3+}} [(C_{60})^{\bullet+}(Cl_4Q^{\bullet-} - Sc^{3+})] \xrightarrow{-Sc^{3+}} C_{60} + Cl_4Q + Sc^{3+}$$

FIGURE 7.9   Formation of the radical intermediate complex $[(C_{60})^{\bullet+}(Cl_4Q^{\bullet-} Sc^{3+})]$.

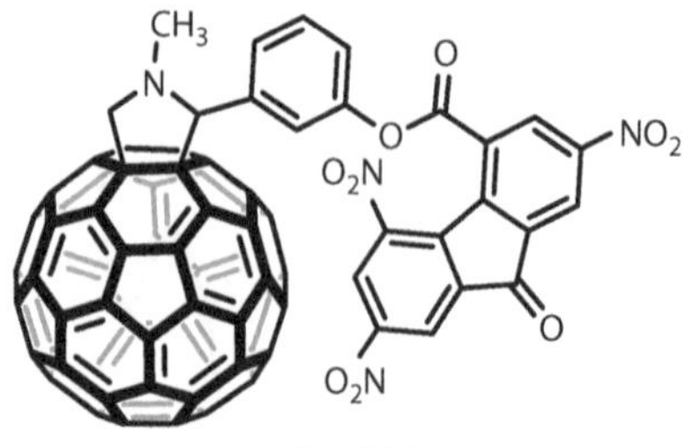

C$_{60}$-TNF

FIGURE 7.10 Chemical structure of C$_{60}$-TNF electron donor–acceptor conjugate. (Data from Hou, H.-Q. et al., *Chem. Phys. Lett.*, 203, 555–559, 1993.)

the triplet excited state by the binding to semiquinone radical anions. The earlier pathway enabled the systematic generation of C$_{60}$·$^+$ and C$_{70}$·$^+$ and the radical cations of their derivatives produced by the Sc(OTf)$_3$ promoted electron transfer from the corresponding triplet excited states to *p*-chloranil [61].

A different pulse radiolysis approach involved the triplet excited state of C$_{60}$, which was obtained following pulse radiolysis of benzene solution of C$_{60}$. The latter reveals electron transfer to either CCl$_4$ or CCl$_3$CF$_3$ with a radiation-induced electron transfer rate constant of 7.8 × 10^6 M^{-1}s^{-1} [69].

Until now, there has been only one example of a covalently linked electron donor–acceptor conjugate demonstrating the role of C$_{60}$ as an excited state electron donor. The photophysics of an electron-accepting trinitrofluorenone (TNF) linked to C$_{60}$–C$_{60}$–TNF conjugate shown in Figure 7.10 gives rise to a photoinduced electron transfer from C$_{60}$ to TNF only in the presence of Sc(OTf)$_3$ upon photoexcitation (Figure 7.11), whereas the triplet excited state of C$_{60}$ is formed in the absence of Sc(OTf)$_3$. Very similarly, replacing TNF with TCNQ lacks in the corresponding C$_{60}$–TCNQ any appreciable electron transfer activity. Instead, the C$_{60}$ singlet excited state intersystem crosses to the corresponding triplet excited state and, finally, to the singlet ground state [70].

Likewise, employing C$_{60}$ as a ground-state electron donor in combination with an electron-accepting perylenebisimide (PDI)—*vide infra*—fails upon photoexcitation of the latter to result in electron transfer. Instead, the photoreactivity of C$_{60}$–PDI is exclusively governed by a cascade of energy transfer processes, namely, shuttling the excited state energy back and forth between PDI and C$_{60}$ [71].

7.1.2.3 Oxidation of Fullerenes—Cosensitized Oxidation

An indirect approach, in light of oxidizing C$_{60}$, has been demonstrated via cosensitization (Figure 7.12). In particular, the photosensitized electron transfer takes place between the singlet excited state of *N*-methylacridinium hexafluorophosphate (MA) and biphenyl (BP). In this process, the primary MA electron acceptor is excited and abstracts an electron from the BP electron donor. BP·$^+$ is the ultimate oxidant for C$_{60}$. The lifetime of the generated BP·$^+$ enables a secondary electron transfer from C$_{60}$ [58].

A similar concept has also been pursued employing pulse radiolysis, rather than photolysis to generate (BP·$^+$) and various polycyclic arene π-radical cations. These

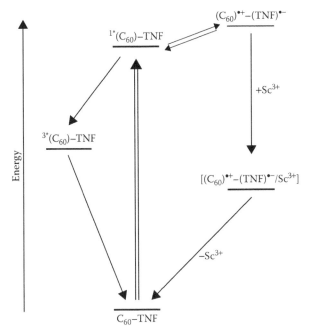

FIGURE 7.11   Energy level diagrams for C_{60}-TNF in benzonitrile reflecting the different pathways of electron transfer on the right.

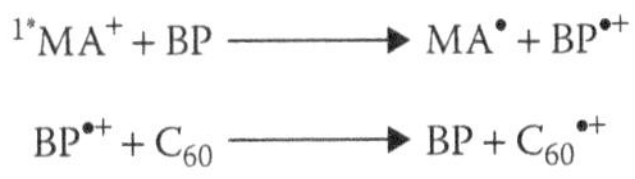

$$^{1*}MA^+ + BP \longrightarrow MA^\bullet + BP^{\bullet +}$$

$$BP^{\bullet +} + C_{60} \longrightarrow BP + C_{60}^{\bullet +}$$

FIGURE 7.12   Cosensitized oxidation of C_{60} with MA and BP.

display fine-tuned oxidation potentials that all enable the subsequent generation of $C_{60}^{\bullet +}$. Varying the oxidation strength of the arene π-radical cations helped to vary the free energy changes. In such charge-shift type of electron transfer reactions, the solvation before and after the electron transfer may be largely canceled out when the free energy change of electron transfer is expected to be rather independent of the solvent polarity. The reaction under investigation gave rise to rate constants for BP, t-stilbene, m-terphenyl, and naphthalene with C_{60} in dichloromethane, which vary between 2.5×10^9 and 7.9×10^9 M^{-1} s^{-1}. The driving forces calculated on the basis of the difference in the respective arenes and C_{60} ionization potentials (IP = 7.59 eV) show no linear correlation with the measured rate constants for the electron transfer reactions. In fact, it is gratifying to note that these results indicate a decrease of the rate constant at higher driving forces [59].

The substantially reduced ionization potentials of C_{76} (D_2) (IP = 7.1 eV) and C_{78} (C_{2v}) (IP = 7.05 eV) relative to C_{60} (I_h) (IP = 7.59 eV) are much more beneficial for following the envisaged intermolecular transfer processes, which now become possible with a significantly increased number of suitable electron-accepting substrates. C_{76} and C_{78} reveal striking parabolic dependencies, that is, a decrease of the rate

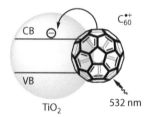

FIGURE 7.13 Photoinduced charge transfer between TiO_2 semiconductors and photoexcited C_{60}. (Data from Guldi, D. M. and Asmus, K.-D., *J. Am. Chem. Soc.*, 119, 5744–5745, 1997.)

constants with increasing free energy. Most interestingly, the pronounced decrease toward the highly exothermic region represents one of the rare confirmations of the existence of the Marcus-inverted region in a truly intermolecular oxidative electron transfer. The ratio between the maximum rate and the rate at the high exothermic end in the set of experiments is 7. From these experiments, an experimental value of approximately 0.6 eV was deduced for the total reorganization energy of C_{76} and C_{78} in oxidative charge-shift processes in dichloromethane [72].

7.1.2.4 Oxidation of Fullerenes—Heterogeneous Approaches

Common to any of the aforementioned approaches is a homogeneous electron transfer. Heterogeneous electron transfer was first reported in 532 nm laser excitation experiments of C_{60}-coated TiO_2 samples (Figure 7.13) [73]. The absorption changes reveal decreased absorptions in the 620 nm region and increased absorptions in the 400 nm region, which prompt to irreversible chemical changes occurring with C_{60} on TiO_2 surface that are insensitive to oxygen. At low surface coverage, the transient absorption spectra of both degassed and air-equilibrated samples of C_{60}/TiO_2 are similar. Some irreversible changes were also seen on an Al_2O_3 surface, but this fraction was small compared to the C_{60}/TiO_2 sample. The higher yields of the oxidation product observed on the TiO_2 surface are attributed to the greater reactivity of the semiconducting support material. The linear dependence of the transient formed on the TiO_2 surface on the square of the laser dose shows that the formation of this oxidation product is a biphotonic process. Notably, the oxidation potential of the singlet excited state of C_{60} is 0.1 V versus normal hydrogen electrode. In other words, one would not expect the singlet excited state of C_{60} to directly inject electrons from its singlet excited state into TiO_2 particles. Therefore, more than a single photon is necessary to induce the photoinjection process. The photoinjection of electrons into TiO_2 particles is likely to be followed by the interaction of $C_{60}^{\bullet+}$ with either the lattice oxygen of TiO_2 or the chemisorbed oxygen that results in the formation of an epoxide-type species. C_{70} reacts essentially similar to C_{60} [73].

7.1.3 ENDOHEDRAL METALLOFULLERENES

7.1.3.1 Oxidation of EMFs—Ground-State Oxidation

EMFs bear, in contrast to empty fullerenes, additional atoms, ions, or clusters within their inner spheres. What renders EMFs particularly appealing is that a significant control over the chemical and physical properties has been realized by just changing the nature and composition of the encapsulated species [35,37]. For instance, $M_2@I_h$-C_{80} (M = Ce or La) reveal, in stark contrast to empty fullerenes, strong electron-accepting and electron-donating features. Specifically, $Lu_3N@C_{80}$ and $La_2@C_{80}$ both exhibit much lower oxidation potentials than C_{60} and, therefore, exhibit much better electron-donating property [74–77].

The redox potentials of EMFs substantially differ from those of empty fullerenes [78]. In general, $La@C_{82}$ and its analogs monometallofullerenes reveal fully reversible oxidations/reductions at very low potentials [37]. In fact, the first oxidation of $La@C_{2v}$-C_{82} is +0.07 V versus ferrocene/ferrocenium [79,80]. The first oxidations of dimetallofullerenes $M_2@C_{80}$ are shifted by about 0.5 V to more positive values relative to $La@C_{2v}$-C_{82} [37]. As a matter of fact, $La_2@I_h$-C_{80} exhibits its first oxidation at +0.56 V [74], whereas trimetallic nitride fullerenes such as $Sc_3N@C_{80}$ exhibit slightly more positive oxidations, which are in the range from +0.6 and +0.7 V [78]. Noteworthy is in this context that different fullerene isomers of the same EMF reveal appreciably different oxidations [81,82]. The first oxidations of the D_{5d} and I_h isomers of $Sc_3N@C_{80}$ differ by 0.27 V. Therefore, selective chemical oxidation of the D_{5d} isomer with tris(*p*-bromophenyl)aminium hexachloroantimonate, which features an oxidation potential between the first oxidations of the I_h and D_{5d} isomers, assists in the purification to isolate isomerically pure $Sc_3N@I_h$-C_{80} [83].

7.1.3.2 Oxidation of EMFs—Excited-State Oxidation

The complementary use of spectroscopy and electrochemistry shed light onto the supramolecular interactions of calixarene scaffold bearing different bisporphyrins as hosts with $Sc_3N@C_{80}$ and $Lu_3N@C_{80}$ as guests. Detailed studies document that an electron transfer from $Sc_3N@C_{80}$ and $Lu_3N@C_{80}$ to the porphyrin is operative in the EMF host–guest complexes [84].

$Ce_2@I_h$-C_{80} has evolved as a versatile building block toward synthesizing novel electron donor–acceptor conjugates, such as $Ce_2@I_h$-C_{80}-ZnP. Detailed spectroscopic analyses revealed unusually strong interactions between $Ce_2@I_h$-C_{80} and ZnP, despite employing a flexible 2-oxy-ethyl butyrate linkage. These interactions are driven by the giant π-system with highly delocalized negative charges. Most impressive is the unprecedented change in charge transfer chemistry in toluene and tetrahydrofuran (THF) versus benzonitrile and dimethylformamide (Figure 7.14). In this context, a systematic investigation of the charge transfer chemistry demonstrated a reductive charge transfer—formation of $(Ce_2@I_h$-$C_{80})^{\bullet-}$-$(ZnP)^{\bullet+}$—in nonpolar media, whereas an oxidative charge transfer—formation of $(Ce_2@I_h$-$C_{80})^{\bullet+}$-$(ZnP)^{\bullet-}$—dominates in polar media. Reduction of the $[Ce_2]^{6+}$ cluster is exothermic in all solvents. Notably weak is the electronic coupling between the $[Ce_2]^{6+}$ cluster and the electron-donating ZnP. The oxidation of $C_{80}{}^{6-}$ and the simultaneous reduction of ZnP, however, necessitates solvent stabilization. In such a case, the strongly exothermic $(Ce_2@I_h$-$C_{80})^{\bullet-}$-$(ZnP)^{\bullet+}$ radical ion pair state formation is compensated by a nonadiabatic charge transfer, and a $C_{80}{}^{6-}$/ZnP electronic matrix element, which exceeds that for $[Ce_2]^{6+}$/ZnP. The corresponding rate constants are 1.2×10^{12} s^{-1} in toluene, 1.3×10^{12} s^{-1} in THF, 1.4×10^{12} s^{-1} in benzonitrile, and 2.3×10^{12} s^{-1} in dimethylformamide. For the charge recombination step, the rate constants 1.8×10^{10} and 8.6×10^9 s^{-1} were measured in benzonitrile and dimethylformamide, respectively [85].

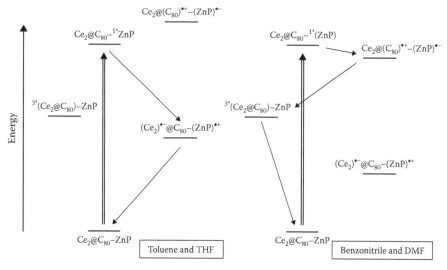

FIGURE 7.14 Energy level diagrams for Ce$_2$@C$_{80}$-ZnP in toluene and THF (left) and benzonitrile and DMF (right) reflecting the different pathways of excited-state deactivation that lead to different charge transfer products.

As the latest development, two stable electron donor–acceptor conjugates with La$_2$@I$_h$-C$_{80}$ and Sc$_3$N@I$_h$-C$_{80}$, on one hand, and ZnP, on the other, have been prepared via [2 + 1] cycloaddition reactions of a diazo precursor. From crystallography and ^{13}C nuclear magnetic resonance (NMR) studies with La$_2$@I$_h$-C$_{80}$-ZnP and Sc$_3$N@I$_h$-C$_{80}$-ZnP, a [6,6]-fulleroid addition pattern was determined. Still, subtly different conformations, that is, a rigid and a comparatively more flexible structure, emerge for La$_2$@I$_h$-C$_{80}$-ZnP and Sc$_3$N@I$_h$-C$_{80}$-ZnP, respectively. In line with this aforementioned difference are the electrochemical measurements, which imply appreciably stronger I$_h$-C$_{80}$/ZnP interactions in La$_2$@I$_h$-C$_{80}$-ZnP. Density functional calculations reveal that in the lowest energy conformations of La$_2$@I$_h$-C$_{80}$-ZnP, La$_2$@I$_h$-C$_{80}$ and ZnP assume van der Waals distances. Such subtle changes also exert an impact on the excited state reactivity. For example, spectroscopic analyses corroborate that the fast deactivation of the ZnP singlet excited state results in a fast charge separation ($>10^{12}$ s^{-1}) process and yields a radical ion pair, whose nature, (La$_2$@I$_h$-C$_{80}$)$^{\bullet-}$-(ZnP)$^{\bullet+}$ versus (La$_2$@I$_h$-C$_{80}$)$^{\bullet+}$-(ZnP)$^{\bullet-}$, varies with solvent polarity. Notably, charge separation evolving from the ZnP singlet excited state (2.0 eV) is thermodynamically feasible in all of the tested solvents, since the La$_2$@I$_h$-C$_{80}$$^{\bullet-}$-ZnP$^{\bullet+}$ radical ion pairs have energy levels of 0.72 eV in toluene, 0.52 eV in THF, 0.46 eV in benzonitrile, and 0.45 eV in dimethylformamide—according to electrochemical measurements. On the contrary, the (La$_2$@I$_h$-C$_{80}$)$^{\bullet+}$-(ZnP)$^{\bullet-}$ radical ion pair state energy is 2.3 eV in toluene, 2.1 eV in THF, 2.0 eV in benzonitrile, and 1.9 eV in dimethylformamide, which renders charge separation from the ZnP singlet excited state thermodynamically feasible only in the more polar solvents, namely, benzonitrile and dimethylformamide. In terms of charge recombination, notable is the lack of any transient that relates to either the ZnP or the La$_2$@I$_h$-C$_{80}$ triplet excited states (i.e., at 840 or 580 nm), which corroborates that the $(La_2@I_h\text{-}C_{80})^{\bullet-}\text{-}(ZnP)^{\bullet+}$ radical ion pair state recombines with 4.3×10^9 s^{-1} in toluene and 5.8×10^9 s^{-1} in THF directly to the ground state. In addition, the time absorption profiles associated with the $(La_2@I_h\text{-}C_{80})^{\bullet+}\text{-}(ZnP)^{\bullet-}$ radical ion pair emerge toward the $La_2@I_h\text{-}C_{80}$ triplet excited state. In fact, multiwavelength analyses provide the rate constants of 9.2×10^9 and 1.6×10^{10} s^{-1} in benzonitrile and dimethylformamide, respectively [86].

In $La_2@C_{80}$-TCAQ, EMF $La_2@C_{80}$ is covalently linked to the strong electron acceptor 11,11,12,12-tetracyano-9,10-anthra-p-quinodimethane (TCAQ). Although weak electronic couplings dictate the interactions between $La_2@C_{80}$ and TCAQ in the ground state, in the excited state the formation of the $(La_2@C_{80})^{\bullet+}\text{-}(TCAQ)^{\bullet-}$ radical ion pair state evolves in nonpolar and polar media. Here, $La_2@C_{80}$ in its singlet excited state (1.4 ± 0.2 eV) powers an electron transfer to yield a spatially separated radical ion pair state (1.15 eV). The one-electron reduced TCAQ and the one-electron oxidized $La_2@C_{80}$ decays with a lifetime of several hundred picoseconds to populate the triplet excited state (1.0 ± 0.1 eV) and, subsequently, the ground state in the absence of molecular oxygen. On one hand, the electron acceptor character of TCAQ and, on the other hand, the electron donor character of $La_2@C_{80}$ are decisive to afford a radical ion pair state in nonpolar as well as polar media [87].

In a covalently linked $Lu_3N@C_{80}$–PDI conjugate (Figure 7.15), electron transfer evolves from the electron-donating $Lu_3N@C_{80}$ to the electron-accepting PDI singlet excited

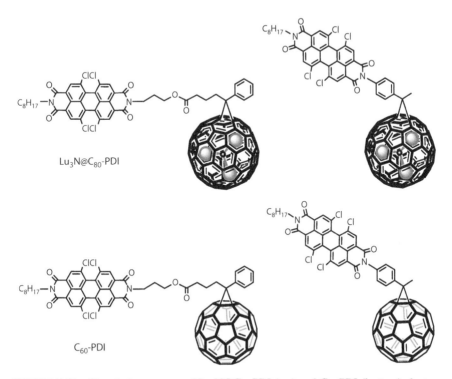

FIGURE 7.15 Chemical structures of $Lu_3N@C_{80}$-PDI (top) and C_{60}-PDI (bottom) electron donor–acceptor conjugates.

state to yield $(Lu_3N@C_{80})^{\bullet+}-(PDI)^{\bullet-}$ (1.43 eV). The latter is metastable and decays with a lifetime of 120 ± 10 ps in toluene, 100 ± 10 ps in chlorobenzene, and 45 ± 5 ps in benzo-nitrile. The product of charge recombination is the PDI triplet excited state [71].

However, in the above-mentioned $Lu_3N@C_{80}$–PDI conjugate, a close-contact geometry turned out to be unfavorable for long-lived charge separation events. As such, a linear $Lu_3N@C_{80}$–PDI conjugate was probed and the outcome is sketched in Figure 7.16. In the excited state, an intramolecular electron transfer commences with the photoexcitation of PDI. The initially formed singlet radical ion pair state, $^1(Lu_3N@I_h\text{-}C_{80})^{\bullet+}-(PDI)^{\bullet-}$, undergoes radical ion pair intersystem crossing induced by electron nuclear hyperfine coupling within the radicals to produce the triplet radical ion pair state, $^3(Lu_3N@I_h\text{-}C_{80})^{\bullet+}-(PDI)^{\bullet-}$. Key to this interconversion is certainly the presence of the Lu_3N cluster. The accordingly formed radical ion pairs recombine much faster from the singlet manifold than from the triplet manifold with life-times as long as 102 ns (THF) and as short as 28 ps (DMF), respectively. Overall, the radical ion pair state lifetime increase is about 1000-fold [88].

7.2 BIOMEDICAL APPLICATIONS OF EMFs

Shasha Zhao and Xing Lu

Because of the presence of metallic species inside the cage, EMFs are potential nano-medicines that show both diagnosis and therapeutic applications. Multifunctional contrast agents and antitumor reagents are particularly attractive and have received wide attention.

7.2.1 X-Ray Contrast Agents

Dorn et al. reported in 2002 that $Lu_3N@C_{80}$ served as X-ray contrast agent because of large cross section of the lutetium (Lu) atoms. $Lu_3N@C_{80}$ shows X-ray contrast when it was distributed onto a nonabsorbing Teflon block and irradiated with X-ray. By contrast, the Teflon blocks without the lutetium species exhibit no contrast (Figure 7.17). The fact that the C_{60}-containing block has no contrast reveals that the contrast cannot be attributed to the fullerene carbon cage.

Compounds encapsulating mixed-metal species of gadolinium/lutetium and holmium/lutetium, for example, $Lu_{3-x}A_xN@C_{80}$ (A = Gd or Ho; $x = 0$–2), may have applications as multifunctional contrast agents for X-ray, magnetic resonance imaging (MRI), and radiopharmaceuticals. The high-contrast ability of these agents that enhance a variety of spectroscopic methods would minimize exposure to patients and simultaneously provide similar contrast images [89]

7.2.2 MRI Contrast Agents

Gadolinium (Gd)–chelate complexes, for example, Gd–diethylenetriamino-pentaacetic acid (DTPA), are the most commercially available MRI contrast agents. However, the application of EMFs as MRI contrast agents was widely reported in recent years because of their high-contrast abilities.

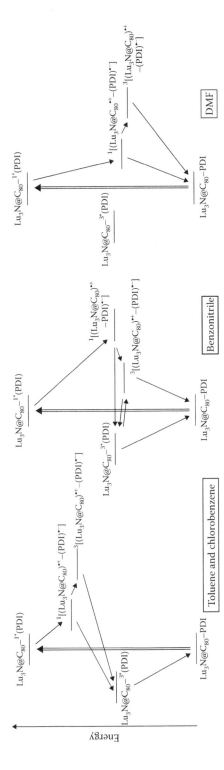

FIGURE 7.16 Energy level diagrams for $Lu_3N@C_{80}$-PDI linear in toluene and chlorobenzene (left), benzonitrile (middle), and DMF (right) reflecting the different pathways of excited-state deactivation that lead to different charge transfer products.

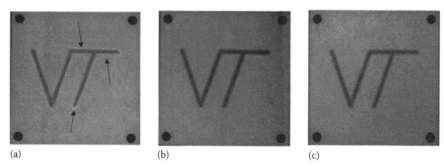

FIGURE 7.17 X-ray photographs of contrast provided by $Lu_3N@C_{80}$ on a Teflon block (a), C_{60} fullerene on a Teflon block (b), and a blank Teflon block (c). The samples were exposed for 3.5 min at 30 kV. (Reproduced with permission from Iezzi, E. B. et al., *Nano Lett.*, 2, 1187–1190, 2002. Copyright 2002 American Chemical Society.)

In 2001, Shinohara et al. synthesized the water-soluble Gd–fullerenol [Gd@ $C_{82}(OH)_{40}$] and measured its proton relaxivities r1 and r2 *in vitro*. The water proton relaxivity r1 (the effect on $1/t1$) (81 mM^{-1} s^{-1}) is significantly higher (more than 20-fold) than Gd–DTPA (3.9 $mM^{-1}s^{-1}$) at 1.0 T, pH = 7.5 [90]. Late in 2002, the same group prepared water-soluble multihydroxyl lanthanoid (La, Ce, Gd, Dy, and Er) metallofullerenols, $M@C_{82}(OH)_n$, the water proton relaxivities r1 (0.8–73 mM^{-1} s^-) and r2 (1.2–80 $mM^{-1}s^-$) of which are higher than those of the corresponding free ions and lanthanoid–DTPA chelate complexes. The significantly high water proton relaxivities were in connection with the dipole–dipole relaxation, as well as a significant decrease of the overall molecular rotational motion of these EMFs [91]. The corresponding results are shown in Figure 7.18.

A large number of $Gd@C_{82}$ derivatives with hydroxyls and other water-soluble groups have been prepared and studied these years. Qu et al. synthesized Gd@ $C_{82}(OH)_{22}$ and studied its imaging efficiency in mice, exploring the relationship between the imaging efficiency and the hydroxyl number in $Gd@C_{82}(OH)_n$ in vivo.

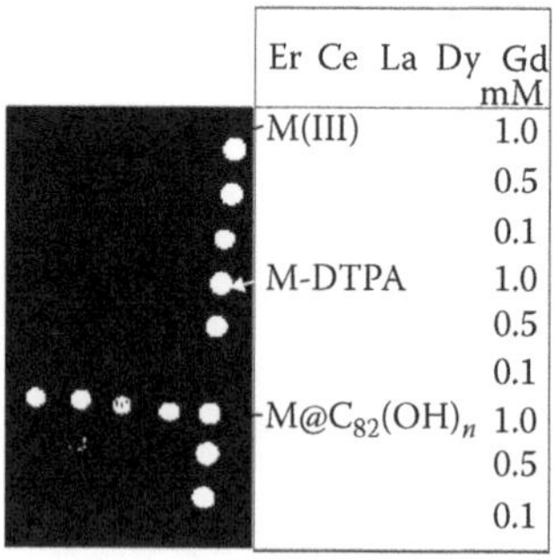

	Er Ce La Dy Gd
	mM
M(III)	1.0
	0.5
	0.1
M-DTPA	1.0
	0.5
	0.1
$M@C_{82}(OH)_n$	1.0
	0.5
	0.1

FIGURE 7.18 Phantom NMR images of various metallofullerenol solutions (together with those of lanthanoid ions and lanthanoid–DTPA complexes) at 1.0, 0.5, and 0.1 mmol metal/L and at 20°C. (Reproduced with permission from Kato, H. et al., *J. Am. Chem. Soc.*, 125, 4391–4397, 2003. Copyright 2003 American Chemical Society.)

It is found that the proton relaxivity of $Gd@C_{82}(OH)_{22}$ is lower than that of $Gd@$ $C_{82}(OH)_{40}$, but the value is still higher than the commercial Gd–DTPA MRI contrast agent [92]. Zhang et al. prepared $Gd@C_{82}(OH)_{16}$ and studied both proton relaxivity and MRI images, which showed high proton relaxivity [93]. Xing et al. synthesized paramagnetic $Gd@C_{82}(OH)_{22\pm2}$ nanoparticles that owned ordered microstructures and they discovered an enhancement of 12 times higher MRI relaxivity r1 *in vitro* or *in vivo* compared to Gd–DTPA. In addition, $Gd@C_{82}(OH)_{22\pm2}$ nanoparticles (65 nm) could successfully escape the reticuloendothelial system (RES) uptake *in vivo* [94]. Many studies were certainly performed on conventional $Gd@C_{82}$ to make its water-soluble derivatives. Shu et al. prepared a series of novel water-soluble $Gd@C_{82}$ derivatives such as $Gd@C_{82}O_6(OH)_{16}$ $(NHC_2H_4CO_2H)_8$ (AAD-EMF$_S$), $Gd@C_{82}O_6(OH)_{16}$ $(NHCH_2CH_2COantiGFP)_5$, and $Gd@C_{82}O_2(OH)_{16}$ $(C(PO_3Et_2)_2)_{10}$ and investigated their MRI properties [95–97].

In 2003, Bolskar et al. prepared $Gd@C_{60}[C(COOCH_2CH_3)_2]_{10}$ and its hydrolyzed water-soluble form, $Gd@C_{60}[C(COOH)_2]_{10}$. They found that it showed the first non-RES-localizing behavior with a relaxivity r1 (4.6 mM^{-1} s^{-1} at 20 MHz and 40°C) *in vivo* MRI biodistribution study (Figure 7.19) [98]. In 2005, Laus et al. measured proton relaxivity for the water-soluble metallofullerenes, $Gd@C_{60}[C(COOH)_2]_{10}$ and $Gd@C_{60}(OH)_{27}$. The ^1H nuclear magnetic relaxation dispersion/dynamic light scattering study showed that aggregation formed in aqueous solution of water-soluble gadofullerenes that can be destroyed by adding salts such as phosphate and sodium halides [99]. Laus et al. recorded multiple-field ^{17}O and ^1H relaxation rates of $Gd@C_{60}(OH)_{27}$ and $Gd@C_{60}[C(COOH_yNa_{1-y})_2]_{10}$ with variable temperature in

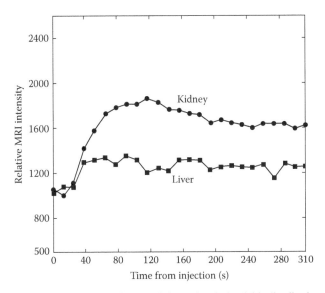

FIGURE 7.19 Representative in vivo MRI intensity-derived biodistribution data showing the $Gd@C_{60}[C(COOH)_2]_{10}$ signal enhancement within the first 5 min of administration, revealing rapid renal uptake with a minimum of liver uptake. (Reproduced with permission from Bolskar, R. D. et al., *J. Am. Chem. Soc.*, 125, 5471–5478, 2003. Copyright 2003 American Chemical Society.)

both aggregated and disaggregated states (monomers) and discovered that the rapid exchange of water molecules with bulk contributes to the high relaxivity of the aggregated gadofullerenes [100].

$Gd_3N@C_{80}$ is certainly more attractive because of the large number of Gd atoms encapsulated. Fatouros et al. functionalized $Gd_3N@C_{80}$ with poly(ethylene glycol) (PEG) units and hydroxylated the carbon cage to improve water solubility and bio-distribution, resulting in $Gd_3N@C_{80}[DiPEG5000(OH)_x]$. In vitro and in vivo imaging studies showed its relaxivity r1 to be 102, 143, and 32 mM^{-1} s^{-1} at 0.35, 2.4, and 9.4 T, respectively, which is markedly higher than that of the commercial reagent [101]. In 2008, MacFarland et al. synthesized a series of novel derivatives of $Gd_3N@C_{80}$ that are called hydrochalarones, namely, $Gd_3N@C_{80}$-hydrochalarone and reported that hydrochalarone-6 exhibits the relaxivity r1 as high as 205 mM^{-1} s^{-1} [102]. In 2009, Dorn et al. synthesized $Gd_3N@C_{80}(OH)_{26}(CH_2CH_2COOM)_{16}$ (M = Na or H) and claimed their longitudinal and transverse relaxivities r1 and r2 to be 207 and 282 mM^{-1} s^{-1} (per C_{80} cage) at 2.4 T, respectively, which are 50 times larger than those of Gd^{3+} poly(aminocarboxylate) complexes, for example, commercial Omniscan and Magnevist [103]. Later in 2010, Fatouros synthesized and charac-terized $Gd_3N@C_{80}[DiPEG(OH)_x]$ and measured its ^1H MRI relaxivity with various molecular weights (300–5000 Da) at 0.35, 2.4, and 9.4 T. It is revealed that $Gd_3N@C_{80}[DiPEG(OH)_x]$ has the highest relaxivities r1 for the 350/750 Da PEG derivatives, 237/232 mM^{-1} s^{-1} for r1, and 460/398 mM^{-1} s^{-1} for r2 at a clinical-range magnetic field of 2.4 T [104].

There is a great need for specific targeted nanoplatforms capable of deliver-ing both therapeutic and imaging agents directly to invasive tumor cells. In 2011, Fillmore et al. functionalized and conjugated $Gd_3N@C_{80}$ with a tumor-specific peptide interleukin (IL)-13 and assessed their ability to bind to glioma cells *in vitro*, leading to the conclusion that IL-13 peptide-conjugated gadolinium metallofuller-enes could serve as a platform to deliver imaging and therapeutic agents to tumor cells [105].

Compared with other Gd-containing EMFs, the abundance of $Sc_xGd_{3-x}N@C_{80}$ (x = 1, 2) makes them very competitive candidates for next-generation high-efficiency MRI contrast agents. In 2007, Wang et al. prepared two water-soluble fullerene derivatives $[Sc_xGd_{3-x}N@C_{80}O_m(OH)_n$ (x = 1, 2; m ≈ 12; n ≈ 26)], the relaxivities r1 of $Sc_2GdN@C_{80}O_{12}(OH)_{26}$ and $ScGd_2N@C_{80}O_{12}(OH)_{26}$ were 20.7 and 17.6 $mM^{-1}s^{-1}$, respectively, which are much higher than that of the commercial MRI contrast agent (Gd–DTPA, 3.2 mM^{-1} s^{-1}) [106]. In 2010, Braun et al. reported the synthesis of a modular system based on $GdSc_2N@C_{80}$ named *Gd-cluster@-BioShuttle*, which con-tains three functional modules: $GdSc_2N@C_{80}$ acting as a contrast agent and cova-lently linking to the nuclear address module and then connecting to the peptide that can facilitate the passage across cell membranes. Gd-cluster@-BioShuttle showed both high proton relaxivity and signal enhancement at very low Gd concentrations in comparison with the commonly used contrast agents such as Gd–DTPA, result-ing in a 500-fold gain of sensitivity. This modularly designed contrast agent repre-sents a new tool for improved monitoring and evaluation of interventions at the gene transcription level. Using the Gd-cluster@-BioShuttle as MRI contrast agent allows an improved evaluation of radiotherapy or chemotherapy-treated tissues [107].

7.2.3 ANTIMICROBIAL AND ANTITUMOR ACTIVITY

Phillips et al. reported the singlet oxygen sensitization by $Sc_3N@C_{80}$ in 2009. They embedded $Sc_3N@C_{80}$ into polystyrene-*block*-polyisoprene-*block*-polystyrene copolymer pressure-sensitive adhesive films, which exhibited an antimicrobial activity because of in situ generated O_2 [108].

In 2005, Chen et al. prepared $Gd@C_{82}(OH)_{22}$ nanoparticles (22 nm) that exhibited a high antineoplastic activity in H22 hepatoma-implanted mice [110]. In addition, the particles can improve immunity and interfere with tumor invasion in normal muscle cells without causing toxicity to cells [109]. In 2008, Yin et al. reported the antitumor activities of $[Gd@C_{82}(OH)_{22}]_n$ nanoparticles in vitro and in vivo. Because reactive oxygen species (ROS) are known to be implicated in the etiology of a wide range of human diseases, including cancer, the present findings demonstrate that the potent inhibition of $[Gd@C_{82}(OH)_{22}]_n$ nanoparticles on tumor growth likely relates with a typical capacity of scavenging ROS [110]. In 2009, Yin et al. found $C_{60}(C(COOH)_2)_2$, $C_{60}(OH)_{22}$, and $Gd@C_{82}(OH)_{22}$ can protect cells against H_2O_2-induced oxidative damage and scavenge all physiologically relevant ROS [111].

In 2009, Liu et al. reported that $Gd@C_{82}(OH)_{22}$ nanoparticles at a high concentration markedly enhanced immune responses and stimulated immune cells to release more cytokines, helping eliminate abnormal cells [112]. In 2010, Liang et al. found that $[Gd@C_{82}(OH)_{22}]_n$ nanoparticles overcome tumor resistance to cisplatin by increasing its intracellular accumulation through the mechanism of restoring defective endocytosis [113]. In 2013, this group also studied the antiangiogenesis activity of the $Gd@C_{82}(OH)_{22}$ fullerenic nanoparticles (f-NPs) as angiogenesis inhibitors and found that the f-NPs can simultaneously downregulate more than 10 angiogenic factors in the messenger ribonucleic acid (mRNA) level (Figure 7.20). Moreover, the efficacy of the treatment using f-NPs in nude mice was comparable to the clinic anti-cancer drug paclitaxel without pronounced side effects [114].

In 2012, the same group developed a novel approach by the use of $Gd@C_{82}(OH)_{22}$ metallofullerenol nanoparticles to inhibit tumor metastasis. Instead of directly killing the cells, the NPs inhibited tumor metastasis mainly through a matrix metalloproteinase (MMP)-inhibitory process. The formation of a thick fibrous cage may serve as a *prison* capable of confining the invasive tumor cells in their primary site [115]. In 2012, Kang et al. reported that $Gd@C_{82}(OH)_{22}$ effectively blocked tumor growth in human pancreatic cancer xenografts in a nude mouse model. Both the suppression of MMP expression and the specific binding mode make $Gd@C_{82}(OH)_{22}$ a potentially more effective nanomedicine for pancreatic cancer than traditional medicines, which usually target the proteolytic sites directly but fail in selective inhibition [116]. In 2012, Kang et al. presented the study on the interaction between $Gd@C_{82}(OH)_{22}$ and a small protein domain, using all-atom explicit solvent molecular dynamics simulations [117].

7.2.4 RADIOTRACERS AND RADIOPHARMACEUTICALS

In 1995, Kobayashi et al. made the first investigation on EMFs used as a radiotracer. They injected the emulsion of $La@C_{82}$ and $La_2@C_{80}$, of which La was labeled as ^{140}La in 5% polyvinylpyrrolidine, into the heart of rats and observed radioactivity in

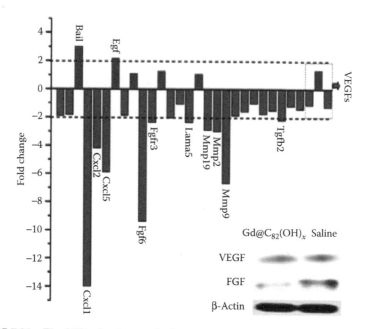

FIGURE 7.20 The f-NPs simultaneously downregulated multiple angiogenic factors on the mRNA level: Analysis of expression of angiogenic factors in tumors with the f-NP treatment by quantitative polymerase chain reaction array. The y-axis indicates the fold changes compared to saline control. Some of the factors, including fibroblast growth factor and vascular endothetial growth factor, were confirmed on the protein level by Western blotting assay. (Reproduced with permission from Meng, H. et al., *ACS Nano*, 4, 2773–2783, 2010. Copyright 2010 American Chemical Society.)

each organ by means of γ-ray spectroscopy. They found large radioactivity in blood and liver, which was due to [140]La-EMFs [118].

In 1999, Wilson et al. hydroxylated a purified $Ho_x@C_{82}$ mixture ($x = 1, 2$) and prepared $^{166}Ho_x@C_{82}(OH)_y$ by a $^{166}Ho[n,\gamma]^{166}Ho$ neutron activation of the fullerenol mixture. Studies of $^{166}Ho_x@C_{82}(OH)_y$ on biodistribution as a radiotracer in BALB/c mice during 48 h indicated a selective localization of the $^{166}Ho_x@C_{82}(OH)_y$ tracer in the liver but with slow clearance, as well as its uptake by bone without clearance. Moreover, a metabolism study of $^{166}Ho_x@C_{82}(OH)_y$ in Fischer rats indicated 20% excretion of intact $^{166}Ho_x@C_{82}(OH)_y$ within 5 days [119].

In 2009, Dorn et al. reported the ^{177}Lu-radiolabeled nitride cluster fullerene ($^{177}Lu_xLu_{3-x}N@C_{80}$) was functionalized and conjugated to a fluorescent tag (tetramethyl-6-carboxyrhodamine [TAMRA])-labeled IL-13 peptide, which was designed to target an overexpressed receptor in glioblastoma multiform tumors. They showed that the radiolabeled ^{177}Lu ions are not readily removed from the fullerene cage and demonstrated the potential targeting with a conjugated IL-13 protein (Figure 7.21) [120].

Shultz et al. reported that a functionalized metallofullerene (f-$Gd_3N@C_{80}$) conjugated to radiolabeled Lu chelated by tetraazacyclododecane tetraacetic acid, resulting in ^{177}Lu-DOTA-f-$Gd_3N@C_{80}$, and it can be applied to effective brachytherapy. Treating

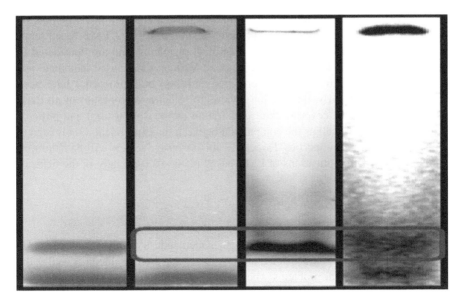

FIGURE 7.21 Polyacrylamide gel electrophoresis images from left to right: Visible lane for the TAMRA-labeled IL-13 peptide alone and visible, fluorescent, and autoradiograph lanes for the $^{177}Lu_xLu_{3-x}N@C_{80}$–TAMRA–IL-13 peptide. The rectangle shows the $^{177}Lu_xLu_{3-x}N@$ C_{80}–TAMRA–IL-13 peptide conjugate. (Reproduced with permission from Shultz, M. D. et al., *J. Am. Chem. Soc.*, 132, 4980–4981, 2010. Copyright 2010 American Chemical Society.)

mice with implanted human glioblastoma cells by ^{177}Lu-DOTA-f-Gd$_3$N@C$_{80}$ gave rise to longitudinal tumor imaging, prolonged intratumoral probe retention, biodistribution, and extended survival in an orthotopic xenograft brain tumor model [121].

Later on, Lu-DOTA-f-Gd$_3$N@C$_{80}$ was infused by *convection-enhanced delivery* (CED) intratumorally to provide brachytherapy for orthotopic xenograft tumor-bearing mice and it was found that a therapeutic dose response and efficacy with in vivo models using two glioblastoma cell lines, U87MG and GBM12 [122].

Horiguchi et al. prepared Gd@C$_{82}$-PEG-*b*-poly(2-(*N*,*N*-diethylamino)ethyl methacrylate) [PAMA]-complexed nanoparticles (GdNPs) using PEG-*b*-PAMA to react with Gd@C$_{82}$ and the toxicity of GdNPs is much reduced. They applied the nanoparticles (GdNPs) to neutron capture therapy (NCT). The neutron irradiation of colon-26 cells in the presence of GdNPs induces cell death, indicating the emission of gamma rays and conversion electrons upon the neutron capture reactions of ^{155}Gd and ^{157}Gd [123].

7.3 ELECTRONIC NANOMATERIALS AND MOLECULAR MATERIALS

Lai Feng

Unlike empty fullerenes that are composed of only carbon atoms, EMFs are hybrid molecules, containing wide compositions of fullerenes and various metals or metallic clusters. Until now, many efforts have been devoted to demonstrating their unique electronic properties, which may arise from the unusual charge transfer interaction between the cage and endohedrals. Furthermore, in recent years, EMF-based semiconducting materials have attracted a great deal of interest and are considered to be novel building blocks of electronic devices. Various EMF-based nanomaterials and molecular materials have been prepared and their bulk properties have been well studied. Particularly, this section reviews the electrical properties of all these EMF-based materials, covering their preparation methods, structural and physical characterizations, as well as the comparison between the experimental tools used to probe their electrical properties. Hopefully, this extensive review will be helpful in guiding the further applications of EMF-based materials especially in the fields of electronics.

7.3.1 INTRODUCTION

Inorganic electronics have been widely used in modern industry, which are mostly made from silicon. However, because of the recently increasing price of silicon as well as the emerging environmental problems when dealing with the scrapped electronics, further development of inorganic electronics was bottlenecked. However, in recent years, organic electronics, which are made from conjugated polymers, macromolecules, or small molecules, have attracted a lot of general interest. They have showed numerous advantages such as low cost, flexibility, ease of fabrication, and environment-friendly, and have been considered to be promising alternatives in the field where inorganic electrics dominated [124–127]. Thus, developing organic electronic devices has been one of the most active research topics in nanoscale science in recent years.

According to the general physical understanding, the performance of any organic device depends on the charge carrier mobility (μ) of the materials, which is the speed (cm/s) at which the charge carriers move in the material in a given direction in the presence of an applied electric field (V/cm):

$$\mu = \frac{cm^2}{Vs}$$

Almost always, higher mobility leads to better device performance. For example, charge carrier mobility in transistors determines how fast the device can be turned on and off. Thus, the key to efficiently utilizing organic electronics is to prepare the organic materials that have charge carrier mobility equivalent to or better than that of Si-based materials (i.e., $\mu = {\sim}5 \times 10^{-1}{-}10^{-3}$ cm^{2}V^{-1}s^{-1}).

Fullerene C$_{60}$ features a well-defined three-dimensional conjugated system [128]. In the last decade, a variety of fullerene materials were well studied and developed to fabricate electronics [129–132]. Typically, because of the popularity of electron carries in fullerene materials, they are usually used as n-channel semiconductors in various applications such as p–n junctions [133,134], organic field-effect transistors (FETs) [135–138], and novel acceptor materials in organic solar cells [139–142]. It is noteworthy that not only pristine fullerenes but also fullerene derivatives that can be solution processed have been reported to exhibit nice electron mobility values (i.e., $\mu = 1{-}10$ cm^{2}V^{-1}s^{-1}), which approach those of amorphous silicon [137,143]. Besides, it was reported that fullerene-based materials also show superiority in air stability [137] and tunable molecular packing and orientation in bulk solid state [144–146]. All of these characters likely make fullerenes promising building blocks in the field of organic electronics.

EMFs are created by filling or trapping metals or metallic clusters in fullerene cavities, and have been considered as special fullerene derivatives [36]. In contrast to non-metallic endohedrals such as $N@C_{60}$ [147] and $He@C_{60}$ [32,148] which simply combine the individual properties of the components, EMFs merge the properties of the components in a way that creates new properties distinct from those of either building block [149]. As a result, EMFs are expected to show novel electrical properties, which have been never reported for their individual components (i.e., fullerenes). In fact, very recently, there has appeared a number of pioneering works, focusing on the electrical properties of EMFs or EMF-based nanomaterials. This section reviews a variety of EMF-based electronic materials. In particular, it describes their preparations, molecular crystals, and supramolecular architectures as well as their unique electrical properties in terms of mainly charge carrier mobilities or conductivities. For example, a cocrystal of $La@C_{82}/Ni^{II}$(octaethylporphyrin) (OEP) exhibits an anisotropic charge carrier transport along each crystal axis on the nanometer scale and its highest electron mobility is determined to be $\mu = 0.9$ $cm^2V^{-1}s^{-1}$, based on the flash-photolysis time-resolved microwave conductivity (TRMC) measurements [150]. However, the crystalline nanorods of $La@C_{82}$-Ad show an ambipolar transport property by using the technique of FET [151].

In addition, the formation of metallofullerene peapods consisting of single-walled carbon nanotubes (SWNTs) and EMF molecules will be described. Important is the approved correlation between the local electronic structure of SWNT and the nested metallofullerene. For instance, when filling SWNT with $Gd@C_{82}$, the band gap of SWNT, which is ~0.5 eV, is reduced to ~0.1 eV at the sites where $Gd@C_{82}$ molecules locate [152]. Thus, encapsulating EMFs inside the inner space of SWNTs can significantly modulate the band gap of SWNTs as well as their electron transport properties. These studies might open up a new way toward the semiconductors with tunable electrical properties.

7.3.2 MONOMETALLOFULLERENES

7.3.2.1 $La@C_{82}$ and Derivatives

Mono-metallofullerene $La@C_{82}$ has a very unique electronic structure, which features the delocalized HOMO and LUMO on the cage of C_{82} as well as a smaller band gap relative to those of empty fullerenes C_{60}, C_{70}, and other metallofullerenes [153,154]. As the intrinsic electrical properties of the molecular materials are highly related to their electronic structures, it might be very possible that $La@C_{82}$-based materials show novel or unusual electrical properties.

The charge carrier mobility and conductivity of the single crystal of $La@C_{82}$(Ad) were first reported by Akasaka et al. [155] in which $La@C_{82}$(Ad) refers to an adamantane carbene adduct of $La@C_{82}$. The single crystal was obtained by a solvent diffusion method and its carrier mobility as well as conductivity was determined using the TRMC measurements. Particularly, TRMC is a recently developed method that

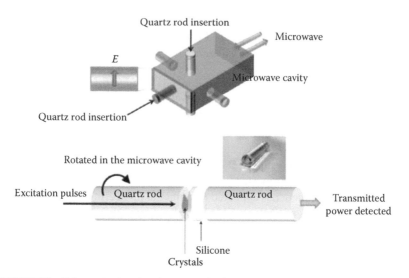

FIGURE 7.22 Schematic showing the setup of anisotropic conductivity measurement based on TRMC measurements. (Reprinted with permission from Sato, S. et al., *J. Am. Chem. Soc.*, 133, 2766–2771, 2011. Copyright 2011 American Chemical Society.)

measures the nanometer-scale mobility of charge carriers under an oscillating microwave electric field with no contact between the semiconductors and the metals [156]. Such electrode–semiconductor separation allows evaluating the intrinsic charge carrier mobility with minimal trapping effects. Thus, this method is very suitable for the electrical measurements of organic crystals, in which a number of charge carriers might be trapped by numerous types of crystal defects, especially, on the surface of crystals. The TRMC experimental setup is presented in Figure 7.22. All crystals are mounted on quartz rods and the quartz rod is rotated in the microwave cavity. Then, the changes in the effective electric field in the crystals by the rotation of the samples are calculated numerically based on the geometry of the crystals captured using a digital charge-coupled device (CCD) camera. The number of photons absorbed by the crystals is estimated through direct measurement of the transmitted power of laser pulses through the (quartz rod)-(crystal with PMMA binder)-(quartz rod) geometry (Figure 7.22) using an Ophir NOVA display laser power meter. Using this technique, the transient conductivity ($\phi\Sigma\mu$) can be measured, where ϕ and $\Sigma\mu$, respectively, denote the photocarrier generation yield (quantum efficiency) and the sum of mobilities for positive and negative charge carriers, respectively. The measured conductivities ($\phi\Sigma\mu$) and the calculated carrier mobilities (μ) are listed in Table 7.1. It is very obvious that the La@C_{82}(Ad) single crystal exhibits anisotropic electron mobilities and conductivities. The highest mobility of 10 $cm^2V^{-1}s^{-1}$ is observed along the c-axis or the long axis of crystal under normal temperatures and pressures in the atmosphere (see Figure 7.23), whereas along other directions (i.e., 45° or 90° relative to the c-axis) lower mobilities (i.e., 2 and 7 $cm^2V^{-1}s^{-1}$, respectively) were observed due to the presence of the Ad group and packed CS_2 molecules, both of which serve as an insulating part. Note that the value of 10 $cm^2V^{-1}s^{-1}$ is higher than previously

TABLE 7.1

Measured transient conductivities ($\Phi\Sigma\mu$) ($cm^2V^{-1}s^{-1}$) and charge carrier mobilities (μ) ($cm^2V^{-1}s^{-1}$) of La@C$_{82}$Ad using TRMC[a] and TOF techniques

	$\Phi\Sigma\mu$ along c-axis	$\Phi\Sigma\mu$ 45° to c-axis	$\Phi\Sigma\mu$ 90° to c-axis	μ along c-axis	μ 45° to c-axis	μ 90° to c-axis
Single crystal of La@C$_{82}$Ad[b]	$3.5 \pm 0.5 \times 10^{-3}$	$1.2 \pm 0.2 \times 10^{-3}$	$6.5 \pm 1.0 \times 10^{-4}$	10 ± 5	2 ± 1	$7 \pm 3 \times 10^{-1}$
Nanorod of La@C$_{82}$Ad[b]	$5.7 \pm 1.0 \times 10^{-3}$	$1.2 \pm 0.2 \times 10^{-3}$	$8.0 \pm 1.0 \times 10^{-4}$	>10	–	–
Cocrystal of La@C$_{82}$/[Ni^{II}(OEP)][b]	3.0×10^{-3}	5.0×10^{-4}	1.0×10^{-3}	0.9	0.1	0.3
Drop-casted film of La@C$_{82}$Ad[b]	$7.0 \pm 1.0 \times 10^{-5}$			$7 \pm 3 \times 10^{-2}$		
Drop-casted film of La@C$_{82}$Ad[c]	–			$8.0 \pm 2.4 \times 10^{-2}$		
Drop-casted film of C$_{60}$Ad[c]	–			$5.0 \pm 1.5 \times 10^{-3}$		

[a] Excitation was conducted at 532 nm: 45 mJ/cm².

[b] Values obtained from TRMC measurements.

[c] Values obtained from TOF measurements.

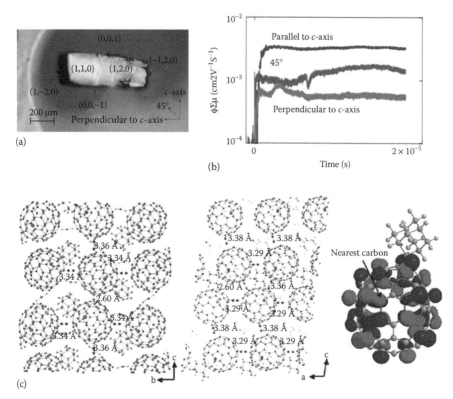

FIGURE 7.23 Photograph (a) and observed transient conductivity (b) for single-crystal La@C_{82}Ad. Excitation was conducted at 532 nm: 10 mJ/cm². (c) Molecular alignment of single-crystal La@C_{82}Ad and the calculated LUMO distribution. (Reprinted with permission from Sato, S. et al., *J. Am. Chem. Soc.*, 133, 2766–2771, 2011. Copyright 2011 American Chemical Society.)

reported organic conductors measured by TRMC, and therefore, the La@C_{82}(Ad) single crystal can be considered as a semiconductor having high conductivity.

For comparison, the TRMC measurement of La@C_{82}Ad nanorods was conducted to examine their electrical properties [155]. It was revealed that La@C_{82}Ad nanorods also exhibit anisotropic electron mobilities and conductivities (see Table 7.1). The values are very close to those of single crystals. The highest electron mobility of $\mu > 10$ cm²V⁻¹s⁻¹ was observed along the long axis of La@C_{82}Ad nanorods, suggesting a similar crystal alignment of La@C_{82}Ad nanorods to that of the La@C_{82}Ad single crystal. This value is also the highest of reported organic semiconductors measured by TRMC. For further shedding light on the origin of such high carrier mobility, analysis based on the *density functional theory* (DFT) calculations was performed and suggested a very small conduction–valence band gap of 0.005 eV for the La@C_{82}Ad single crystal at the applied computational level. It appears that the valence electrons of La@C_{82}Ad single crystal would be facilely excited to the conduction band, thus providing sufficient carriers in the crystal with very low-energy power source. This might account for the observed high electron mobility of the crystal in this work. In addition, X-ray analysis revealed a very short intermolecular distance of 2.60 Å in La@C_{82}(Ad). As the LUMO of La@C_{82}Ad has a large distribution on the nearest carbon atoms, such close molecular packing results in almost continuous LUMO distribution in the crystal and contributes to the electron mobility.

For further comparison, the charge carrier mobility of the drop-casted La@C_{82}Ad film was measured using time-of-flight (TOF) techniques that can probe the charge transport in the bulk ultrapure organic crystals at small carrier densities, consequently minimizing the trapping effect as well [157]. Thus, the measured conductivity of the drop-casted La@C_{82}Ad film is nearly 1 order of magnitude higher than that of drop-casted C_{60}Ad. The related *scanning electron microscopy* (SEM) and *atomic force microscopy* (AFM) characterizations revealed better crystalline nature of the drop-casted La@C_{82}Ad relative to that of C_{60}Ad, which might account for the higher electron mobility of the drop-casted La@C_{82}Ad relative to that of C_{60}Ad and highlight the importance of endohedral La doping in enhancing the mobility. However, the conductivity of drop-casted La@C_{82}Ad film is 2–3 orders of magnitude lower than those of La@C_{82}Ad single crystals and nanorods probably due to the lower crystalline nature of the former.

Moreover, the intrinsic carrier mobility of the cocrystal of La@C_{82}/[Ni^{II}(OEP)] was similarly investigated by TRMC measurements [150]. Black crystals of La@C_{82}/[Ni^{II}(OEP)] suitable for measurements were obtained by diffusion of a benzene solution of the La@C_{82} into a $CHCl_3$ solution of [Ni^{II}(OEP)]. X-ray analysis has revealed the formation of an interdigitated network of La@C_{82} and [Ni^{II}(OEP)], in which the closest intermolecular distances ranging from 3.02 to 3.23 Å are found between La@C_{82} units, shorter than the sum of van der Waals radii of the carbon atoms (3.35 Å). This might suggest that the presence of [Ni^{II}(OEP)] moiety contributes to the close packing of La@C_{82} molecules and therefore enhances the electron transport between them. The highest electron mobility of this cocrystal along the direction pointing to the shortest contact between La@C_{82} molecules (i.e., along the c-axis) was determined to be $\mu = 0.9$ $cm^2V^{-1}s^{-1}$, whereas the highest values in the other directions are much lower. For instance, the mobilities observed along the a-axis or the directions rotated 30° and 45° to the *a*-axis are, respectively, 0.3, 0.3, and 0.1 $cm^2V^{-1}s^{-1}$. Detailed X-ray analysis revealed that such anisotropic electron mobility is probably due to the presence of the interstitially packed benzene molecules that may block the electron carrier transport by acting as an insulator between the La@C_{82} molecules. Note that all these reported electron mobilities of cocrystal La@C_{82}/[Ni^{II}(OEP)] are higher than that of single-crystal [Ni^{II}(OEP)] ($\mu = 0.07$ $cm^2V^{-1}s^{-1}$), indicating that the aligned La@C_{82} molecules shall be responsible for providing highly conductive pathways for the electron transport.

Besides, in another study, the charge transport properties of La@C_{82}(Ad) nanorod and La@C_{82}(Ad) thin film were investigated via FET measurements [151]. Unlike the TRMC measurements that are electrodeless, the fabrication of FET requires electrodes (i.e., source and drain electrodes) that shall be bottom layered or top layered on the surface of the semiconductors. In a typical measurement, a SiO_2/doped Si substrate with thickness of an oxidized layer of 200 nm was used and a fullerene nanorod or thin film was deposited on it. The source and drain electrodes were fabricated by Ag paste. As shown in Figure 7.24a, the La@C_{82}(Ad) nanorod FETs show completely a p-type behavior, whereas the analog FETs fabricated by C_{60}(Ad) nanorod

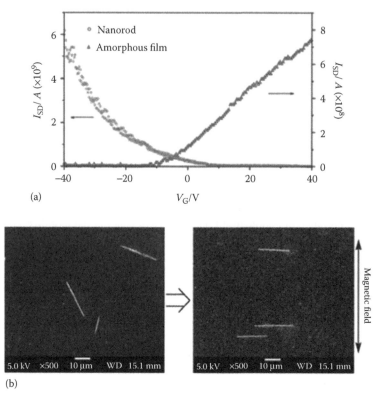

(a)

(b)

FIGURE 7.24 (a) I_{SD} curve of nanorod and amorphous film of La@C$_{82}$(Ad). (b) SEM observation for the magnetic orientation of nanorods of La@C$_{82}$(Ad). (These orientations of nanorods were observed in the magnetic field at the region of ≥ 3 T.) (Reprinted with permission from Sato, S. et al., *J. Am. Chem. Soc.*, 133, 2766–2771, 2011. Copyright 2008 American Chemical Society.)

and the thin film of La@C$_{82}$(Ad) or C$_{60}$(Ad) show n-type behaviors in general. Thus, the p-type characteristic of La@C$_{82}$(Ad) nanorod is unique within fullerene-based FETs. Nevertheless, the mechanism of the p-type behavior of La@C$_{82}$(Ad) nanorod is not very clear and has been considered to relate with the nanointerfacial contact between the electrodes and La@C$_{82}$(Ad) nanorod or the special electronic states such as the lowering of work function in the rod-shaped La@C$_{82}$(Ad). Anyway, this result suggests that La@C$_{82}$(Ad) nanorods are p-type FET channels, which might be an important candidate for high-performance fullerene FET devices. In addition, these La@C$_{82}$(Ad) nanorods display unusual response toward magnetic field. As shown in Figure 7.24b, the La@C$_{82}$(Ad) nanorods orientate perpendicularly to the magnetic field. This is in contrast to the C$_{60}$-based nanorods, which usually have orientations parallel to the magnetic field. This observation indicates that the nanorods of La@ C$_{82}$(Ad) have a negative magnetic anisotropy (i.e., the magnetic susceptibility in the direction parallel to the rod axis is smaller than that in the direction perpendicular to the rod axis). However, so far the mechanism of this anisotropy has remained unknown and further studies are necessary for disclosing this mechanism.

7.3.2.2 Dy@C$_{82}$

In early years, there appeared a number of pioneering works focusing on the transporting properties of mono-metallofullerenes. For instance, in 2003, Kubozino et al. reported the fabrication of a thin-film FET of Dy@C$_{82}$ [158]. The related I_D–V_{DS} and I_D–V_G plots are shown in Figure 7.25, revealing that the I_D increases linearly with increasing V_{DS} as well as V_G. It is interesting to note that enough high I_D (i.e., I_D = 760 nA at V_{DS} = 110 V and V_G = 0) was detected even when no carrier was induced into the Dy@C$_{82}$ interface from the dielectric gate (i.e., V_G = 0 V). The authors proposed that this detectable I_D at V_G = 0 V originates from the intrinsic bulk current of Dy@C$_{82}$, which is different from the case of the C$_{60}$ and C$_{70}$ FETs, in which I_D = 0 nA at V_G = 0 V. These results imply that the Dy@C$_{82}$ FET shows an n-channel normally-on property, in contrast to the C$_{60}$ or C$_{70}$ FET that exhibits an n-channel normally-off property. In particular, in Dy@C$_{82}$, a three-electron transfer occurs from the Dy atom to the C$_{82}$ cage, which renders Dy@C$_{82}$ FET high carrier concentration due to the presence of delocalized electrons on C$_{82}$ cage donated from Dy and small band-gap semiconductor behavior (i.e., E_g = 0.2 eV). These might account for the presence of the intrinsic bulk current in Dy@C$_{82}$ FET as well as its normally-on property. In addition, the carrier mobility of Dy@C$_{82}$ FET was estimated to be 8.9 × 10^{-5} cm^{2}V^{-1}s^{-1}, much lower than that of similarly fabricated C$_{60}$ FET (i.e., 1.8 × 10^{-2} cm^{2}V^{-1}s^{-1}), which was mainly attributed to the low crystallinity of thus fabricated Dy@C$_{82}$ thin film.

7.3.2.3 Ce@C$_{82}$ and Gd@C$_{82}$

A single fullerene molecule with the size of ~1 nm has been considered as an ideal building block for the fabrication of a single molecule junction showing a conductance value that is both large and fixed. The related topics are of great interest because of the anticipation that the employment of such molecular junction might greatly improve electronic device performance.

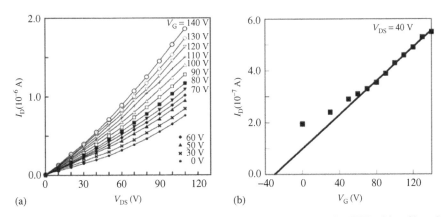

FIGURE 7.25 Plots of I_D vs. V_{DS} (a) and I_D vs. V_G (b) for the Dy@C$_{82}$ FET with a film of 120 nm. (Reprinted with permission from Kanbara, T. et al., *Chem. Phys. Lett.*, 379, 223–229, 2003. Copyright 2003 Elsevier.)

In a recent study, a single molecular junction was fabricated by directly binding a fullerene molecule (i.e., C_{60}, $Ce@C_{82}$) to a metallic electrode without the use of anchoring groups [159]. Figure 7.26a and b shows the experiment setup for conductivity measurement and the schematic view of the single $Ce@C_{82}$ molecule junction. As shown in Figure 7.26c and f, after the introduction of $Ce@C_{82}$ to Ag contacts, the conductance of $1G_0$ plateau ($G_0 = 2e^2/h$, where e is the charge of an electron and h is Planck's constant) was elongated with increasing stretch length, and the $1G_0$ peak was enhanced in the conductance histograms. A new sequence of steps appeared in the conductance traces at the lower conductance region (Figure 7.26c). The conductance value of the steps was an integer multiple of $0.2G_0$–$0.3G_0$. The corresponding conductance histogram showed a peak around $0.3G_0$ (Figure 7.26d). The conductance data obtained from the repeated measurements is summarized in Table 7.2. When using Ag as electrodes, the single $Ce@C_{82}$ molecule junction had a conductance of $0.28(\pm 0.05)G_0$, which is much larger than that of single-molecule junctions having anchoring groups (i.e., $<0.01G_0$), but only half that of the single C_{60} molecule junction (i.e., $0.5G_0$). This result demonstrates that the conductance of $Ce@C_{82}$ is reduced compared with that of C_{60} despite the fact that the HOMO–LUMO gap of $Ce@C_{82}$ is smaller than that of C_{60}. It appears to be contradictory to the general

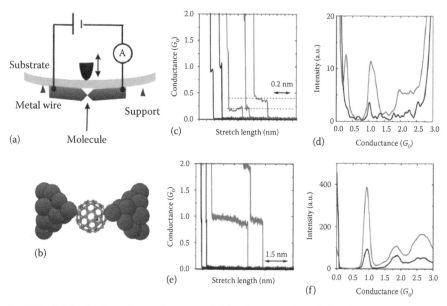

FIGURE 7.26 (a) Experimental setup and (b) schematic view of the single $Ce@C_{82}$ molecule junction. (c, e) Typical conductance traces and (d, f) histograms of Ag and Au contacts after the introduction of $Ce@C_{82}$. The metal electrodes were (c, d) Ag and (e, f) Au. The black traces and histograms indicate the values obtained for the clean Ag or Au contacts. The intensity of the conductance histograms was normalized with the number of the conductance traces. The conductance histograms were constructed without data selection from 1000 conductance traces of breaking metal contacts. The bin size was $0.004G_0$. (Reprinted with permission from Kaneko, S. et al., *Phys. Rev. B*, 86, 2012. Copyright 2012 American Physical Society.)

TABLE 7.2

Summary of the conductance for the single C_{60} and Ce@C_{82} molecule junctions

System	Au/Ce@C_{82}/Au	Ag/Ce@C_{82}/Ag	Au/C_{60}/Au	Ag/C_{60}/Ag
Conductance	–	0.28($\pm$0.05)G_0	0.3($\pm$0.1)G_0	0.5($\pm$0.1)G_0

understanding that the conductance of single-molecule junction increases with decreasing energy difference between the Fermi level and the conduction orbital relating to the HOMO–LUMO gap. However, the related theoretical calculations revealed that the LUMO and LUMO+1 of Ce@C_{82} are more localized than those of C_{60} due to the electron transfer interaction between the endohedral Ce atoms residing at an off-center position and the cage carbons nearby. Therefore, it was proposed that such LUMO or LUMO+1 distribution reduces some of the conducting channels for the electron transport, causing a reduction in the conductance of Ce@C_{82} species relative to empty C_{60} fullerene. By contrast, in the case of Ag/C_{60}/Ag, the conductive π-orbital of C_{60} is delocalized only on the cage and directly hybridizes with the orbital of the metal electrodes, resulting in high conductance of 0.5G_0. This value is almost constant due to the spherical molecular shape of fullerene in the break junction process.

Moreover, in an early study, theoretical calculations predicted reduced conduction property of Gd@C_{82} relative to that of empty C_{82} fullerene [160]. Detailed calculations revealed a significant overlap of the electron distribution between Gd and the C_{82} cage, with the transferred Gd electron density localized mainly on the nearest carbon atoms. The authors proposed that the conduction occurs primarily through the C_{82} cage and that charge donation from the Gd atom to the cage disrupts these conduction channels. This proposal is fully consistent with the experimental results concerning the reduced conductivity of a single-molecule junction Ag/Ce@C_{82}/Ag relative to that of Ag/C_{60}/Ag.

7.3.2.4   Gd@C_{82}(OH)$_x$ (x = 12 or 20)

Metallofullerenols such as Gd@C_{82}(OH)$_x$ have been widely studied due to their promising applications in the fields of nanomaterials and biomedical sciences [109,161]. Nevertheless, their electronic properties have been seldom reported except for a work of Yuliang Zhao et al. [162]. In that work, the electronic properties of Gd@C_{82}(OH)$_x$ (x = 12, 20) were studied using synchrotron radiation ultraviolet photoelectron spectroscopy (UPS), which is known to be a very sensitive tool to probe the difference of the density of states (DOS) near the Fermi edge (E_F). The measurements were first performed at room temperature. As shown in Figure 7.27, when the number of OH changes from 0, to 12, to 20, the photoelectron onset energy of Gd@C_{82}(OH)$_x$ shifts systematically from 0.3, to 1.9, to 3.5 eV, respectively. The higher onset energy corresponds to a wider band gap between the HOMO and LUMO levels. If E_F (set to 0) is located in the middle between the valence and conduction bands, the band gap E_g for Gd@C_{82}(OH)$_{12}$ and Gd@C_{82}(OH)$_{20}$ was estimated to be about 3.8 and 7.0 eV, respectively, suggesting their insulating properties. However, after 1-week storage of the Gd@C_{82}(OH)$_x$ film at 4°C, it was found that the photoelectron onset

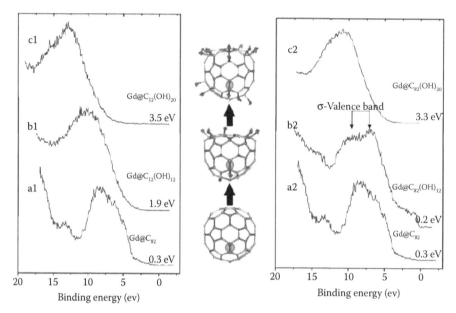

FIGURE 7.27 UPS of Gd@C_{82}(OH)$_x$ ($x = 0$, 12, 20) samples prepared at room temperature (a1–c1) and 4°C (a2–c2). The middle column shows the schematic drawings for hydroxyl addition and structural information of metallofullerenols. (Reprinted with permission from Tang, J. et al., *J. Phys. Chem. B,* 111, 11929–11934, 2007. Copyright 2007 American Chemical Society.)

energy of Gd@C_{82}(OH)$_{12}$ was largely reduced to 0.2 eV that is even smaller than that of Gd@C_{82}, indicating its semiconductor character. Thus, the electrical property of metallofullerenols is tunable as a function of temperature. The following *transmission electron microscopy* (TEM) studies further revealed a better microcrystalline structure of the sample Gd@C_{82}(OH)$_{12}$ stored at low temperature. The authors proposed that the stronger electric dipole moment of Gd@C_{82}(OH)$_{12}$ induced unique intermolecular interactions and the formation of self-assembled microcrystalline structures at low temperature, which may cause reconstruction of the upper valence band formed by σ-like electrons and hence the deeper valence band formed by σ-like electrons. Nevertheless, such switching from insulator to semiconductor was not observed for Gd@C_{82}(OH)$_{20}$. It appears that the more uniform dispersion of hydroxyl groups on the cage of C_{82} results in weaker electric dipole moment of Gd@C_{82}(OH)$_{20}$ and the formation of microcrystalline Gd@C_{82}(OH)$_{20}$ was unseen in the related TEM studies.

7.3.3 ENDOHEDRAL CLUSTER FULLERENES

7.3.3.1 M_x@C_{80} (M_x = La$_2$, Sc$_3$N, and Sc$_3$C$_2$)

A fullerene cage can encapsulate not only a metal atom but also a metallic cluster such as dimetallic cluster, metallic nitride cluster, metallic oxide cluster, and metallic sulfide cluster. Similar to that of mono-metallofullerenes, the endohedral cluster

TABLE 7.3

Electrochemical properties, conductivity transient ($\phi\Sigma\mu$), charge carrier generation yield (ϕ), and charge carrier mobility (μ) of thin films of La$_2$@C$_{80}$, Sc$_3$N@C$_{80}$, and Sc$_3$C$_2$@C$_{80}$

	E_1^{ox} (V)[a]	E_2^{ox} (V)[a]	$\phi\Sigma\mu$ (cm^{2}V^{-1}s^{-1})	Φ (%)	μ (cm^{2}V^{-1}s^{-1})
La$_2$@C$_{80}$	0.56 [164]	−0.31 [164]	1.0×10^{-5}	0.2	5.0×10^{-3}
Sc$_3$N@C$_{80}$	0.62 [74]	−1.22 [74]	4.0×10^{-5}	0.70	5.7×10^{-3}
Sc$_3$C$_2$@C$_{80}$	−0.03 [165]	−0.50 [165]	1.0×10^{-3}	0.75	0.13
La@C$_{82}$	0.07 [166]	−0.42 [166]	2.0×10^{-4}	0.15 [150]	0.13

[a] Volts versus Fc/Fc$^+$.

usually donates a number of electrons to the fullerene cage, giving rise to a *core shell*-like electronic structure that is different from those of empty fullerenes. In a recent study, the charge carrier mobilities of a series cluster fullerenes including La$_2$@C$_{80}$, Sc$_3$N@C$_{80}$, and Sc$_3$C$_2$@C$_{80}$ have been measured using the TRMC technique that enables electrodeless measurements to be conducted for determination of the intrinsic charge carrier mobility [163]. Typically, TRMC measurements were performed on the casted thin films of cluster fullerenes. Upon exposure to a laser pulse with an excitation wavelength of 355 nm, the transient conductivities of all samples ($\phi\Sigma\mu$) were measured. The ϕ values were determined using conventional dc current integration with semitransparent Au electrodes as counter electrodes under excitation at 355 nm. It is noteworthy that, for all films, the current transients were observed in negative bias mode, suggesting that electrons are the major charge carrier species. The measured and calculated values of $\phi\Sigma\mu$, ϕ, and μ are summarized in Table 7.3 [163]. It can be seen that the μ-value of Sc$_3$C$_2$@C$_{80}$ is the same as that reported for La@C$_{82}$ and 2 orders of magnitude higher than those of La$_2$@C$_{80}$ and Sc$_3$N@C$_{80}$. Detailed X-ray diffraction (XRD) studies revealed that all the casted films are composed of crystallites with a base-centered monoclinic lattice as shown in Figure 7.28. However, the charge carrier mobility of Sc$_3$N@C$_{80}$ thin film was determined to be lower than that of Sc$_3$C$_2$@C$_{80}$, despite the fact that Sc$_3$N@C$_{80}$ has better crystalline quality. Furthermore, DFT calculations revealed the semiconductor character of these crystalline cluster fullerenes that are La$_2$@C$_{80}$, Sc$_3$N@C$_{80}$, and Sc$_3$C$_2$@C$_{80}$ with band gaps of 0.41, 1.23, and 0.07 eV, respectively (see Figure 7.28). As the band gap (E_g) of Sc$_3$C$_2$@C$_{80}$ is only 1/6 and 1/17 of those of La$_2$@C$_{80}$ and Sc$_3$N@C$_{80}$, respectively, the intrinsic charge carrier concentration (n) of Sc$_3$C$_2$@C$_{80}$ shall be over 10^3 times higher than those of the other two species at room temperature ($T = 300$ K) according to the following formula, which might account for the better charge carrier mobility of crystalline Sc$_3$C$_2$@C$_{80}$ relative to others:

$$\sigma = ne(\mu_e + \mu_h)$$

$$= 2\left(\frac{k_B T}{2\pi h^2}\right)^{3/2} (m_e * m_h *)^{3/4} e^{-E_g/2k_B T} e(\mu_e + \mu_h) \tag{7.1}$$

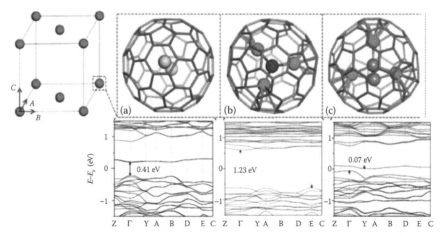

FIGURE 7.28 (Left) Schematic structure of the base-centered monoclinic lattice. (Right) Geometrical structures and band structures of $La_2@C_{80}$ (a), $Sc_3N@C_{80}$ (b), and $Sc_3C_2@C_{80}$ (c) in the respective optimized crystals. The corresponding band gap is labeled in each panel. (Reprinted with permission from Sato, S. et al., *J. Am. Chem. Soc.*, 134, 11681–11686, 2012. Copyright 2012 American Chemical Society.)

where:

   σ is the conductivity

   n is the intrinsic carrier concentration

   μ_e and μ_h are the electron and hole carrier mobilities, respectively

   m_e^* and m_h^* are the effective masses of the conduction band maximum and the
        valence band minimum, respectively

   k_B is the Boltzmann constant

7.3.3.2 $La_2@C_{80}$

In another work, the electron transport property of $La_2@C_{80}$ thin film was alternatively characterized using the FET technique [167]. In a typical measurement, as shown in Figure 7.29, a layer of $La_2@C_{80}$-based thin film was fabricated on the drain and source electrodes. The plots of drain–source current against drain voltage for various gate voltage values suggested an n-channel normally-on character of $La_2@C_{80}$. The electron mobility (μ) of this thin-film transistor was estimated to be $\mu = 1.1 \times 10^{-4}$ cm²/Vs, which is comparable to the value ($\mu = 5.0 \times 10^{-3}$ cm²/Vs) obtained by means of the TRMC technique as well as the value of $Dy@C_{82}$ FET ($\mu = 8.9 \times 10^{-5}$ cm²/Vs) but much lower than that of C_{60} or C_{70} FET [135,168]. According to the authors' proposal, such a low carrier mobility of $La_2@C_{80}$ was ascribed to the conduction mechanism through encapsulated metal ions. From a viewpoint of quantum theory, because the LUMO of $La_2@C_{80}$ is dominated by the orbital of encapsulated La, the intermolecular overlap of LUMOs is very small, which is unfavorable for the diffusion of electron carriers and results in very low mobility of $La_2@C_{80}$-based thin film.

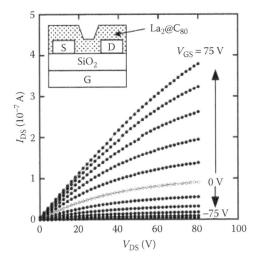

FIGURE 7.29 Drain–source current (I_{DS}) is plotted as a function of drain–source voltage (V_{DS}) for several values of gate voltage (V_{GS}). Data sets are displayed for every 15 V from V_{GS} (–75 to 75 V). Inset shows a schematic diagram of La$_2$@C$_{80}$ thin-film transistor. S, D, and G denote the source, the drain, and the gate, respectively. (Reprinted with permission from Kobayashi, S. et al., *J. Am. Chem. Soc.*, 125, 8116–8117, 2003. Copyright 2003 American Chemical Society.)

7.3.3.3   Ce$_2$@C$_{80}$

In a recent study, the technique of low-temperature scanning tunneling microscopy (LT-STM) has been used for characterizing the electrical property of metallofullerenes for the first time [169]. This method allows constructing atomic-size contacts between a sharp tip and a flat substrate covered with fullerenes or metallofullerenes. Also, the imaging capability of STM allows selecting the desired contact geometry and modifying the contact properties by including different single atoms in the contact area. Thus, the authors measured the conductance of Ce$_2$@C$_{80}$ and C$_{60}$ under the exactly same condition. As shown in Figure 7.30a, in the contact regime the conductance of Ce$_2$@C$_{80}$ (i.e., 0.1G_0) is ~5 times lower than that of C$_{60}$. Furthermore, the DFT calculation demonstrated that the low-lying unoccupied states of closely contacted units of Ce$_2$@C$_{80}$/Cu(111) are dominated by Ce and Cu contributions with surprisingly low cage contributions (see Figure 7.30c), which is at 0.306 eV above E_f. However, the energy of delocalized cage–Cu orbitals that are forming upon adsorption as well is at 1.573 eV above E_f. Thus, when the conductance measurements were performed (arrow in Figure 7.30f), only the separated Ce/Cu orbitals are available as conducting channels for electrons with energy up to 0.3 eV. Therefore, the lack of sufficient overlap between the low-energy Ce/Cu orbitals may account for the low conductance of Ce$_2$@C$_{80}$ absorbed on Cu(111), which correlates well with the previously reported results [163,167], suggesting the poor transport property of M$_2$@C$_{80}$ (M = La, Ce) either in bulk state or at single molecular level.

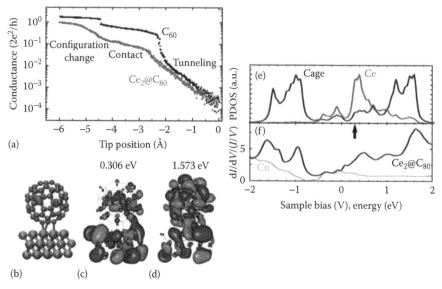

FIGURE 7.30 (a) Conductance traces measured with the same STM tip on C_{60} and $Ce_2@C_{80}$ as a function of tip-molecule distance. In the contact regime, the conductance of $Ce_2@C_{80}$ is ~5 times lower than that of C_{60}. (b–e) DFT calculations of $Ce_2@C_{80}$ on Cu(111). (b) Minimum energy configuration. Localized (c) and delocalized orbitals (d) at 0.306 and 1.573 eV above E_f. (e) Partial density of states for Ce and cage C atoms. At energy around 0.3 eV where the conductance measurements are performed [arrow in (f)], electron states localized on Ce dominate the spectrum. Electron states delocalized on the cage have energy ≥1 eV. (f) DOS measured by scanning tunneling spectroscopy on $Ce_2@C_{80}$ and the clean substrate. (Reprinted with permission from Strozecka, A. et al., *Appl. Phys. Lett.*, 95, 2009. Copyright 2009 AIP Publishing LLC.)

7.3.4 METALLOFULLERENE PEAPODS

In the past 10 years, SWNTs have attracted a great deal of interest because of their potential applications in the fields of electronics and material science. In particular, SWNTs exhibit either a metallic or a semiconductor character depending on their chirality. Furthermore, their band gaps can be varied between 0 and 1.5 eV by varying their geometrical structure. However, encapsulating atoms or molecules such as fullerenes or EMFs into their inner space can alternatively change their electronic structures. For instance, in 2002, Kahng et al. first reported the electronic measurements of C_{60}@SWNT, which were performed with an STM and demonstrated that the encapsulated C_{60} molecules modify the local electronic structure of the nanotube [170]. It was believed that the strong interaction between C_{60} and SWNT is the origin of this modified electronic structure of SWNT. As such an interaction varies depending on the nature of endohedrals, it is expected that replacing C_{60} with other species would give rise to the peapods having different electronic features.

In 2002–2003, Shinohara et al. reported that filling SWNTs with $Gd@C_{82}$ can modulate the band gap of SWNTs [152,167,171–173]. Specifically, the band gap of SWNT (i.e., ~0.5 eV) is reduced to ~0.1 eV at the sites where $Gd@C_{82}$ molecules locate [152]. Thus, the $Gd@C_{82}$ peapod can be considered as an array of quantum dots with small band gaps. It was believed that such a modulation originated from a combination of elastic strain and charge transfer between $Gd@C_{82}$ and SWNT. In particular, the charge transfer interaction between EMF and SWNT was confirmed by Warner et al. They revealed a charge transfer between the spin-active $Sc@C_{82}$ and SWNT for $C_{60}:Sc@C_{82}$ peapod, which is expressed by the loss of measurable hyperfine structure of Sc^{3+}-related electron spin resonance signal [174]. These results might be important specifically for the further advancement of solid-state architectures of the aligned quantum dots and spin-dependent transport studies.

More recently, a variety of EMF peapods have been prepared and well studied, in which not only various pristine EMFs such as $M@C_{82}$ (M = Sc [175], La [176], Ce [177], Pr [178], Gd [179,180], Sm [181], $La_2@C_{80}$ [182], and $Sc_3C_2@C_{80}$51) but also the functionalized EMF (i.e., pyrrolidine adduct of $Sc_3N@C_{80}$) [183] can be trapped into SWNTs. Interestingly, these endohedrally nested EMFs show different dynamic behaviors inside SWNTs, which have been never reported for C_{60} or C_{70} peapods. For instance, a series of TEM studies reported a continuous spinning for $Sm@C_{82}$ in SWNTs [181], a tumbling motion for $La_2@C_{80}$ in SWNTs [182], and a mixed dynamic behavior of $Gd@C_{82}$ in SWNTs where some cages are stationary and some are spinning [179,180]. Herein, the different electrostatic interactions between the highly charged EMF cages are considered to account for these behaviors. Moreover, in the case of $Sc_3C_2@C_{80}$ peapod, a dynamic conformational change of SWNT was reported. As shown in Figure 7.31, the TEM images showed that $Sc_3C_2@C_{80}$ molecules adopt a zigzag packing probably for reducing the intermolecular Coulomb repulsive forces [175]. Surprisingly, the slow expansion and contraction of the diameter of the SWNT in $Sc_3C_2@C_{80}$ peapods were observed with the rotation of the zigzag fullerene chain. This entirely new phenomenon was attributed to the higher HOMO level and the smaller ionization potential of $Sc_3C_2@C_{80}$ relative to other fullerenes, which result in easier ionization of $Sc_3C_2@C_{80}$ under electron beam

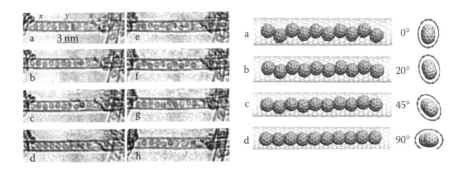

FIGURE 7.31 TEM images of $Sc_3C_2@C_{80}$@SWNT peapods as a function of time. (a–h) Each image has 2 s acquisition and 10 s between frames. Three sections of the peapod are labeled as x, y, and z for future reference (left panel). Schematic representation of the rotation of a zigzag chain of $Sc_3C_2@C_{80}$ inside an SWNT causing structural deformation that leads to an elliptical cross section (right panel). (Reprinted with permission from Warner, J. H. et al., *Nano. Lett.*, 8, 2328–2335, 2008. Copyright 2008 American Chemical Society.)

irradiation and therefore more significant interaction between SWNT and highly charged fullerene species. The authors believed that this result is very exciting and might provide possibilities for nanoactuation.

7.3.5 CONCLUDING REMARKS

This chapter showed that EMF-based nanomaterials and molecular materials exhibit various electrical properties depending on their molecular electronic structures as well as the crystalline quality of the bulk materials. Recent advances also suggest that some EMFs, such as the small band-gap species $La@C_{82}$ and $Sc_3C_2@C_{80}$, are promising building blocks for the fabrication of semiconductors showing excellent electron transport properties. The next imperative challenge is to further stimulate innovations based on molecular and material designs that can realize more efficient charge carrier transport in EMF-based semiconductors, thereby widening the scope for their potential applications.

ACKNOWLEDGMENTS

This work was supported by the National Natural Science Foundation of China (51372158, 21241004) and the Natural Science Foundation of Jiangsu Province (BK2012611).

REFERENCES

1. H. W. Kroto, J. R. Heath, S. C. O'Brien, R. F. Curl, R. E. Smalley, *Nature* **1985**, *318*, 162–163.
2. J. Ginsberg, A. C. Society, *The Discovery of Fullerenes: A National Historic Chemical Landmark, October 11, 2009*, American Chemical Society, Office of Communications, National Historic Chemical Landmark Program, **2010**.
3. R. Ehrhardt, Wikimedia Commons, licensed under CreativeCommons Attribution 2.0 Generic lizence, URL: http://creativecommons.org/licenses/by/2.0/deed.en.
4. W. Kraetschmer, K. Fostiropoulos, D. R. Huffman, *Astrophys. Space Sci. L.* **1990**, *165*, 89–93.
5. F. Diederich, R. Ettl, Y. Rubin, R. L. Whetten, R. Beck, M. Alvarez, S. Anz et al., *Science* **1991**, *252*, 548–551.
6. H. Murayama, S. Tomonoh, J. M. Alford, M. E. Karpuk, *Fuller. Nanotub. Car. N.* **2005**, *12*, 1–9.
7. R. F. Bunshah, S. Jou, S. Prakash, H. J. Doerr, L. Isaacs, A. Wehrsig, C. Yeretzian, H. Cynn, F. Diederich, *J. Phys. Chem.* **1992**, *96*, 6866–6869.
8. M. Ozaki, A. Taahashi, *Chem. Phys. Lett.* **1986**, *127*, 242–244.
9. K. Hedberg, L. Hedberg, D. S. Bethune, C. A. Brown, H. C. Dorn, R. D. Johnson, M. de Vries, *Science* **1991**, *254*, 410–412.
10. G. E. Scuseria, *Chem. Phys. Lett.* **1991**, *176*, 423–427.
11. C. S. Yannoni, P. P. Bernier, D. S. Bethune, G. Meijer, J. R. Salem, *J. Am. Chem. Soc.* **1991**, *113*, 3190–3192.
12. S. E. Lyshevski, *Nano- and Micro-Electromechanical Systems: Fundamentals of Nano- and Microengineering*, 2nd edition, Taylor & Francis Group, Boca Raton, FL, **2005**.
13. D. M. Guldi, M. Prato, *Acc. Chem. Res.* **2000**, *33*, 695–703.

14. W. S. Shin, Y. M. Hwang, W.-W. So, S. C. Yoon, C. J. Lee, S.-J. Moon, *Mol. Cryst. Liq. Cryst.* **2008**, *491*, 331–338.

15. L. M. Andersson, Y.-T. Hsu, K. Vandewal, A. B. Sieval, M. R. Andersson, O. Inganäs, *Org. Electron.* **2012**, *13*, 2856–2864.

16. P. Morvillo, E. Bobeico, S. Esposito, R. Diana, *Energy Procedia* **2012**, *31*, 69–73.

17. R. V. Bensasson, T. Hill, C. Lambert, E. J. Land, S. Leach, T. G. Truscott, *Chem. Phys. Lett.* **1993**, *201*, 326–335.

18. C. S. Foote, *Top. Curr. Chem.* **1994**, *169*, 347–363.

19. C. Bingel, *Chem. Ber.* **1993**, *126*, 1957–1959.

20. M. Maggini, G. Scorrano, M. Prato, *J. Am. Chem. Soc.* **1993**, *115*, 9798–9799.

21. R. Tenne, *Adv. Mater.* **1995**, *7*, 965–995.

22. J. R. Heath, S. C. O'Brien, Q. Zhang, Y. Liu, R. F. Curl, F. K. Tittel, R. E. Smalley, *J. Am. Chem. Soc.* **1985**, *107*, 7779–7780.

23. W. Kratschmer, L. D. Lamb, K. Fostiropoulos, D. R. Huffman, *Nature* **1990**, *347*, 354–358.

24. R. D. Johnson, M. S. de Vries, J. Salem, D. S. Bethune, C. S. Yannoni, *Nature* **1992**, *355*, 239–240.

25. D. S. Bethune, R. D. Johnson, J. R. Salem, M. S. de Vries, C. S. Yannoni, *Nature* **1993**, *366*, 123–128.

26. K. Kikuchi, S. Suzuki, Y. Nakao, N. Nakahara, T. Wakabayashi, H. Shiromaru, K. Saito, I. Ikemoto, Y. Achiba, *Chem. Phys. Lett.* **1993**, *216*, 67–71.

27. L. Dunsch, M. Krause, J. Noack, P. Georgi, *J. Phys. Chem. Solids* **2004**, *65*, 309–315.

28. S. Stevenson, K. Harich, H. Yu, R. R. Stephen, D. Heaps, C. Coumbe, J. P. Phillips, *J. Am. Chem. Soc.* **2006**, *128*, 8829–8835.

29. S. Stevenson, M. A. Mackey, J. E. Pickens, M. A. Stuart, B. S. Confait, J. P. Phillips, *Inorg. Chem.* **2009**, *48*, 11685–11690.

30. Z. Ge, J. C. Duchamp, T. Cai, H. W. Gibson, H. C. Dorn, *J. Am. Chem. Soc.* **2005**, *127*, 16292–16298.

31. B. Pietzak, M. Waiblinger, T. A. Murphy, A. Weidinger, M. Höhne, E. Dietel, A. Hirsch, *Carbon* **1998**, *36*, 613–615.

32. E. Shabtai, A. Weitz, R. C. Haddon, R. E. Hoffman, M. Rabinovitz, A. Khong, R. J. Cross, M. Saunders, P.-C. Cheng, L. T. Scott, *J. Am. Chem. Soc.* **1998**, *120*, 6389–6393.

33. K. Komatsu, M. Murata, Y. Murata, *Science* **2005**, *307*, 238–240.

34. M. Murata, Y. Murata, K. Komatsu, *J. Am. Chem. Soc.* **2006**, *128*, 8024–8033.

35. T. Akasaka, S. Nagase (Eds.), *Endofullerenes: A New Family of Carbon Clusters*, Kluwer, Dordrecht, The Netherlands, **2002**.

36. H. Shinohara, *Rep. Prog. Phys.* **2000**, *63*, 843–892.

37. M. Rudolf, S. Wolfrum, D. M. Guldi, L. Feng, T. Tsuchiya, T. Akasaka, L. Echegoyen, *Chem. Eur. J.* **2012**, *18*, 5136–5148.

38. K. Kobayashi, S. Nagase, *Chem. Phys. Lett.* **1996**, *262*, 227–232.

39. H. Shinohara, *Nature* **1992**, *357*, 52–54.

40. S. Stevenson, G. Rice, T. Glass, K. Harich, F. Cromer, M. R. Jordan, J. Craft et al., *Nature* **1999**, *401*, 55–57.

41. L. Dunsch, S. Yang, *Small* **2007**, *3*, 1298–1320.

42. M. Krause, F. Ziegs, A. A. Popov, L. Dunsch, *ChemPhysChem* **2007**, *8*, 537–540.

43. S. Stevenson, M. A. Mackey, M. A. Stuart, J. P. Phillips, M. L. Easterling, C. J. Chancellor, M. M. Olmstead, A. L. Balch, *J. Am. Chem. Soc.* **2008**, *130*, 11844–11845.

44. C.-R. Wang, T. Kai, T. Tomiyama, T. Yoshida, Y. Kobayashi, E. Nishibori, M. Takata, M. Sakata, H. Shinohara, *Angew. Chem. Int. Ed.* **2001**, *40*, 397–399.

45. T.-S. Wang, L. Feng, J.-Y. Wu, W. Xu, J.-F. Xiang, K. Tan, Y.-H. Ma et al., *J. Am. Chem. Soc.* **2010**, *132*, 16362–16364.

46. L. Dunsch, S. Yang, L. Zhang, A. Svitova, S. Oswald, A. A. Popov, *J. Am. Chem. Soc.* **2010**, *132*, 5413–5421.

47. P. M. Allemand, A. Koch, F. Wudl, Y. Rubin, F. Diederich, M. M. Alvarez, S. J. Anz, R. L. Whetten, *J. Am. Chem. Soc.* **1991**, *113*, 1050–1051.
48. D. Dubois, K. M. Kadish, S. Flanagan, R. E. Haufler, L. P. F. Chibante, L. J. Wilson, *J. Am. Chem. Soc.* **1991**, *113*, 4364–4366.
49. D. Dubois, K. M. Kadish, S. Flanagan, L. J. Wilson, *J. Am. Chem. Soc.* **1991**, *113*, 7773–7774.
50. Y. Ohsawa, T. Saji, *J. Chem. Soc. Chem. Commun.* 1992, 781.
51. Q. Xie, E. Perez-Cordero, L. Echegoyen, *J. Am. Chem. Soc.* **1992**, *114*, 3978–3980.
52. F. Zhou, C. Jehoulet, A. J. Bard, *J. Am. Chem. Soc.* **1992**, *114*, 11004–11006.
53. C. Bruno, I. Doubitski, M. Marcaccio, F. Paolucci, D. Paolucci, A. Zaopo, *J. Am. Chem. Soc.* **2003**, *125*, 15738–15739.
54. Q. Xie, F. Arias, L. Echegoyen, *J. Am. Chem. Soc.* **1993**, *115*, 9818–9819.
55. R. D. Webster, G. A. Heath, *PCCP* **2001**, *3*, 2588–2594.
56. T. Kato, T. Kodama, T. Shida, T. Nakagawa, Y. Matsui, S. Suzuki, H. Shiromaru, K. Yamauchi, Y. Achiba, *Chem. Phys. Lett.* **1991**, *180*, 446–450.
57. D. M. Guldi, H. Hungerbühler, E. Janata, K. D. Asmus, *J. Phys. Chem.* **1992**, *97*(43), 11258–11264.
58. S. Nonell, J. W. Arbogast, C. S. Foote, *J. Phys. Chem.* **1992**, *96*, 4169–4170.
59. D. M. Guldi, P. Neta, K.-D. Asmus, *J. Phys. Chem.* **1994**, *98*, 4617–4621.
60. M. Fujitsuka, A. Watanabe, O. Ito, K. Yamamoto, H. Funasaka, *J. Phys. Chem. A* **1997**, *101*, 7960–7964.
61. S. Fukuzumi, H. Mori, H. Imahori, T. Suenobu, Y. Araki, O. Ito, K. M. Kadish, *J. Am. Chem. Soc.* **2001**, *123*, 12458–12465.
62. R. D. Bolskar, Dissertation thesis, University of Southern California, Los Angeles, CA, **1997**.
63. Z. Gasyna, L. Andrews, P. N. Schatz, *J. Phys. Chem.* **1992**, *96*, 1525–1527.
64. C. A. Reed, R. D. Bolskar, *Chem. Rev.* **2000**, *100*, 1075–1120.
65. H. S. Shin, H. Lim, H. J. Song, H.-J. Shin, S.-M. Park, H. C. Choi, *J. Mater. Chem.* **2010**, *20*, 7183.
66. C. A. Reed, *Science* **2000**, *289*, 101–104.
67. V. A. Nadtochenko, N. N. Denisov, I. V. Rubtsov, A. S. Lobach, A. P. Moravskii, *Chem. Phys. Lett.* **1993**, *208*, 431–435.
68. S. Michaeli, V. Meiklyar, M. Schulz, K. Moebius, H. Levanon, *J. Phys. Chem.* **1994**, *98*, 7444–7447.
69. H.-Q. Hou, C. Luo, Z.-X. Liu, D.-M. Mao, Q.-Z. Qin, Z.-R. Lian, S.-D. Yao, W.-F. Wang, J.-S. Zhang, N.-Y. Lin, *Chem. Phys. Lett.* **1993**, *203*, 555–559.
70. K. Ohkubo, J. Ortiz, L. Martin-Gomis, F. Fernandez-Lazaro, A. Sastre-Santos, S. Fukuzumi, *Chem. Commun.* **2007**, 589–591.
71. L. Feng, M. Rudolf, S. Wolfrum, A. Troeger, Z. Slanina, T. Akasaka, S. Nagase et al., *J. Am. Chem. Soc.* **2012**, *134*, 12190–12197.
72. D. M. Guldi, K.-D. Asmus, *J. Am. Chem. Soc.* **1997**, *119*, 5744–5745.
73. P. V. Kamat, M. Gevaert, K. Vinodgopal, *J. Phys. Chem. B* **1997**, *101*, 4422–4427.
74. Y. Iiduka, O. Ikenaga, A. Sakuraba, T. Wakahara, T. Tsuchiya, Y. Maeda, T. Nakahodo, T. Akasaka, M. Kako, N. Mizorogi, S. Nagase, *J. Am. Chem. Soc.* **2005**, *127*, 9956 9957.
75. T. Cai, L. Xu, M. R. Anderson, Z. Ge, T. Zuo, X. Wang, M. M. Olmstead, A. L. Balch, H. W. Gibson, H. C. Dorn, *J. Am. Chem. Soc.* **2006**, *128*, 8581–8589.
76. M. N. Chaur, F. Melin, A. L. Ortiz, L. Echegoyen, *Angew. Chem.* **2009**, *121*, 7650–7675.
77. M. N. Chaur, F. Melin, A. L. Ortiz, L. Echegoyen, *Angew. Chem. Int. Ed.* **2009**, *48*, 7514–7538.
78. A. A. Popov, S. Yang, L. Dunsch, *Chem. Rev.* **2013**, *113*, 5989–6113.
79. T. Akasaka, T. Wakahara, S. Nagase, K. Kobayashi, M. Waelchli, K. Yamamoto, M. Kondo et al., *J. Am. Chem. Soc.* **2000**, *122*, 9316–9317.
80. K. Yamamoto, H. Funasaka, T. Takahasi, T. Akasaka, T. Suzuki, Y. Maruyama, *J. Phys. Chem.* **1994**, *98*, 12831–12833.

81. Y. Yang, F. Arias, L. Echegoyen, L. P. F. Chibante, S. Flanagan, A. Robertson, L. J. Wilson, *J. Am. Chem. Soc.* **1995**, *117*, 7801–7804.
82. J. Xu, M. Li, Z. Shi, Z. Gu, *Chem. Eur. J.* **2006**, *12*, 562–567.
83. B. Elliott, L. Yu, L. Echegoyen, *J. Am. Chem. Soc.* **2005**, *127*, 10885–10888.
84. B. Grimm, J. Schornbaum, C. M. Cardona, J. D. van Paauwe, P. D. W. Boyd, D. M. Guldi, *Chem. Sci.* **2011**, *2*, 1530–1537.
85. D. M. Guldi, L. Feng, S. G. Radhakrishnan, H. Nikawa, M. Yamada, N. Mizorogi, T. Tsuchiya et al., *J. Am. Chem. Soc.* **2010**, *132*, 9078–9086.
86. L. Feng, S. Gayathri Radhakrishnan, N. Mizorogi, Z. Slanina, H. Nikawa, T. Tsuchiya, T. Akasaka, S. Nagase, N. Martín, D. M. Guldi, *J. Am. Chem. Soc.* **2011**, *133*, 7608–7618.
87. Y. Takano, S. Obuchi, N. Mizorogi, R. García, M. Á. Herranz, M. Rudolf, D. M. Guldi, N. Martín, S. Nagase, T. Akasaka, *J. Am. Chem. Soc.* **2012**, *134*, 19401–19408.
88. M. Rudolf, L. Feng, Z. Slanina, T. Akasaka, S. Nagase, D. M. Guldi, *J. Am. Chem. Soc.* **2013**, *135*, 11165–11174.
89. E. B. Iezzi, J. C. Duchamp, K. R. Fletcher, T. E. Glass, H. C. Dorn, *Nano Lett.*, **2002**, *2*(11), 1187–1190.
90. M. Mikawa, H. Kato, M. Okumura, M. Narazaki, Y. Kanazawa, N. Miwa, H. Shinohara, *Bioconjugate Chem.* **2001**, *12*(4), 510–514.
91. H. Kato, Y. Kanazawa, M. Okumura, A. Taninaka, T. Yokawa, H. Shinohara, *J. Am. Chem. Soc.* **2003**, *125*(14), 4391–4397.
92. L. Qu, W. B.Cao, G. M. Xing, J. Zhang, H. Yuan, J. Tang, Y. Cheng, B. Zhang, Y. L. Zhao, H. Lei, *J. Alloys Compd.* **2006**, *408*, 400–404.
93. J. Zhang, K. M. Liu, G. M. Xing, T. X. Ren, S. K. Wang, *J. Radioanal. Nucl. Chem.* **2007**, *272*(3), 605–609.
94. G. M. Xing, H. Yuan, R. He, X. Y. Gao, L. Jing, F. Zhao, Z. F. Chai, Y. L. Zhao, *J. Phys. Chem. B.* **2008**, *112*(20), 6288–6291.
95. C. Y. Shu, C. R. Wang, J. F. Zhang, H. W. Gibson, H. C. Dorn, F. D. Corwin, P. P. Fatouros, T. J. S. Dennis, *Chem. Mater.* **2008**, *20*(6), 2106–2109.
96. C. Y. Shu, X. Y. Ma, J. F. Zhang, F. D. Corwin, J. H. Sim, E. Y. Zhang, H. C. Dorn et al., *Bioconjugate Chem.* **2008**, *19*(3), 651–655.
97. C. Y. Shu, L. H. Gan, C. R. Wang, X. L. Pei, H. B. Han, *Carbon* **2006**, *44*(3), 496–500.
98. R. D. Bolskar, A. F. Benedetto, L. O. Husebo, R. E. Price, E. F. Jackson, S. Wallace, L. J. Wilson, J. M. Alford, *J. Am. Chem. Soc.* **2003**, *125*(18), 5471–5478.
99. S. Laus, B. Sitharaman, V. Toth, R. D. Bolskar, L. Helm, S. Asokan, M. S. Wong, L. J. Wilson, A. E. Merbach, *J. Am. Chem. Soc.* **2005**, *127*(26), 9368–9369.
100. S. Laus, B. Sitharaman, E. Toth, R. D. Bolskar, L. Helm, L. J. Wilson, A. E. Merbach, *J. Phys. Chem. C.* **2007**, *111*(15), 5633–5639.
101. P. P. Fatouros, F. D. Corwin, Z. J. Chen, W. C Broaddus, J. L. Tatum, B. Kettenmann, Z. Ge et al., Radiology, **2006**, *240*(3), 756–764.
102. D. K. MacFarland, K. L. Walker, R. P. Lenk, S. R. Wilson, K. Kumar, C. L. Kepley, J. R. Garbow, *J. Med. Chem.* **2008**, *51*(13), 3681–3683.
103. C. Y. Shu, F. D. Corwin, J. F. Zhang, Z. J. Chen, J. E. Reid, M. H. Sun, W. Xu et al., *Bioconjugate Chem.* **2009**, *20*(6), 1186–1193.
104. J. F. Zhang, P. P. Fatouros, C. Y. Shu, J. Reid, L. S. Owens, T. Cai, H. W. Gibson et al., *Bioconjugate Chem.* **2010**, *21*(4), 610–615.
105. H. L. Fillmore, M. D. Shultz, S. C. Henderson, P. Cooper, W. C. Broaddus, Z. J. Chen, C. Y. Shu et al., Nanomedicine, **2011**, *6*(3), 449–458.
106. E. Y. Zhang, C. Y. Shu, L. Feng, C. R. Wang *J. Phys. Chem. B.* **2007**, *111*(51), 14223–14226.
107. K. Braun, L. Dunsch, R. Pipkorn, M. Bock, T. Baeuerle, S. Yang, W. Waldeck, M. Wiessler, *Int. J. Med. Sci.* **2010**, *7*(3), 136–146.

108. D. M. McCluskey, T. N. Smith, P. K. Madasu, C. E. Coumbe, M. A. Mackey, P. A. Fulmer, J. H. Wynne, S. Stevenson, J. P. Phillips, *ACS Appl. Mater. Interfaces* **2009**, *1*, 882–887.

109. C. Y. Chen, G. M. Xing, J. X. Wang, Y. L. Zhao, B. Li, J. Tang, G. Jia et al., *Nano Lett.*, **2005**, *5*(10), 2050–2057.

110. J. J. Yin, F. Lao, J. Meng, P. P. Fu, Y. L. Zhao, G. M. Xing, X. Y. Gao et al., *Mol. Pharmacol.* **2008**, *74*(4), 1132–1140.

111. J. J. Yin, F. Lao, P. P. Fu, W. G. Wamer, Y. L. Zhao, P. C. Wang, Y. Qiu et al., *Biomaterials*, **2009**, *30*(4), 611–621.

112. Y. Liu, F. Jiao, Y. Qiu, W. Li, F. Lao, G. Q. Zhou, B. Y. Sun et al., *Biomaterials*, **2009**, *30*(23-24), 3934–3945.

113. X. J. Liang, H. Meng, Y. Z. Wang, H. Y. He, J. Meng, J. Lu, P. C. Wang et al., *Proc. Natl. Acad. Sci. USA.* **2010**, *107*(16), 7449–7454.

114. H. Meng, G. M. Xing, B. Y. Sun, F. Zhao, H. Lei, W. Li, Y. Song et al., *ACS Nano*, **2010**, *4*(5), 2773–2783.

115. H. Meng, G. Xing, E. Blanco, Y. Song, L. Zhao, B. Sun, X. Li et al., *Nanomed. Nanotechnol. Biol. Med.* **2012**, *8*, 136–146.

116. S.-G. Kang, G. Zhou, P. Yang, Y. Liu, B. Sun, T. Huynh, H. Meng et al., *Proc. Natl. Acad. Sci. USA.* **2012**, *109*, 15431–15436.

117. S.-G. Kang, T. Huynh, R. Zhou, *Sci. Rep.* **2012**, *2*, 957.

118. K. K. M. Kobayashi, K. Sueki, K. Kikuchi, Y. Achiba, H. K. N. Nakahara, M. Watanabe, K. Tomura, *J. Radioanal. Nucl. Chem.* **1995**, *192*, 81–89.

119. D. W. Cagle, S. J. Kennel, S. Mirzadeh, J. M. Alford, L. J. Wilson, *Proc. Natl. Acad. Sci. USA.* **1999**, *96*(9), 5182–5187.

120. M. D. Shultz, J. C. Duchamp, J. D. Wilson, C. Y. Shu, J. C. Ge, J. Y. Zhang, H. W. Gibson et al., *J. Am. Chem. Soc.* **2010**, *132*(14), 4980–4981.

121. M. D. Shultz, J. D. Wilson, C. E. Fuller, J. Y. Zhang, H. C. Dorn, P. P. Fatouros, *Radiology,* **2012**, *261*(1), 136–143.

122. J. D. Wilson, W. C. Broaddus, H. C. Dorn, P. P. Fatouros, C. E. Chalfant, M. D. Shultz, *Bioconjugate Chem.* **2012**, *23*, 1873–1880.

123. Y. Horiguchi, S. Kudo, Y. Nagasaki, *Sci. Technol. Adv. Mater.* **2011**, *12*.

124. C. K. Chiang, C. R. Fincher, Y. W. Park, A. J. Heeger, H. Shirakawa, E. J. Louis, S. C. Gau, A. G. Macdiarmid, *Phys. Rev. Lett.* **1977**, *39*, 1098–1101.

125. J. L. Bredas, J. E. Norton, J. Cornil, V. Coropceanu, *Accounts. Chem. Res.* **2009**, *42*, 1691–1699.

126. Y. Y. Liang, L. P. Yu, *Accounts. Chem. Res.* **2010**, *43*, 1227–1236.

127. K. Takimiya, S. Shinamura, I. Osaka, E. Miyazaki, *Adv. Mater.* **2011**, *23*, 4347–4370.

128. D. M. Guldi, *Chem. Commun.* **2000**, *5*, 321–327.

129. T. Akasaka, S. Nagase, K. Kobayashi, T. Suzuki, T. Kato, K. Yamamoto, H. Funasaka, T. Takahashi, *J. Chem. Soc. Chem. Comm.* **1995**, *13*, 1343–1344.

130. X. X. Liao, T. S. Wang, J. Z. Wang, J. C. Zheng, C. R. Wang, V. W. W. Yam, *ACS. Appl. Mater. Inter.* **2013**, *5*, 9579–9584.

131. T. Kasuya, M. Sera, T. Suzuki, *J. Phys. Soc. Jpn.* **1993**, *62*, 3364–3367.

132. J. Miller, G. F. Krebs, J. Panetta, L. S. Schroeder, P. N. Kirk, Z. F. Wang, W. Bauer et al., *Phys. Lett. B* **1993**, *314*, 7–12.

133. Y. G. Zhen, N. Obata, Y. Matsuo, E. Nakamura, *Chem-Asian. J.* **2012**, *7*, 2644–2649.

134. S. J. Noever, S. Fischer, B. Nickel, *Adv. Mater.* **2013**, *25*, 2147–2151.

135. R. C. Haddon, A. S. Perel, R. C. Morris, T. T. M. Palstra, A. F. Hebard, R. M. Fleming, *Appl. Phys. Lett.* **1995**, *67*, 121–123.

136. W. Kang, M. Kitamura, Y. Arakawa, *Org. Electron.* **2013**, *14*, 644–648.

137. H. Yu, H. H. Cho, C. H. Cho, K. H. Kim, D. Y. Kim, B. J. Kim, J. H. Oh, *ACS. Appl. Mater. Inter.* **2013**, *5*, 4865–4871.

138. J. M. Ball, R. K. M. Bouwer, F. B. Kooistra, J. M. Frost, Y. B. Qi, E. B. Domingo, J. Smith et al., *J. Appl. Phys.* **2011**, *110*, 014506.
139. G. Yu, J. Gao, J. C. Hummelen, F. Wudl, A. J. Heeger, *Science* **1995**, *270*, 1789–1791.
140. Y. Zhang, Y. Matsuo, C. Z. Li, H. Tanaka, E. Nakamura, *J. Am. Chem. Soc.* **2011**, *133*, 8086–8089.
141. R. B. Ross, C. M. Cardona, D. M. Guldi, S. G. Sankaranarayanan, M. O. Reese, N. Kopidakis, J. Peet et al., *Nat. Mater.* **2009**, *8*, 208–212.
142. Y. J. He, G. J. Zhao, B. Peng, Y. F. Li, *Adv. Funct. Mater.* **2010**, *20*, 3383–3389.
143. J. H. Schon, C. Kloc, B. Batlogg, *Science,* **2002**, *298*, 961–961.
144. M. Capone, M. Fabrizio, C. Castellani, E. Tosatti, *Science,* **2002**, *296*, 2364–2366.
145. R. E. Dinnebier, O. Gunnarsson, H. Brumm, E. Koch, P. W. Stephens, A. Huq, M. Jansen, *Science,* **2002**, *296*, 109–113.
146. L. T. Scott, M. M. Boorum, B. J. McMahon, S. Hagen, J. Mack, J. Blank, H. Wegner, A. de Meijere, *Science,* **2002**, *295*, 1500–1503.
147. J. Lu, X. W. Zhang, X. G. Zhao, *Chem. Phys. Lett.* **1999**, *312*, 85–90.
148. Y. Morinaka, S. Sato, A. Wakamiya, H. Nikawa, N. Mizorogi, F. Tanabe, M. Murata et al., *Nat. Commun.* **2013**, *4*, 1554/1–1554/5.
149. X. Lu, L. Feng, T. Akasaka, S. Nagase, *Chem. Soc. Rev.* **2012**, *41*, 7723–7760.
150. S. Sato, H. Nikawa, S. Seki, L. Wang, G. F. Luo, J. Lu, M. Haranaka, T. Tsuchiya, S. Nagase, T. Akasaka, *Angew. Chem. Int. Ed.* **2012**, *51*, 1589–1591.
151. T. Tsuchiya, R. Kumashiro, K. Tanigaki, Y. Matsunaga, M. O. Ishitsuka, T. Wakahara, Y. Maeda et al., *J. Am. Chem. Soc.* **2008**, *130*, 450–451.
152. J. Lee, H. Kim, S. J. Kahng, G. Kim, Y. W. Son, J. Ihm, H. Kato et al., *Nature,* **2002**, *415*, 1005–1008.
153. K. Kobayashi, S. Nagase, *Chem Phys Lett* **1998**, *282*, 325–329.
154. Z. Slanina, K. Kobayashi, S. Nagase, *Chem. Phys. Lett.* **2004**, *388*, 74–78.
155. S. Sato, S. Seki, Y. Honsho, L. Wang, H. Nikawa, G. Luo, J. Lu et al., *J. Am. Chem. Soc.* **2011**, *133*, 2766–2771.
156. A. Saeki, Y. Koizumi, T. Aida, S. Seki, *Accounts. Chem. Res.* **2012**, *45*, 1193–1202.
157. V. Podzorov, E. Menard, A. Borissov, V. Kiryukhin, J. A. Rogers, M. E. Gershenson, *Phys. Rev. Lett.* **2004**, *93*, 1739–1741.
158. T. Kanbara, K. Shibata, S. Fujiki, Y. Kubozono, S. Kashino, T. Urisu, M. Sakai, A. Fujiwara, R. Kumashiro, K. Tanigaki, *Chem. Phys. Lett.* **2003**, *379*, 223–229.
159. S. Kaneko, L. Wang, G. F. Luo, J. Lu, S. Nagase, S. Sato, M. Yamada, Z. Slanina, T. Akasaka, M. Kiguchi, *Phys. Rev. B* **2012**, *86*, 155406.
160. L. Senapati, J. Schrier, K. B. Whaley, *Nano. Lett.* **2004**, *4*, 2073–2078.
161. J. X. Wang, C. Y. Chen, B. Li, H. W. Yu, Y. L. Zhao, J. Sun, Y. F. Li et al., *Biochem. Pharmacol.* **2006**, *71*, 872–881.
162. J. Tang, G. M. Xing, Y. L. Zhao, L. Jing, H. Yuan, F. Zhao, X. Y. Gao et al., *J. Phys. Chem. B* **2007**, *111*, 11929–11934.
163. S. Sato, S. Seki, G. F. Luo, M. Suzuki, J. Lu, S. Nagase, T. Akasaka, *J. Am. Chem. Soc.* **2012**, *134*, 11681–11686.
164. T. Suzuki, Y. Maruyama, T. Kato, K. Kikuchi, Y. Nakao, Y. Achiba, K. Kobayashi, S. Nagase, *Angew. Chem. Int. Ed. Engl.* **1995**, *34*, 1094–1096.
165. Y. Iiduka, T. Wakahara, T. Nakahodo, T. Tsuchiya, A. Sakuraba, Y. Maeda, T. Akasaka et al., *J. Am. Chem. Soc.* **2005**, *127*, 12500–12501.
166. T. Suzuki, Y. Maruyama, T. Kato, K. Kikuchi, Y. Achiba, *J. Am. Chem. Soc.* **1993**, *115*, 11006–11007.
167. S. Kobayashi, S. Mori, S. Iida, H. Ando, T. Takenobu, Y. Taguchi, A. Fujiwara, A. Taninaka, H. Shinohara, Y. Iwasa, *J. Am. Chem. Soc.* **2003**, *125*, 8116–8117.
168. R. C. Haddon, *J. Am. Chem. Soc.* **1996**, *118*, 3041–3042.

169. A. Strozecka, K. Muthukumar, A. Dybek, T. J. Dennis, J. A. Larsson, J. Myslivecek, B. Voigtlander, *Appl. Phys. Lett.* **2009**, *95*, 133118.
170. D. J. Hornbaker, S. J. Kahng, S. Misra, B. W. Smith, A. T. Johnson, E. J. Mele, D. E. Luzzi, A. Yazdani, *Science*, **2002**, *295*, 828–831.
171. T. Shimada, T. Okazaki, R. Taniguchi, T. Sugai, H. Shinohara, K. Suenaga, Y. Ohno, S. Mizuno, S. Kishimoto, T. Mizutani, *Appl. Phys. Lett.* **2002**, *81*, 4067–4069.
172. K. Kimura, N. Ikeda, Y. Maruyama, T. Okazaki, H. Shinohara, S. Bandow, S. Iijima, *Chem. Phys. Lett.* **2003**, *379*, 340–344.
173. T. Okazaki, T. Shimada, K. Suenaga, Y. Ohno, T. Mizutani, J. Lee, Y. Kuk, H. Shinohara, *Appl. Phys. a-Mater.* **2003**, *76*, 475–478.
174. M. Zaka, J. H. Warner, Y. Ito, J. J. L. Morton, M. H. Rummeli, T. Pichler, A. Ardavan, H. Shinohara, G. A. D. Briggs, *Phys. Rev. B* **2010**, *81*, 075424.
175. J. H. Warner, Y. Ito, M. Zaka, L. Ge, T. Akachi, H. Okimoto, K. Porfyrakis, A. A. R. Watt, H. Shinohara, G. A. D. Briggs, *Nano. Lett.* **2008**, *8*, 2328–2335.
176. K. Suenaga, T. Okazaki, K. Hirahara, S. Bandow, H. Kato, A. Taninaka, H. Shinohara, S. Iijima, *Appl. Phys. a-Mater. Sci. Process.* **2003**, *76*, 445–447.
177. A. N. Khlobystov, K. Porfyrakis, M. Kanai, D. A. Britz, A. Ardavan, H. Shinohara, T. J. S. Dennis, G. A. D. Briggs, *Angew. Chem. Int. Ed.* **2004**, *43*, 1386–1389.
178. Y. Iijima, K. Ohashi, N. Imazu, R. Kitaura, K. Kanazawa, A. Taninaka, O. Takeuchi, H. Shigekawa, H. Shinohara, *J. Phys. Chem. C* **2013**, *117*, 6966–6971.
179. K. Suenaga, T. Tence, C. Mory, C. Colliex, H. Kato, T. Okazaki, H. Shinohara, K. Hirahara, S. Bandow, S. Iijima, *Science,* **2000**, *290*, 2280.
180. K. Hirahara, K. Suenaga, S. Bandow, H. Kato, T. Okazaki, H. Shinohara, S. Iijima, *Phys. Rev. Lett.* **2000**, *85*, 5384–5387.
181. T. Okazaki, K. Suenaga, K. Hirahara, S. Bandow, S. Iijima, I. E. Shinohara, *J. Am. Chem. Soc.* **2001**, *123*, 9673–9674.
182. B. W. Smith, D. E. Luzzi, Y. Achiba, *Chem. Phys. Lett.* **2000**, *331*, 137–142.
183. M. D. Gimenez-Lopez, A. Chuvilin, U. Kaiser, A. N. Khlobystov, *Chem. Commun.* **2011**, *47*, 2116–2118.

8 Concluding Remarks and Perspectives

Xing Lu

Carbon is viewed as the most attractive element in the periodical table. Its unique valence electron distributions and hybridization styles provide variable valence states and sequentially enable the covalent connection with nearly all the other elements in the periodic table. The unique bonding characteristics of carbon also allow the formation of many types of compounds with different geometrical molecular structures, including linear, triangular, tetrahedral, and more complicated structures of the compounds.

For pure carbon allotropes, the research community has expended greatly during the last three decades and several new allotropes have been discovered.[1] Among these novel *artificial carbon allotropes*, such as fullerenes, carbon nanotubes, and graphene (Figure 8.1), fullerenes are the only soluble form having clear molecular structures, which ensures the evaluation of the effect of structures on the properties.[2] It is believed that the research on fullerenes is helpful to understanding the formation mechanism of related carbon allotropes, especially carbon nanotubes. The advent of fullerenes provides the third allotrope of carbon, and more importantly, it creates a new concept for molecular construction because the endohedral chemistry of fullerenes is a revolutionary strategy for chemical derivatization.[3] Indeed, endohedral metallofullerenes (EMFs) have shown unique structures and interesting properties, which are much different from conventional hydrocarbon materials.[4]

Although research on EMFs has expanded greatly during recent years, many unresolved difficulties still exist in this area.[5] The biggest problem is the availability of pure samples. The currently applicable arc-discharge method is only effective to give quantities sufficient for scientific research. The efficiency of this method should be improved to realize the large-scale industrial production. To this end, it is still desirable that new synthetic methods could be invented in the near future. Meanwhile, the many isomers present in the raw soot hinder further investigations. Selective production of certain isomers of EMFs by, for instance, adding certain kinds of catalysts or using different kinds of starting materials, gaseous, liquid, or solid, should be the final goal of EMF production. In addition, extraction from the raw soot and isolation of pure isomers of EMFs are additional blocks on the road to EMFs research. Although several non-chromatographic methods have been invented to isolate typical EMFs, none of them has shown a general power for the isolation of all kinds of EMFs.[6,7] Accordingly, more efficient works have to be done in this field.

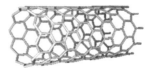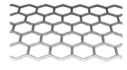

FIGURE 8.1 Schematic illustration of typical *artificial* carbon allotropes. From left to right: fullerene, endohedral metallofullerene, carbon nanotube, and graphene.

Secondly, the applications of EMFs have not been intensively explored. Taking into account the low yield of EMFs, we may have to search for some new application fields that are not *replaceable* by other materials. For example, one may utilize the magnetic or optical properties of the internal metallic species in combination with the functionalizable carbon cages to create some new materials useful in biology or medical therapy or diagnosis.[8] This should be one of the most brilliant potential applications of EMFs. In addition, recent results have revealed that the electron mobility of crystalline nanorods of some EMFs and their derivatives can be rather high, which is promising and exciting for their future applications in electronics.[9–11] As to the photovoltaic applications, devices embedded with $Lu_3N@C_{80}$-PCBH as an accepter material have shown superior performances than the C_{60} analogues.[12,13] From the perspective of chemistry, exohedral functionalization of EMFs is an effective way to realize the application of EMFs.[14] However, due to the complexity of the composition of ENFs and the unclear chemical properties which are strongly dependent on both the cage structure and the internal composition, many difficulties exist in the research.

Moreover, the unique structure-related spinning properties of some EMFs, such as $La@C_{82}$ and $Sc_3C_2@C_{80}$, could be used to make practical tools in quantum information systems or as molecular positioning indicators. For example, Wang and coworkers found that exohedral modification of $Sc_3C_2@C_{80}$ is an effective way to alter the spin density distributions of the three scandium atoms inside the fullerene cage and thus the paramagnetic properties of the whole molecule, which makes them likely candidates for future application as novel molecular devices.[15] In addition, the paramagnetic properties of $La@C_{82}$ can also be tuned by chemical modification. When cycloaddition takes place, paramagnetism remains in the adduct but the hyperfine structure of the lanthanum nuclear changed upon modification.[16,17] In contrast, when a singly-bonded addend is attached onto the fullerene surface, the paramagnetism of $La@C_{82}$ vanished completely.[18,19] Such magnetic properties are unique for EMFs and can be utilized in many applications.

Finally, the emergence of EMFs has not only generated a new class of hybrid materials but also provided some new concepts in coordination chemistry.[20] For instance, the motion of the two lanthanum atoms in $La_2@C_{80}$ is beyond the scope of modern chemistry because in conventional metallic complexes, the metal cores are invariably fixed relative to the coordinative ligand.[21] The mutual influence between the motion, location, and composition of the internal metallic clusters and the structures and properties of the cage is still of great importance. It is still anticipated that the hybridization of metals will give some new properties to EMFs, although the current status of research does not satisfy our anticipations yet.

More fascinating than other pure carbon allotropes because of the presence of the metallic species and the strong metal–cage interactions, EMFs deserve to gain their due attention from the scientific community and will afford real products that will benefit humanity in the near future.

REFERENCES

1. Hirsch, A. The era of carbon allotropes; *Nat. Mater.* **2010**, *9*, 868–871.
2. Kadish, K. M.; Ruoff, R. *Fullerenes: Chemistry, Physics, and Technology*; Wiley-VCH: New York, 2000.
3. Heath, J. R.; Obrien, S. C.; Zhang, Q.; Liu, Y.; Curl, R. F.; Kroto, H. W.; Tittel, F. K.; Smalley, R. E. Lanthanum complexes of spheroidal carbon shells; *J. Am. Chem. Soc.* **1985**, *107*, 7779–7780.
4. Popov, A.; Yang, S.; Dunsch, L. Endohedral fullerenes; *Chem. Rev.* **2013**, *113*, 5989–6113.
5. Lu, X.; Feng, L.; Akasaka, T.; Nagase, S. Current status and future developments of endohedral metallofullerenes; *Chem. Soc. Rev.* **2012**, *41*, 7723–7760.
6. Akiyama, K.; Hamano, T.; Nakanishi, Y.; Takeuchi, E.; Noda, S.; Wang, Z. Y.; Kubuki, S.; Shinohara, H. Non-HPLC rapid separation of metallofullerenes and empty cages with TiCl4 Lewis acid; *J. Am. Chem. Soc.* **2012**, *134*, 9762–9767.
7. Stevenson, S.; Harich, K.; Yu, H.; Stephen, R. R.; Heaps, D.; Coumbe, C.; Phillips, J. P. Nonchromatographic "stir and filter approach" (SAFA) for isolating $Sc_3N@C_{80}$ metallofullerenes; *J. Am. Chem. Soc.* **2006**, *128*, 8829–8835.
8. Bolskar, R. D. Gadofullerene MRI contrast agents; *Nanomedicine* **2008**, *3*, 201–213.
9. Sato, S.; Nikawa, H.; Seki, S.; Wang, L.; Luo, G. F.; Lu, J.; Haranaka, M.; Tsuchiya, T.; Nagase, S.; Akasaka, T. A Co-crystal composed of the paramagnetic endohedral metallofullerene $La@C_{82}$ and a Nickel Porphyrin with high electron mobility; *Angew. Chem. Int. Ed.* **2012**, *51*, 1589–1591.
10. Sato, S.; Seki, S.; Honsho, Y.; Wang, L.; Nikawa, H.; Luo, G.; Lu, J. et al. Semi-metallic single-component crystal of soluble La@C-82 derivative with high electron mobility; *J. Am. Chem. Soc.* **2011**, *133*, 2766–2771.
11. Sato, S.; Seki, S.; Luo, G.; Suzuki, M.; Lu, J.; Nagase, S.; Akasaka, T. Tunable charge-transport properties of I-h-C-80 endohedral metallofullerenes: Investigation of La-2@C-80, Sc3N@C-80, and Sc3C2@C-80; *J. Am. Chem. Soc.* **2012**, *134*, 11681–11686.
12. Ross, R. B.; Cardona, C. M.; Guldi, D. M.; Sankaranarayanan, S. G.; Reese, M. O.; Kopidakis, N.; Peet, J. et al. Endohedral fullerenes for organic photovoltaic devices; *Nat. Mater.* **2009**, *8*, 208–212.
13. Ross, R. B.; Cardona, C. M.; Swain, F. B.; Guldi, D. M.; Sankaranarayanan, S. G.; Van Keuren, E.; Holloway, B. C.; Drees, M. Tuning conversion efficiency in Metallo Endohedral fullerene-based organic photovoltaic devices; *Adv. Funct. Mater.* **2009**, *19*, 2332–2337.
14. Lu, X.; Akasaka, T.; Nagase, S. Chemistry of endohedral metallofullerenes: The role of metals; *Chem. Commun.* **2011**, *47*, 5942–5957.
15. Wang, T. S.; Wu, J. Y.; Xu, W.; Xiang, J. F.; Lu, X.; Li, B.; Jiang, L.; Shu, C. Y.; Wang, C. R. Spin divergence induced by exohedral modification: ESR study of $Sc_3C_2@C_{80}$ fullero-pyrrolidine; *Angew. Chem. Int. Ed.* **2010**, *49*, 1786–1789.
16. Maeda, Y.; Matsunaga, Y.; Wakahara, T.; Takahashi, S.; Tsuchiya, T.; Ishitsuka, M. O.; Hasegawa, T. et al. Isolation and characterization of a carbene derivative of La@C82; *J. Am. Chem. Soc.* **2004**, *126*, 6858–6859.
17. Feng, L.; Slanina, Z.; Sato, S.; Yoza, K.; Tsuchiya, T.; Mizorogi, N.; Akasaka, T.; Nagase, S.; Martin, N.; Guldi, D. Covalently linked Porphyrin-La@C82 hybrids: Structural elucidation and investigation of intramolecular interactions; *Angew. Chem. Int. Ed.* **2011**, *50*, 5909–5912.

18. Takano, Y.; Yomogida, A.; Nikawa, H.; Yamada, M.; Wakahara, T.; Tsuchiya, T.; Ishitsuka, M. O. et al. Radical coupling reaction of paramagnetic endohedral metallofullerene La@C$_{82}$; *J. Am. Chem. Soc.* **2008**, *130*, 16224–16230.

19. Lu, X.; Nikawa, H.; Tsuchiya, T.; Akasaka, T.; Toki, M.; Sawa, H.; Mizorogi, N.; Nagase, S. Nitrated Benzyne derivatives of La@C-82: Addition of NO2 and its positional directing effect on the subsequent addition of Benzynes; *Angew. Chem. Int. Ed.* **2010**, *49*, 594–597.

20. Xie, Y. -P.; Lu, X.; Akasaka, T.; Nagase, S. New features in coordination chemistry: Valuable hints from X-ray analyses of endohedral metallofullerenes; *Polyhedron* **2013**, *52*, 3–9.

21. Akasaka, T.; Nagase, S.; Kobayashi, K.; Walchli, M.; Yamamoto, K.; Funasaka, H.; Kako, M.; Hoshino, T.; Erata, T. ^{13}C and ^{139}La NMR studies of La$_2$@C$_{80}$: First evidence for circular motion of metal atoms in endohedral dimetallofullerenes; *Angew. Chem. Int. Ed.* **1997**, *36*, 1643–1645.

Index

Note: Locators followed by "*f*" and "*t*" denote figures and tables in the text.